Structural Steelwork Connections

Graham W. Owens BSc(Eng) PhD DIC CEng MICE MWeldI

Brian D. Cheal BSc(Eng) CEng MICE MWeldI

Butterworths

London Boston Singapore Sydney Toronto Wellington

PART OF REED INTERNATIONAL P.L.C.

First published 1989

British Library Cataloguing in Publication Data

Owens, Graham W.
Structural steelwork connections.
1. Steel structures. Structural connections. Design
I. Title II. Cheal, Brian
624.1'82

ISBN 0-408-01214-5

Library of Congress Cataloging-in-Publication Data

Owens, Graham, BscENG.
Structural steelwork connections / Graham W. Owens, Brian Cheal.
330 p.24.6cm.
Bibliography.
Includes index.
ISBN 0-408-01214-5:
1. Building, Iron and steel – Joints. 2. Steel, Structural. I. Cheal, Brian. II. Title. III. Title: Structural steel work connections.
TA684.094 1989
624.1'821 – dc19 88-39116

Typeset by Activity Ltd, Salisbury, Wilts.
Printed and bound in Great Britain by Courier International Ltd, Tiptree, Essex

Preface

This book provides a rational and up-to-date guide for the design of structural steelwork connections, combining scientific principles with practical application in a single volume. It concentrates on two themes. Firstly, having established an appropriate simple method of analysis, it insists on the consideration of *all* the components within the connection. Thus all the elements of each load path are checked, thereby ensuring that there can be no weak links anywhere in the connection. Secondly, it gives the background principles and reasons behind all the design checks that are put forward. Only if designers understand the underlying reasons can they apply design rules with confidence and safety.

An introductory chapter develops and discusses an overall philosophy for connection design, illustrating its application by some simple examples. Chapters 2 to 9 provide the background information necessary for informed design, covering welding, bolts, and bolting, weld behaviour, bolt behaviour, fatigue resistance of connections, other components within the connection, analysis and practicalities of construction. In all cases research and other information has been summarised and presented in a form that is of greatest use to the designer. Chapters 10 to 16 give general descriptions of the most commonly occurring types of connection and detailed design examples that demonstrate the application of the overall design approach and the detailed information in the earlier part of the book.

Much of the material here had its origins in the connections course which forms part of the MSc in Structural Steel Design at Imperial College. It has been refined by exposure to ten generations of post-graduate students, who collectively have several hundred years' of design experience. It has drawn substantially on the working practices and experience of W.S. Atkins and Partners, a leading firm of consultants with particular expertise in heavy steelwork.

The book does *not* list detailed design sequences for every kind of connection that commonly occurs. That task requires a several volume text and is being addressed by the SCI/BCSA Connections Group, on which the first author serves. However, the authors believe that this book, with its emphasis on the need for completeness in design and its presentation of the background reasons to design rules, is an important contribution to the development of improved detailed design methods for connections. They also believe it will be of considerable use to practising connection designers as they strive to achieve simplicity, economy and safety. The principles and general methods put forward should enable a designer to tackle any connection, irrespective of scale and complexity, with confidence and safety.

Graham W. Owens
Brian D. Cheal

Contents

1

Introduction: a rational basis for connection design

1.1 Engineering uncertainty

Structural engineering can never be an exact science and design philosophies should recognize and accommodate this uncertainty. Even in a carefully controlled laboratory experiment, perfect correlation between behaviour and analysis will not be achieved. In a practical structure the divergence between behaviour and prediction is generally greater. In addition to environment, the degree of uncertainty also depends on the type of structure or structural element.

Thus in a laboratory experiment on a steel beam or column with well-defined boundary conditions and known material properties a research worker would feel disappointed if theoretical and experimental deflections and strains did not agree to within 10%. Only in a situation of high imperfection sensitivity would greater divergence be acceptable. Even on complete structures, close agreement can still be achieved with care. In a major experimental study of multispan bridge behaviour[1] the first author achieved agreement to within 12% for deflections and 20% for significant stresses. In a full-scale study of industrial building structures[2] agreements were within 16% on deflections and 11% on strength.

1.2 Uncertainties and complexities of practical connection behaviour

Uncertainties in connection behaviour are frequently much greater, even in the laboratory. In a study of short end-plate beam/column connections[3] measurements were taken of the prying forces that the design method predicted would develop between the end-plate and the face of the column. In one specimen no such forces developed – an error of 100%. This error was due to bad fit.

In some tests on beam splices with high-strength friction-grip (HSFG) bolts and splice plates[4] the relative movements of the web plates to the web were carefully monitored. Figure 1.1 shows some results of this study as well as the positions of the two possible theoretical pairs of centres of rotation. These relative movements are not just of academic interest. As shown in Chapter 8, any analysis of a bolt group under eccentric loading has to assume a centre of rotation, either explicitly as in plastic design or implicitly as in elastic analysis. The position of this centre of rotation is a function of the eccentricity of loading. Clearly, the conventional design method for beam splices, which assumes that the web splice is subject to a shear force with a given eccentricity, cannot be modelling the true behaviour of this connection, whichever method of analysis is used. In practice, there must be a complex interaction of moments and shears between flanges and web as differing parts of the connection reach their limiting capacity.

There are several reasons why connection behaviour is more uncertain and more complex than that of other steel elements.

Geometric imperfections and lacks of fit

All steel components contain geometric imperfections and lacks of fit but the differing degree of uncertainty between elements and connections is mirrored in their varying degree of imperfection. The significant imperfection in a beam or column is a bow or twist with a maximum permitted amplitude of length/1000. This should be contrasted with the lack of fit permitted in a bolted connection using

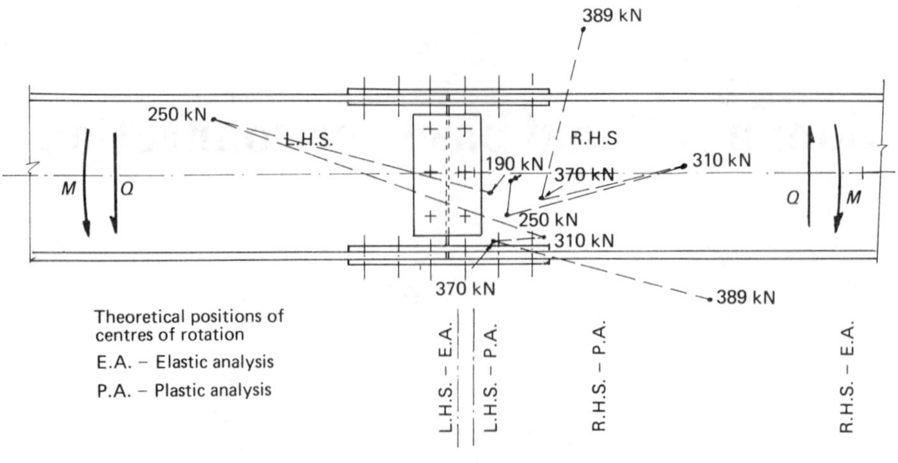

Figure 1.1 Positions of centres of rotation for relative movement between web splice plates and web in beam connection subject to shear and moment. Experimental measurements of paths of centres of rotation are shown thus – – –. Loads are indicated for specific centres of rotation

black bolts in clearance holes that is illustrated in Figure 1.2.

In the former case the imperfection will produce a secondary system of bending stresses which may, in extreme circumstances, attain values of the same order as the average stress. (These are, of course, accounted for in the column strength curves, where average stress decreases with increasing slenderness.) In the latter many of the bolts may not be contributing to the load resistance of the connection (for example, bolt 2 if the nearer plate is loaded to the left) until other bolts have sustained deformations of up to 4 mm (i.e. bolt 3). The influence of this on behaviour can be seen in Figure 1.3.

Alternatively, the 'eye-straight' imperfections that are acceptable for the steel elements may be compared with the gaps that may exist between a beam end plate and the face of the column to which it is connected.[5] Due to weld shrinkage, the end plate is likely to be distorted from its ideal plane; as anyone who has ever inspected site steelwork will know, gaps are likely to exist over much of the

Figure 1.2 Part of bolted connection showing hole misalignment

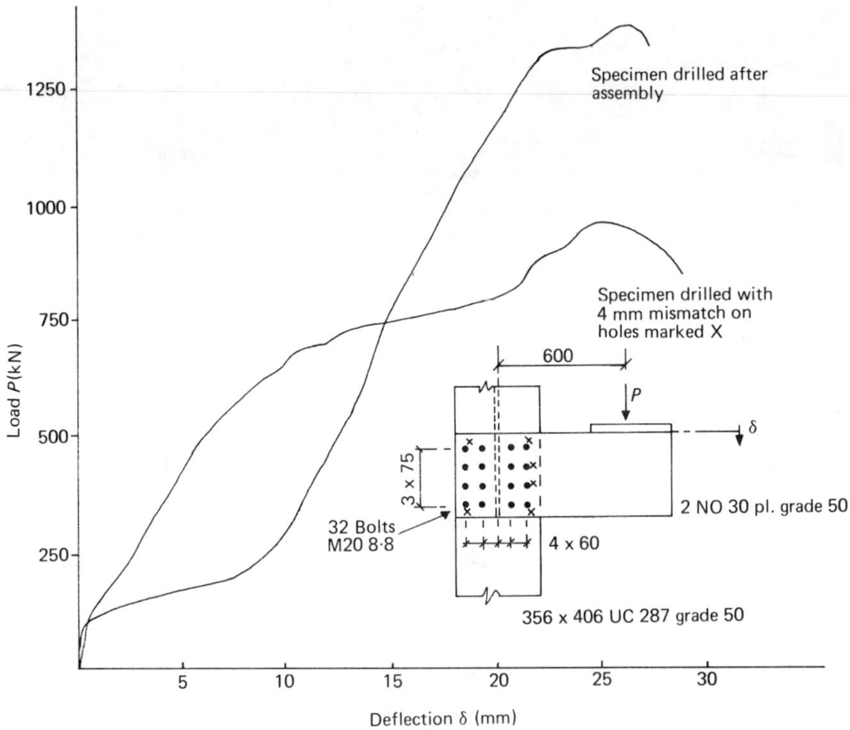

Figure 1.3 Load/deflection response of bolted column bracket, showing influence of lack of alignment of bolt holes

nominal area of contact. (It was this welding distortion that was responsible for the bad fit described earlier.)

Residual stresses and strains

Almost all steelwork contains residual stresses, that is, sets of self-equilibrating stresses that are locked in during manufacture and fabrication. A stiffened plate panel is likely to have compressive residual stresses of $0.2 \times \sigma_y$ and tensile residual stresses of σ_y. A hot rolled section may have residual stresses of $0.3 \times \sigma_y$ in compression and tension.[6] These two situations correspond to elastic strain incompatibilities of 0.15% and 0.08%, respectively. These figures should be contrasted with the mechanical properties that it has been found necessary to specify for the Hyzed range of steels.[7] These steels may be used in major welded connections with a high degree of restraint in order to eliminate lamellar tearing. The through-thickness properties of the three grades of Hyzed steel have specified minimum reductions of area of 25%, 15% and 10%, corresponding to local strain capacities of approximately 33%, 18% and 11%. These figures illustrate the potential magnitude of the plastic strains in a heavily welded connection. Their amplitude, and that of the associated distortion, in a particular situation depends on many factors, including connection geometry, weld process and sequence of fabrication. Despite considerable research effort, it has not proved possible to predict their magnitude accurately.[8]

Geometric complexity

It is axiomatic that there is greater geometric complexity within a connection than along the length of a structural member. This complexity has two important influences on connection behaviour. First, it causes considerable elastic stress concentrations within the connection; Figure 1.4 presents stress concentration factors (s.c.f.) for two standard situations.[9] Thus any bolted connection must have a s.c.f. of more than 2.2 and any connection with a sharp discontinuity (it is difficult to think of one that does not have such complexity) will have one of more than 3. Two important points should be noted in relation to stress concentration factors in connections:

1. The figures presented above are *macroscopic* s.c.f.: much greater *microscopic* s.c.f. will exist in the presence of any local weld defects or other surface irregularities.
2. Stress concentration factors of much larger magnitude (up to 20) exist in welded tubular connections.

The second effect of geometric complexity on connection behaviour is that simple theories do not

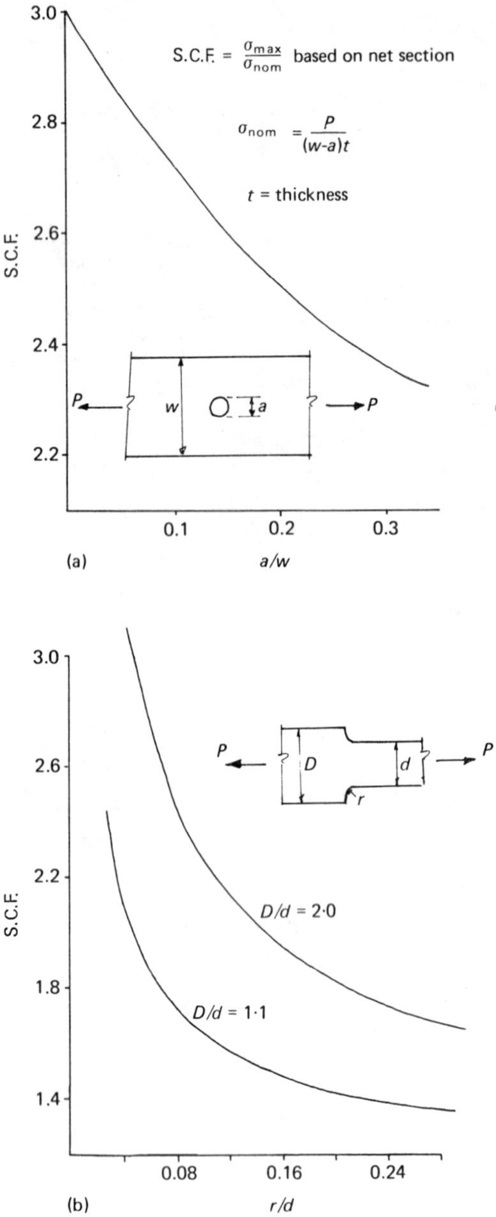

Figure 1.4 Elastic stress concentration factors for (a) axial load on a finite-width plate with a transverse hole and (b) tension across a reducing section – plane stress

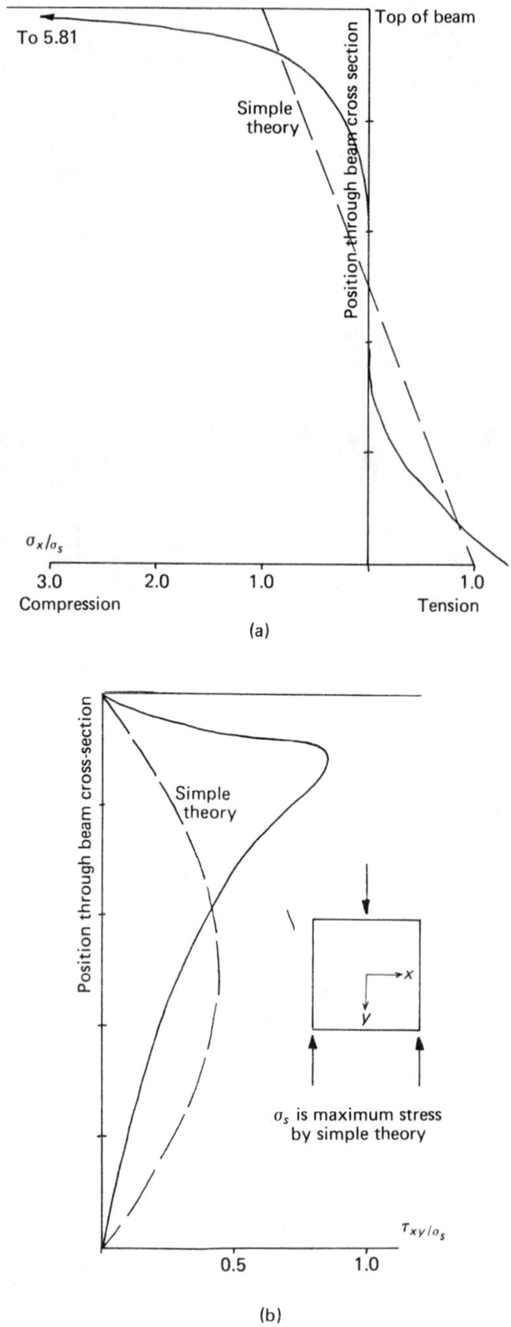

Figure 1.5 Elastic stress distribution in a deep beam with depth/span ratio of 1.0. (a) σ_x/σ_s at centreline; (b) τ_{xy}/σ_s at span/8 from centreline

predict accurately the distribution of stress within the connection. For example, gusset plates and stiffeners have such proportions that, when subject to bending moments, plane sections do not remain plane and engineers' simple bending theory gives a poor estimate of maximum stresses. Figure 1.5 shows the true stress distribution that exists in a deep beam with the proportions that can occur in a connection element.[10]

1.3 Shortcomings of traditional methods of analysis

The derivation of traditional methods of analysis is presented in Chapter 8. However, it is appropriate to examine critically some of the underlying assumptions in the search for a rational design philosophy.

In general, the connected parts are considered to be rigid, that is, their deformations may be ignored when compared with those of the connectors themselves. Figure 1.6[3] shows an example where this is manifestly not correct. In such connections it is clear that the deformations of the plates are greater than those of the connectors themselves and must have a significant influence on the distribution of the connector forces.

The other set of assumptions underlying traditional analytical methods relates to connector response. The most common assumption is that connectors behave in a linear-elastic manner, although rigid-plastic methods are becoming increasingly popular.

Figure 1.7 shows the range of responses that are possible with the most common type of bolt, the untorqued bolt loaded in shear. Because of the uncertainty of alignment of the holes with their 2 mm clearance on the bolt diameter there is the possibility of up to approximately 4 mm of movement at low load before the bolt is bearing against the side of the hole, hence the variation in initial response. As failure is approached the connector exhibits only limited ductility because high strains are very localized. Any method of analysis which represents this behaviour by either a linear or a bilinear response can clearly only be regarded as an approximate model of true behaviour.

Figure 1.6 Short end-plate connection for an I-beam where yielding of beam web has permitted overall bending of end plate

Figure 1.8 shows the response of a single HSFG bolt connection to shear loading.[11] Prior to slip there is very little connector deformation and so, in a

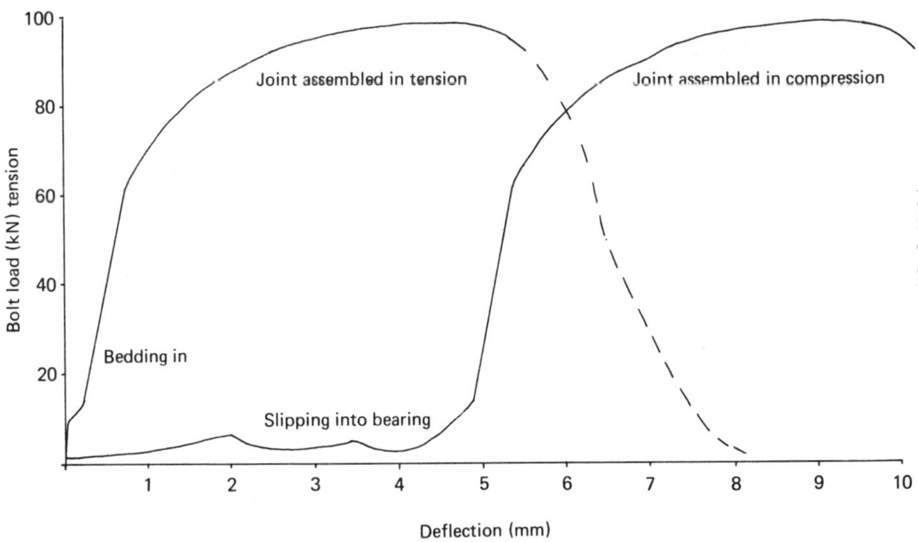

Figure 1.7 Load/deformation response of a single 20 mm diameter Grade 4.6 bolt in single shear

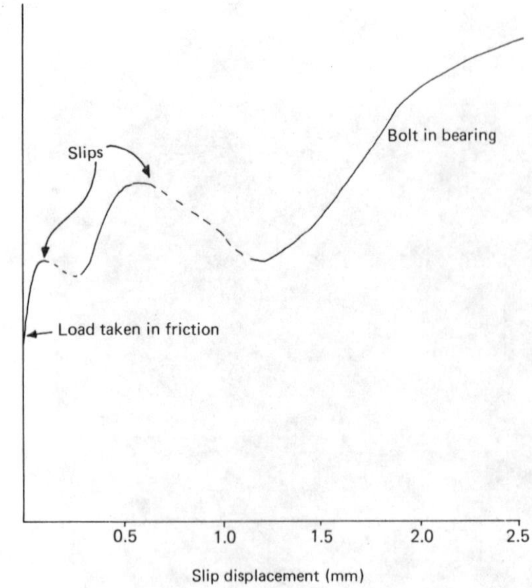

Figure 1.8 Load/deformation response of a single 20 mm diameter general grade HSFG bolt in double shear

multibolt connection, it is the deformation of the connected parts that has the predominant influence on the distribution of connector loads. The magnitude of the slip is governed by the alignment of the clearance holes and the post-slip behaviour is similar to the latter stages of that of the untorqued bolt connection. In a connection with such bolts an elastic analysis that only considers the flexibility of the bolts can only achieve limited accuracy if the connected parts are relatively flexible. Plastic analysis should give a reasonable estimate of overall slip load, at least for compact connections – this is probably where theory and behaviour are in closest agreement. (However, even here agreement is not always good in practice; in symmetric connections plastic analysis tends to overestimate slip capacity because dynamic friction is less than static friction and an 'adjusting' slip can trigger a major slip. Under eccentric load even plastic methods tend to underestimate capacity for reasons that are discussed in Chapter 8.)

Figure 1.9 describes the behaviour of fillet welds under varying directions of loading.[12] There is a substantial variation in both strength and stiffness with the direction of the loading vector. End fillets are both stiffer and stronger than side fillets. However, the former have considerably less ductility than the latter – end-fillet welds will rupture at less than 1 mm of local deformation. Both types of weld show a relatively early departure from linearity because of both the high stress concentration at the root of the weld and the presence of high residual stresses. Once again, complexity of behaviour calls into question the accuracy of our methods of analysis. Elastic analysis based on a single value of weld stiffness cannot be accurate; the limited ductility of the weld prevents the use of simple plastic analysis.

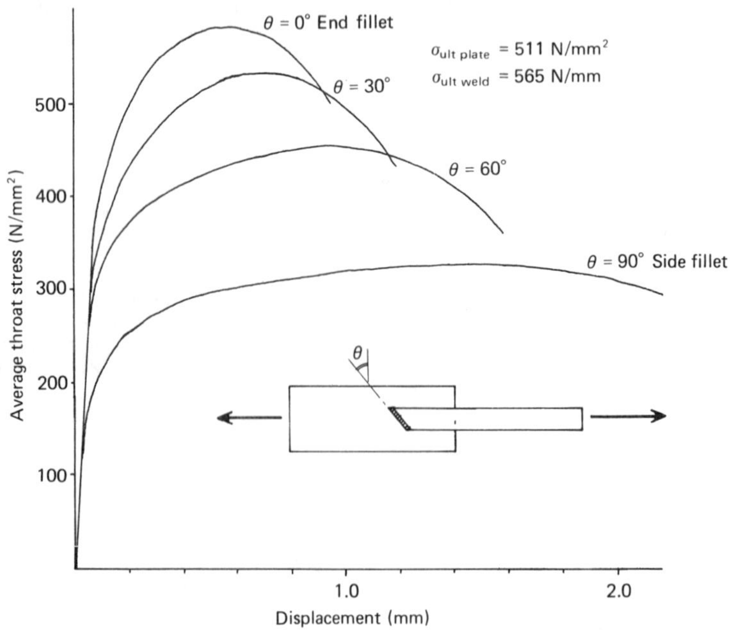

Figure 1.9 Load/deformation responses for 8 mm leg fillet welds at varying angles θ to load vector. Weld length 50 mm; plate thickness 19 mm

1.4 An appropriate design philosophy for connections

Any rational design philosophy has to take account of both the complexity and variability of practical connection behaviour. It is the uncertainty that presents the greater difficulty. If connection behaviour was merely complex it would be possible for designers to make a suitable choice for a particular situation. Detailed analysis could be used that took account of the complexity, thus permitting economic design with small load factors. Alternatively, simple analysis could be used in conjunction with higher load factors to account for possible variations between analysis and behaviour. Because of the *variability* of behaviour the first alternative cannot lead to economic and safe design. However refined the analysis, it would not be appropriate to reduce the load factors significantly.

The following design philosophy, based on simple analysis, would seem to be the most appropriate way of dealing with the problem. This approach has been developed over many years, based on the authors' experience of connection design and research into connection behaviour. It has been tested on successive generations of postgraduate students, many of whom themselves had substantial design experience. As put forward here in its most general form it is widely applicable. The only connections that have been found to be outside its scope are certain classes of tubular connections that present particular difficulty.

It should not, of course, be regarded as some sort of panacea for all the designer's problems. It cannot ensure that an inexperienced designer can immediately tackle any connection and produce an economical and satisfactory solution. Nor can it eliminate a proper consideration of the complexities of certain classes of connection. However, it should provide an overall framework within which the art of successful connection design can be developed, and it should enable the designer who is experienced in one class of connections to translate that expertise to some other structural form with greater confidence and certainty.

The method is presented below and its application in outline is demonstrated in the following section. Its detailed application requires an understanding of the connector and element behaviour, and this is presented in Chapters 2–7. Chapters 10–16 demonstrate its detailed application to the more common classes of connection.

1. Taking account of overall connection behaviour, carry out an appropriate simple analysis to determine a realistic distribution of forces within the connection.

For many connections this analysis should be based on the concept of 'force paths'. Here the overall loads acting on the connection are replaced by equivalent systems of forces which can then be assigned specific paths through the connection. In carrying out this analysis take account of:

1. The distribution of forces in the elements to be connected. For example, if the connection involves a beam carrying shear and moment, then remote from the connection the shear will be concentrated in the beam web and the flanges will carry most of the moment. In many instances this will be a satisfactory basis for the analysis of the forces within the connection. Indeed, it is a common simplifying assumption that the flanges carry all the moment; this is quite satisfactory provided it does not lead to an overstress in the flanges.
2. The flexibility of the components of the connection. It is the most flexible components that will govern the distribution of forces. For example, in an end-plate connection, if the bolts are of small diameter and the end plate is thick, it is the bolt flexibility that will govern the distribution of forces – as is indicated by conventional analysis. However, if the bolts are stiff compared to the end plates it is the flexural action of the latter that will primarily govern the distribution of forces in the connection, including the distribution of forces in the bolts.

It follows from the above that the conventional methods of analysis may be used in the context of this overall philosophy. They are most appropriate when the dominant flexibility is that of the connectors themselves.

It is most important to ensure that the analysis is consistent throughout the connection. In general, this is achieved by carrying out a single analysis of the most critical part of the connection and using that to determine the distribution of forces in other parts of the connection. Surprisingly, it is not uncommon to see designs where serious inconsistencies in analysis have occurred. These most commonly arise when more than one analysis has been used to determine the distribution of forces. For example, in the end-plate bracket connection shown in Figure 1.10 it would *not* be correct to use separate conventional analyses to determine both the distribution of forces in the bolts and in the weld attaching the end plate to the beam. Such separate analyses would assign different proportions of the tensile and compressive forces to different levels in the beam and imply instantaneous redistributions of these forces at the plane of contact between the weld and the end plate, as shown in Figure 1.10(c).

2. Ensure that each *component of* each *force path has sufficient strength to transmit the required force.*

This is self-evident, and yet it is surprising how frequently designers leave a weak link somewhere in

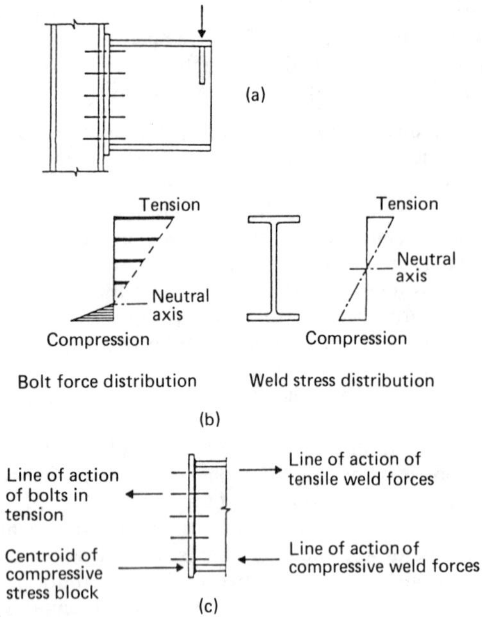

Figure 1.10 Inconsistency in connection analysis. (a) Bracket connection; (b) conventional elastic analyses; (c) stress resultants

a connection. A major disadvantage of traditional methods of analysis is that they concentrate on the distribution of forces in the connectors. Many codes of practice only give guidance on connector strength. Unwary or inexperienced designers are thus beguiled into thinking that, provided they have checked the bolts and/or welds, the connection is satisfactory. In reality, more design effort has frequently to be devoted to the other components than to the connectors themselves.

The only way to be certain that a design is satisfactory is for designers to have a clear understanding of how they wish the connection to behave and for them to ensure that all the components and critical sections have the capacity for this mode of behaviour.

3. Recognizing that the above procedure can only give a connection where equilibrium is capable of being achieved but where compatibility is unlikely to be satisfied, ensure that the components are capable of ductile behaviour.

This may be expressed alternatively as follows. Steps 1 and 2 have ensured that there is a reasonable way in which the connection *can* behave and hence have adequate strength. Ductility must now be ensured so that it *will* attain this condition without any premature rupture or buckling.

The incompatibilities may arise either from simplifications of the analysis or because of some lack of fit; their cause is not important. However, because of their possible presence, it is *essential* that the connection is capable of sufficiently ductile behaviour for plastic deformation to remove them prior to failure. Provided that this precaution is taken, even if it has not been possible to predict elastic response accurately, the connection will redistribute forces until it is acting in the way that was assumed in design. Fortunately, it is usually a straightforward matter to ensure that the components can achieve the necessary ductility.

4. Recognizing that the preceding steps only relate to static ultimate capacity, ensure that the connection will achieve satisfactory serviceability, fatigue resistance, etc.

For connections in buildings that have been designed by conventional elastic approaches this step may generally be omitted. However, in the following cases further calculations will be necessary:

1. Where either overall analysis or individual component design has been based on simple rigid-plastic methods (for example, using yield line analysis for an end plate – see Chapter 7) it will be necessary to ensure that only limited plastic deformation has taken place at working-load levels.
2. Where the connection is subject to significant repeated loading, a separate assessment of fatigue resistance should be carried out. This can create considerable difficulties because it requires both detailed consideration of the elastic response of the connection and an evaluation of important stress concentration factors. In extreme circumstances (for example, the tubular connections referred to previously) design for fatigue resistance should govern the overall design procedure and the sequence outlined above should be reversed. In less extreme circumstances (for example, cross-girder connections in bridges) static strength should still govern overall design but the connection layout should be arranged to minimize stress concentrations because of the importance of fatigue considerations.

1.5 Application of the design philosophy

Most connection design is very straightforward and satisfies the preceding criteria by implication rather than by specific calculation. For example, conventional design of the simple beam-to-column connection shown in Figure 1.11(a) is carried out in a few lines of calculation (or by means of design tables) as a designer/detailer determines the number of bolts necessary to resist the eccentric load shown on the basis of their shear and bearing capacity. Here the

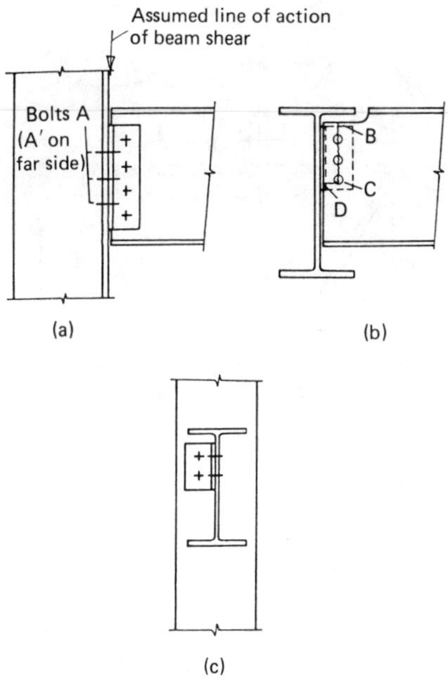

Figure 1.11 'Simple' beam-to-column and beam-to-beam connections. (a) Conventional beam-to-column connection with double-web cleat; (b) beam-to-beam grillage connection with double-web cleats; (c) single-web cleat connection

simple analysis is based on the assumption that the eccentricities between the lines of bolts A and A′ may be discounted because of symmetry in a connection of normal proportions with two web cleats. Experience indicates that the only critical checks are the bolts in shear and the cleats and beam web in bearing. Other 'weak links' are likely be designed out by minimum pitch criteria and practical edge distance requirements. Experience has also indicated that such connections have sufficient ductility to accommodate both lack of fit within the connection and the beam end rotation as it takes up its deflected profile.

If all connection design were as straightforward there would be little point in the explicit procedure presented in the preceding section. However, it is only necessary to vary some details in this simple example to illustrate the importance of a sound appreciation of connection design. Figure 1.11(b) shows the beam end detail that results if the top flange has to be notched to accommodate a flange in the supporting beam. Line BCD becomes a very important critical section to be checked in shear and tension. If a single short web cleat were to be used, as shown in Figure 1.11(c), the same local moments that were reasonably ignored in the double-cleat connection may at least cause unserviceability as the beam twists and could lead to an unacceptable reduction in strength. Similarly, the same simple approach to analysis could lead to an unacceptable reduction in strength if a fillet welded beam end connection were used, because of lack of ductility in the weld.

The direct use of the design philosophy outlined in the previous section is well demonstrated on a beam splice with HSFG bolts. Figure 1.12(a) shows the straightforward set of forces that can replace the applied moment and shear. (This simplification with its implied redistribution is only applicable for building structures. In bridges such redistribution is not usually permitted, primarily because of general concern over fatigue.) Figure 1.12(b) enumerates the checks that have to be carried out to ensure that there are no weak links within the connections, and these checks are as follows:

1. The capacity of the flange to resist the tensile force. The critical section is the vertical net section through the first line of holes, in tension, together with the horizontal web/flange intersection in shear. In addition, the effective section of the flange through the first line of holes should be checked under the flange stress resultant, without any redistribution of bending moment to the flanges. The effective section is defined as the net area times a coefficient greater than unity which recognizes that nominal stresses slightly above yield may occur on a net section without detriment;

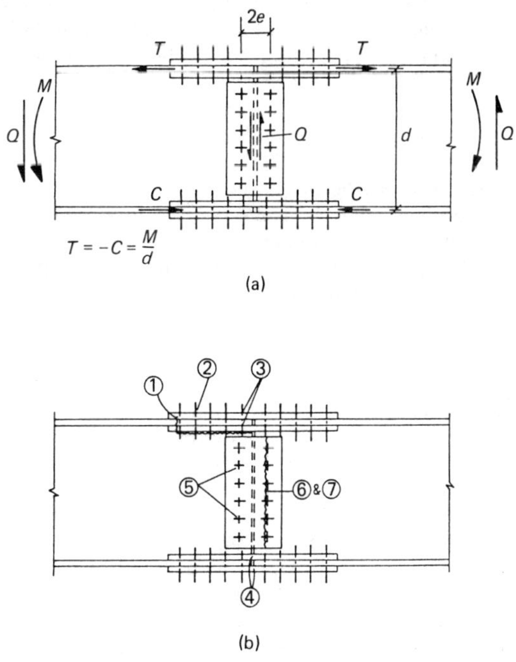

Figure 1.12 Analysis and strength assessment of beam splice. (a) Conversion of applied loading to equivalent system of forces; (b) strength checks required to demonstrate adequacy of connection

2. The frictional capacity of the HSFG bolts;
3. The net sections of the splice plates, assuming that the force T is equally divided between the pair of cover plates.

Note that because of the symmetry of the connection these checks also demonstrate the adequacy of most of the compressive flange splice. The only additional checks are:

4. The compressive capacity of the splice plates, free to buckle vertically (this is satisfied by maximum pitch criteria);
5. The frictional capacity of each web bolt group under a load Q at an eccentricity e;
6. and 7. The capacity of the net section of the web and splice plates. (These are most unlikely to be critical unless the shear is a very high proportion of the beam capacity.)

Figure 1.13 shows the application of the design philosophy to an exterior beam-to-column connection subject to moment and shear, one of the most complex connections that commonly occurs in practice. (A fuller discussion of this connection is presented in Chapter 12; some aspects of behaviour are simplified here in order not to confuse this general example.)

Figure 1.13(a) shows the analytical procedure that is appropriate for this connection for values of the factored applied moment up to approximately 70% of the plastic moment capacity of the beam. Above that value it will be necessary to mobilize some of the bending capacity of the web in order not to overstress the flanges.

Remote from the connection, the beam and column flanges will be making the greatest contribution to the moment resistance of the respective members and the beam shear will be concentrated primarily in the web. These distributions are used as the basis of the simple analysis. Thus the forces within the connection transmitting the bending moments are given by:

$$F_b = \frac{M}{d_b} \quad \text{and} \quad F_c = \frac{M}{2d_c}$$

where d_b and d_c are the distance between flange centroids of the beam and column, respectively. The shear P is assumed to remain in the web. An alternative analysis could be based on some notional centre of rotation at the end-plate/column face contact surface and an assumed linear response of the bolts, in accordance with traditional elastic analysis. This method is not used here because it requires more computation and is unlikely to give greater accuracy for the reasons outlined in Section 1.3.

The next step is to trace the paths of these forces through the connection. The bending forces F_b and F_c may be considered to continue on their original lines of action until they reach the panel ABCD, whose corners are defined by the intersections of these lines of action, and thus has dimensions $d_c \times d_b$. Overall equilibrium of these forces is achieved by shear on this panel. The shear flow on AB and CD is thus $F_b/d_c = M/d_b.d_c$. The vertical (complementary) shear flow is $2F_c/d_b = M/d_c.d_b$. The beam shear is simply transferred straight through to the near column flange and hence down the column. It is customary to ignore the eccentricity between the column face and the column centreline. This leads to an apparent lack of equilibrium, which arises because, in the earlier calculation of bending forces, the applied moment M is taken as the value

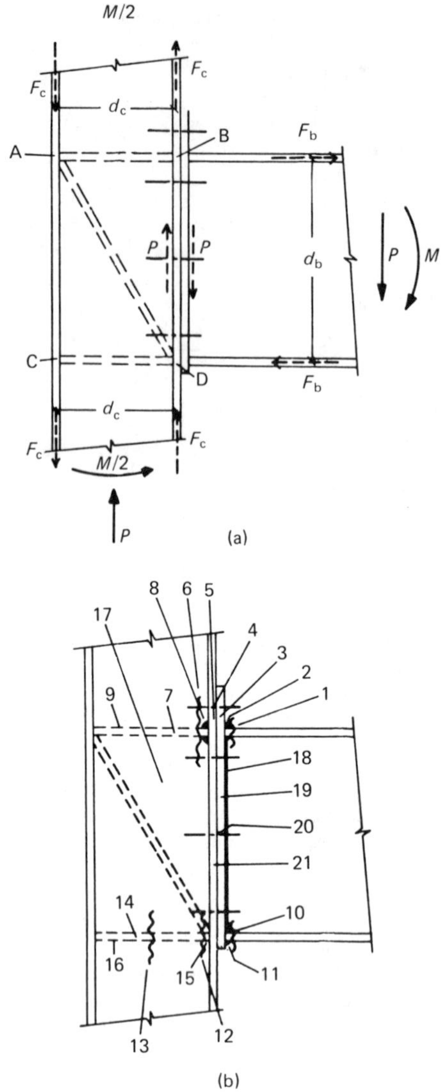

Figure 1.13 Analysis and strength assessment of an exterior beam-to-column connection

at the intersection of the column and beam centrelines rather the value at the column face. (Precisely minded readers who find this unacceptable are invited to base F_b on this latter moment and take separate account of the moment arising from the eccentricity of the shear force path! The resultant overall loading on panel ABCD will then be the same as that arising from the simpler analysis given above.)

It is now possible to check the elements on the various force paths. Thus, starting with the path of the tensile force F_b in the top flange of the beam and tracing it through to the panel ABCD, it is necessary to check:

1. The tension capacity of the flange adjacent to the connection. The effective width of the flange may be limited by the bolt layout;
2. The welds attaching the beam flange to the end plate;
3. The end plate in flexure as it disperses F_b into the bolts;
4. The bolts in tension. Account should be taken of prying action as appropriate;
5. The column flange in flexure as it transfers the bolt loads into the column web. If this is inadequate, extra capacity may be obtained by providing a stiffener along AB. In extreme cases the column flange may still be inadequate with a stiffener;
6. The tensile capacity of the effective length of the column web. If this is inadequate a stiffener has to be provided;
7. The tensile capacity of stiffener AB if this has been provided to satisfy either 5 or 6;
8. The welds connecting the column flange to the stiffener, in tension;
9. The welds connecting the column web to the stiffener, in shear.

The detailed check is then continued by examining the path of the compressive force F_b from the beam bottom flange into the bottom of panel ABCD:

10. The compressive capacity of the beam flange. This is affected, in the absence of stiffener CD, by the stress concentration caused by the column web. If the flange is overstressed, a stiffener should be provided;
11. The flange/end-plate welds;
12. Local crushing of the column web. If there is an overstress a stiffener should be provided;
13. Buckling of the column web. If there is an overstress a stiffener should be provided;
14. The compressive capacity of stiffener CD, if this has been provided to satisfy any of 10, 12 or 13;
15. The stiffener/column flange welds, in compression;
16. The stiffener/column web welds, in shear.

Finally, to complete the force paths associated with the moments:

17. Panel ABCD in shear. If this is inadequate a diagonal stiffener should be provided along AD. Both this stiffener and its welds should be checked in accordance with 14–16 above.

The shear force path requires less attention for this particular connection. It is only necessary to check:

18. The beam web/end-plate welds in shear;
19. The end plate in shear and bearing;
20. The bolts in shear;
21. The column flange in bearing.

The design procedure is presented as a straightforward sequence. In practice, there may be an interaction between sections; for example, if a stiffener has to be provided along AD one will not generally be necessary along CD. This change in geometry will affect the force paths and should be acknowledged by a modification to the overall analysis.

When carrying out these detailed checks it is important to keep in mind the third part of this design procedure, namely that the components of the connection should be ductile. For this connection this means complying with certain criteria for stiffener proportions, minimum weld sizes and bolt/end-plate thickness ratios.

This example looks very daunting; apparently twenty-one separate calculations have to be carried out! However, it should be appreciated that many of the checks are very straightforward and may either be satisfied by inspection or by very simple calculation. Experienced designers will recognize that for a connection of conventional proportions, only checks 3, 4, 5, 13, 17 and 20 require significant computation. Checks 1, 6, 7, 10, 11, 12 and 14 can be dealt with by a single line of calculation; the remainder are either governed by ductility criteria or can be satisfied by inspection. However, the principle of considering each link in the chain explicitly would seem to be a very sound discipline for all connections. It is certainly essential for connections of unusual proportions.

References

1. Owens, G. W., Dowling, P. J. and Hargreaves, A. C., 'Experimental behaviour of a composite box girder bridge', Int. Conf. on Short and Medium Span Bridges, Ontario, 1982.
2. Dowling, P. J., Mears, T. F., Owens, G. W. and Raven, G. K., 'A development in the automated design and fabrication of portal framed industrial buildings', *The Structural Engineer*, October, **60A**, No. 10 (1982).

3. Hargreaves, A. C., MSc thesis, University of London, September 1974.
4. Owens, G. W., MSc thesis, University of London, September 1970.
5. Morris, L. J. and Mann, A. P., 'Lack of fit in steel structures', *CIRIA Report 87*, London, 1981.
6. Conference on Residual Stresses in Welded Connections and their Effects, The Welding Institute, London, 1977.
7. BSC, *Hyzed Steel Plates with High Through-thickness Ductility*, BSC Plates, Glasgow, 1978.
8. *Control of Distortion in Welded Fabrications*, 2nd edn, The Welding Institute, Abington, Cambridge, 1982.
9. Peterson, R. E., *Stress Concentration Design Factors*, Wiley, New York, 1953.
10. Saad, S. and Hendry, A. W., 'Stresses in a deep beam with a central concentrated load', *Proc. Soc. Exper. Stress Analysis*, **18**, No. 1 (1961).
11. Harding, J. E., MSc thesis, University of London, September 1971.
12. Clarke, A., MSc thesis, University of London, September 1970.

2

Basic welding technology

2.1 Scope

A full coverage of the topics covered in this chapter is inappropriate in a general book on connection design; the presentation is therefore limited to those aspects that are of greatest importance to the designer. A bibliography is provided at the end of the chapter for the reader who wishes to pursue a topic in greater detail, and a list of books is given for each of the main headings of the text. The order within each list indicates the relative standard of the references; more general and elementary texts are listed first, followed by more advanced ones.

2.2 Welding processes in structural engineering

2.2.1 General

Most electrical fusion welding processes have several features in common:

1. The electrode and workpiece are connected to opposite sides of the power supply; an arc is struck between the electrode and the workpiece, releasing heat energy.
2. This heat energy melts the surface of the workpiece, the tip of the electrode if it is consumable and any flux that may be used. The metallic components of these molten elements form a weld pool which is held together by electromagnetic and surface tension forces.
3. The edges of this molten weld pool are on the point of solidification. As the electrode is moved it draws the arc centre and weld pool with it. Semi-solid metal on the boundary of the weld pool remains behind, fusing with the parent metal and forming a weld bead. The rate of build-up of metal (the *deposition rate*) is measured in g/min.
4. The arc, weld pool and hot weld bead must be protected from the atmosphere to prevent oxidation of the weld metal. In some processes this is achieved by providing an inert gaseous atmosphere; in others a flux is used, and this is a mixture of compounds that has to fulfil several functions. When heated, parts of it form a gaseous envelope to protect the arc and weld pool; some parts may be drawn into the weld pool to provide necessary alloying additions and the remainder of the melted components will form a slag over the deposited weld bead. This serves both to protect the bead and control its shape.

A wide variety of fluxes is available, and the most common types are briefly described in a later section.

2.2.2 Manual metal arc (MMA) welding

In this process, shown in Figure 2.1, the electrode is hand held and fed by hand into the weld pool. No shielding gas is used but a flux is incorporated as a coating to the electrode. This is probably still the most common process in structural fabrication. It is certainly the most flexible, as it can be used in all welding positions and the hand-held electrode can be manipulated in many situations where other processes could not be used. In addition, there is no time-consuming initial set-up process.

Quality of welding is closely related to operator skill. A good 'stick' welder has both a sound understanding of the process and a high level of manipulative skill. Because good welders command high rates of pay and deposition rates are relatively

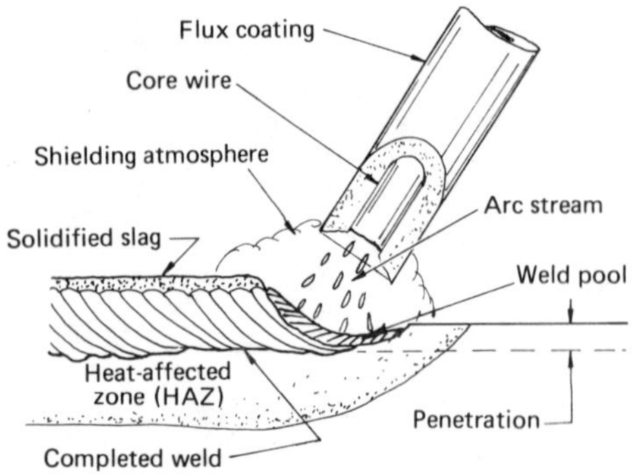

Figure 2.1 Schematic representation of manual metal arc welding

low this can be an expensive process. It was traditionally regarded as the method that was capable of producing the highest quality welds, but this reputation is now threatened by recent advances in more efficient processes.

Typical electrode diameters for structural work vary between 2.5 and 6 mm. Open-circuit voltages are usually 50–90 V, depending on electrode; as soon as the arc is struck the voltage falls to between 20 and 35 V. Typical currents range from 50 to 400 amps, deposition rates from 20 to 100 g/min.

2.2.3 Automatic welding with continuous coated electrodes

This is the automatic equivalent of MMA welding; both electrodes have similar cross-sections. However, instead of being hand held, the electrode in this process is mounted on a drum and incorporated into a continuous feed system in an automatic welding plant. To prevent the flux from breaking away from the electrode 'spin' wires are helically wound around the outside of the flux. These fulfil the additional role of acting as the electrode to transmit the current to the weld arc. The central wire is thus not energized, since it is insulated by the flux.

2.2.4 Gas-shielded welding

This is sometimes called metal active gas (MAG), CO_2 or metal inert gas (MIG), although the last term is properly only applied to welding where the shielding gas is argon or helium (these gases are used for non-ferrous metals.)

In this process the bare electrode, welding arc and weld pool are protected only by the gas shield; there is generally no flux. A flexible hose supplies the welding gun, shown in Figure 2.2, with electrode wire from a drum, shielding gas and electric current, all these consumables being automatically controlled. The shielding gas is usually carbon dioxide, sometimes with a small addition of argon, for carbon and carbon/manganese steels; it emerges from an annular opening directly around the electrode.

A common variation of this method is to use flux-cored wire[4] where the electrode wire has a tubular cross-section and surrounds a central core of flux. The flux contains arc stabilizers, deoxidants and alloying elements. Sometimes it also has constituents that emit carbon dioxide, in which case it is possible to omit the shielding gas. Such processes are particularly suitable for site use because they are less susceptible to wind, although protection should still be provided.

The mode of metal transfer from electrode to weld pool varies with welding current. At low currents the electrode will dip into the weld pool causing an intermittent short circuit and consequent melting of the wire tip; this is called dip transfer. At higher currents the arc is continuous and the plasma stream from electrode to workpiece transfers a spray of molten metal; this is called spray transfer. Recently, pulsed current welding plant has been developed which controls molten metal transfer more closely, one or two metal droplets being transferred with each pulse of high current.

When the gas-shielded welding gun is hand held the method is considered to be semi-automatic, because although the welding consumables are automatically controlled, manipulative skill is still required. However, the method is readily incorporated into fully automatic processes. Its particular advantage as an automatic process is that it can be used for many types of positional welding, and it is finding increasing application in production engineering with robotic systems.

In its early development this process had only a mediocre reputation for quality. Weld appearance was not good, with a lot of spatter; because the process was semi-automatic it was assumed that

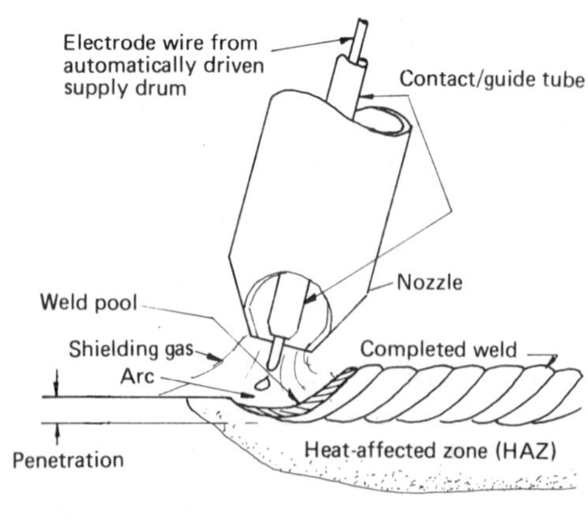

Figure 2.2 Schematic representation of gas-shielded welding

operator skill was less important than with traditional manual metal arc welding. However, with skilled operators and modern plant, particularly those with pulsed current, weld quality should be comparable with MMA. Deposition rates are certainly higher, leading to greater economy, and this process is growing in popularity. For example, in the fabrication of offshore structures, in 1970 all site welding was by MMA but by 1986 the use of gas-shielded welding had risen to approximately 50% of this high-quality work.

Typical electrode diameters for structural work vary between 0.75 and 2.00 mm. Welding voltages vary between 20 and 30 V. Typical currents are from 50 to 200 A for dip transfer and 150 to 500 A for spray transfer. Deposition rates can be as high as 150 g/min for downhand welding with spray transfer. Because the process is essentially continuous, duty cycles are generally significantly higher than for MMA welding.

2.2.5 Submerged arc welding

In this process the arc is entirely submerged in a granular flux. Because of this, high currents can be used without danger of air entrainment or spatter. Figure 2.3 shows a diagrammatic representation of a simple, single-wire, submerged arc welding process.

Deposition rates for this process are considerably higher than for either MMA or CO_2 welding. They can be increased still further by using two- or three-electrode wires for the same weld pool, sometimes with a further 'hot wire' addition. This is a wire which is not electrically charged but is heated and introduced into the weld pool independently of the electrode wires.

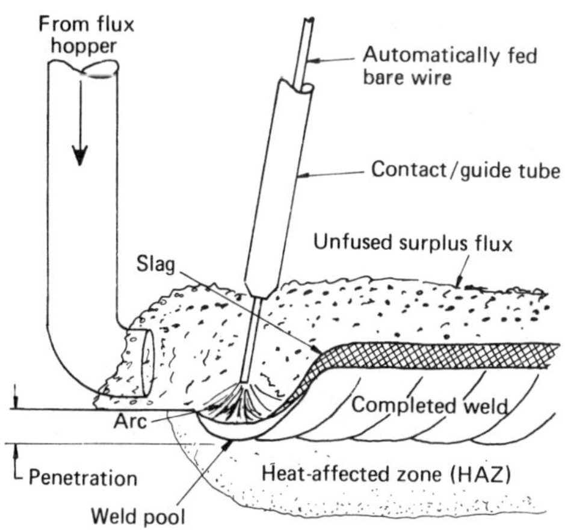

Figure 2.3 Schematic representation of submerged arc welding

The process is normally incorporated into some fully automatic system where either the workpiece is moved, as in the circumferential welding of pipes, or the welding plant is mounted on a tractor unit, as in butt welding between flat plates.

A recent development of this process is the use of narrow-gap submerged arc welding.[5] This is only suitable for situations where there can be tight process control during preparation, machining and welding. However, in those circumstances it can achieve considerable economies on thicker materials (say, above 70 mm).

The weld quality can be extremely high because there is no longer any reliance on manipulative skill. Weld appearance is good, partly because of the automatic nature of the process and partly because of the smoothing and containing nature of the fused slag. The high current also produces greater penetration than other processes, in a consistent manner which can be recognized in design. The two principal disadvantages of the process are that first, because of the granular flux, it can only be used in the flat and horizontal/vertical positions; and second, at very high deposition rates, weld metals of low toughness may be produced because the associated low cooling rates lead to a large-grained microstructure. However, post-weld heat treatments can alleviate this second disadvantage.

Typical electrode diameters for structural work vary between 2 and 5 mm. Welding voltages vary between 30 and 40 V. Currents for single-wire welding can be as high as 1200 A. Deposition rates for single-wire welding can reach 300 g/min and five times that value for sophisticated multiwire sets.

2.2.6 Electroslag welding

This automatic process is only used for the butt welding of thick plates that are in the vertical or nearly vertical position. As shown in Figure 2.4, the plates are simply positioned with a gap of around 30 mm between them; no special weld preparation is required and the plates are cut square.

A very large weld pool is created, contained on two sides by the plates being welded and on the other sides by water-cooled copper shoes. One or more electrodes are fed automatically from the top of the joint into the weld pool. The flux is conductive in its molten state; once it has melted, the process ceases strictly to be an arc welding process. The arc is extinguished and the heat input arises from electrical resistance through the electrode and the conductive slag.

Originally the electrode guides were linked to the copper shoes and the entire unit made to climb as the weld pool rose. Consumable guides have now been developed that do not move and are permitted to melt away as the weld pool rises; this has considerably simplified the process mechanization.

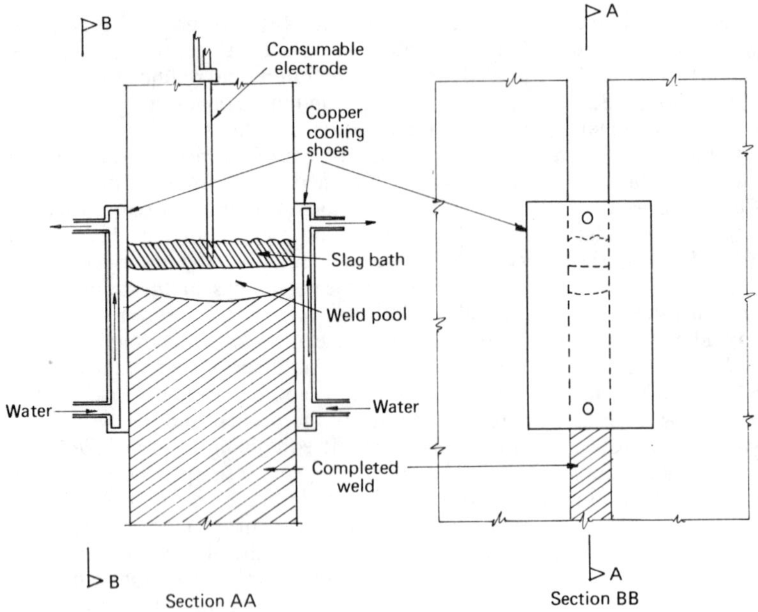

Figure 2.4 Schematic sections of electroslag welding

Because of the high heat input and large weld volume, cooling rates are low. A coarse-grained microstructure with poor fracture toughness is generally produced which may require post-weld heat treatment.

2.3 Welding fluxes and electrode classification

Fluxes have to be used for manual metal arc, submerged arc, flux cored wire and electroslag welding. Their compositions and functions vary considerably from process to process, but it is possible to categorize their constituents in the following way:

1. *Gas formers* that decompose to provide shielding gases when heated. Carbonates produce CO_2, fluorides develop fluoride shielding and organic materials produce hydrocarbons and CO_2;
2. *Slag formers* that produce the slag crust which protects the weld metal after solidification. These include calcium and manganese carbonates, titanium, silicon, manganese and iron oxides, silicates and clays;
3. *Arc initiators and stabilizers* – these are either metallic, such as powdered nickel or iron, or elements producing sodium and potassium ions, such as feldspar and clay;
4. *Fluxing agents* – these decrease impurities in the weld, and they are usually carbonates or oxides;
5. *Deoxidizers* – reducing agents such as ferrosilicon, ferromanganese and iron powder are added to reduce the oxygen in the weld pool;
6. *Minerals controlling physical properties of the flux* – these influence bead profile and slag detachability, and include many of the oxides and fluorides required for other functions;
7. *Metallic additions* – iron powder and ferroalloys can be added to improve the deposition rate;
8. *Binders* such as mica, sodium silicate and organic binders are added to improve the flux strength.

It is also possible to categorize fluxes according to the principal properties of their main constituents. Part of this classification relates to the basicity of the oxide constituents. (In order of increasing basicity, from the most acidic, the common oxides are SiO_2, Al_2O_3, TiO_2, Z_nO_2, FeO, MgO, MnO_2, CaO, NaO, K_2O.) A broadly similar system can be used for manual metal arc, submerged arc and gas-shielded flux-cored wire welding.

1. *Acid or high silica fluxes* – these are good cleaning fluxes and produce a weld of excellent profile. However, the weld will often have high oxygen, hydrogen and silicon contents with consequent low strength and toughness;
2. *Neutral fluxes* produce welds of only moderate strength and toughness;
3. *Basic or low-hydrogen fluxes* produce welds with low oxygen and hydrogen contents, provided that the electrodes are kept oven-dry before welding. These produce welds with good

mechanical properties that are not susceptible to cracking;

4. *Cellulosic fluxes* contain up to 30% of organic material. Good penetration is achieved but at the expense of increased hydrogen levels and greater likelihood of cracking;
5. *Rutile fluxes* are those with a significant proportion of titanium dioxide (rutile) in place of the manganese and iron oxides of acidic fluxes. This improves metal transfer and arc stability, and these electrodes are therefore easy to manipulate. Mechanical properties are moderately good and these are considered to be a good general-purpose electrode.
6. *Iron powder fluxes* – these manual metal arc electrodes have coatings with a considerable proportion of iron powder in a basic or rutile flux to give an improved deposition rate.

Flux-cored wire fluxes for use without any shielding gas do not follow the classification outlined above. They all contain calcium fluoride to give fluoride gas protection and some contain calcium carbonate to give additional carbon dioxide gas. In some, deoxidation is carried out by aluminium, in others by iron powder. Rutile is sometimes used for reasons similar to those outlined above.

In electroslag welding there is considerably more time available for interaction between the molten flux and the weld pool, and this interaction can significantly alter the weld metal composition. Apart from alloying additions associated with this interaction, electroslag fluxes have many similarities to those for submerged arc welding. In addition, they must have current-carrying properties because there is no arc once the process is established. The molten flux must have sufficient conductivity to transmit the current to the weld pool but not so high a conductivity that current demand will be excessive.

Electrode and flux classifications for the various welding processes are governed by different standards, and those in most common use are listed below:

Manual metal arc welding – BS 639 AWS A5.1
Flux-cored wire welding – AWS A5.20
Submerged arc welding – DIN 8557 and AWS A5.17
Electroslag welding – these fluxes have not been classified.

To illustrate the range of parameters considered in these classification systems an interpretation of a single electrode coding, namely E5133B/12029H10, by the BS 639 system is as follows:

E: indicates a covered electrode for manual metal arc welding
51: a tensile strength of 510–650 N/mm^2 and a minimum yield stress of 360 N/mm^2
3: a minimum elongation of 20% and an impact value of 28 J at −20°C
3: a minimum elongation of 22% and an impact value of 47 J at −20°C
B: a basic type of coating

The symbols above are compulsory; those below are optional:

120: nominal electrode efficiency, i.e. 120% minimum metal recovery
2: electrode is suitable for all welding positions apart from vertical down
9: should be used with positive polarity for DC welding and a minimum open circuit voltage of 90 V for AC welding
H10: less than 10 ml H_2 per 100 g weld

2.4 Weld preparations

Fillet welds do not require weld preparations. In butt welds, it is generally necessary to prepare the surfaces of the elements being joined. The ideal weld preparation is one that:

1. *Provides satisfactory access throughout the depth of the weld.* This is essential to ensure that sound weld metal is deposited and that it is properly fused to the parent metal and earlier weld deposits. In addition, it must be possible to clean the weld between runs. This is particularly important for processes where the weld metal is covered by fused slag after welding.
2. *Minimizes the volume of deposited weld metal.* Deposited weld metal is obviously very expensive, and this is an important requirement for economy.
3. *Minimizes the cost of weld preparation.* Simple plane weld preparations can be gas cut, which is a considerably cheaper operation than the machining that is required for more complex preparations, such as those for U and J butt welds.
4. *Does not lead to notches or discontinuities.* This is particularly important where corrosion is severe or where fatigue is a design consideration.
5. *Can accommodate the variations in fit that are likely to be encountered in practice.* Practical tolerances should vary with the situation. Thus in a butt weld between two unstiffened plates a tolerance on fit of ±1–2 mm should be achieved with reasonable care. In a joint where the elements are less flexible and the geometry is more complex (for example, a node in an offshore structure) the cost of achieving a similar fit would be prodigious. In such a situation it is clearly appropriate to adopt a more tolerant welding procedure and preparation.

There are certain general principles that underly the design of weld preparations. The most critical welding pass is the first (root) pass and the most critical part of the weld preparation is the root preparation, as it has to accommodate variations in fit. The root gap must not be so narrow that the weld pool cannot penetrate to the far side, and it must not be so wide that a weld pool cannot be established.

The root face must be sufficiently deep to prevent the arc from burning through but not so deep that there is incomplete penetration. The sizes of the root gap and root face depend on the choice of welding process, welding variables and welding position. For a MMA or CO_2 semi-automatic weld, typical values are 2–3 mm and 1–2 mm, respectively; for a submerged arc weld typical values might be 0–2 mm and 4–6 mm, respectively, because of the greater penetration achieved with this process.

Unless the root geometry can be closely controlled the weld metal of the root runs will not be entirely sound. If there are doubts about its integrity it should be removed by back gouging and replaced from the other side. If, because of overall structural tolerances, the root gap cannot be maintained within suitable limits, a metal backing strip can be used in conjunction with a large root gap, to support the weld pool. The principal disadvantage of the backing strip is that it introduces a crevice into the weld geometry. An alternative system without this disadvantage uses temporary ceramic tiles to support the weld pool during the early passes; these are held in place by magnetic clamps.

Above the root region of the weld the preparation is modified so that the filler passes can be efficiently carried out. The side-wall slopes must be such that the weld arc can be directed against them so that proper side-wall fusion is achieved. For shallow welds it is appropriate to use plane preparations, such as those shown in Figure 2.5(a), because of their lower cost. With deeper welds it becomes more economic to use the more complex weld preparations shown in Figure 2.5(b); the extra preparation cost is more than offset by the saving in weld volume.

The choice between single or double preparations depends on access, the ease with which the structure may be turned, the plate thickness and the means by which distortion is being controlled.

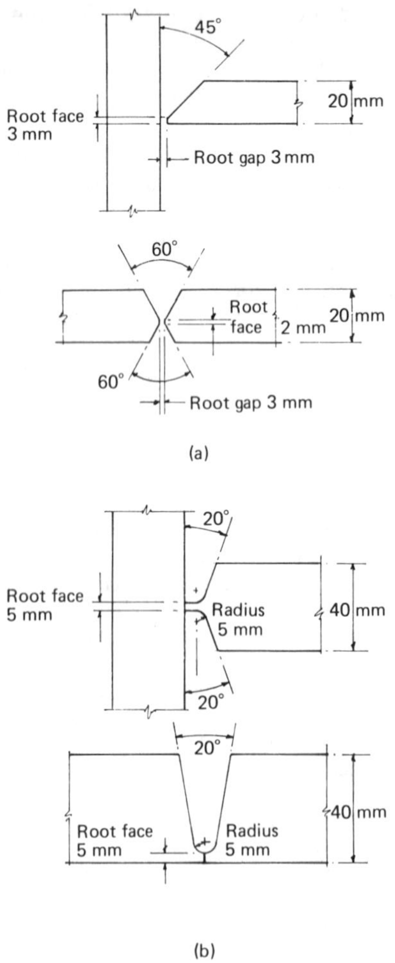

Figure 2.5 Typical weld preparations for butt welds. (a) 20 mm plates; (b) 40 mm plates

2.5 Control of distortion

Once a molten weld bead has been deposited and starts to cool it will solidify and attempt to contract, both along and transverse to its axis. This tendency to contract will induce tensile residual stresses, probably accompanied by tensile yielding where it is resisted by the surrounding structure, and distortions where the surrounding structure is less than fully rigid (which is almost always the case in practice).

Longitudinal shrinkage can cause slender outstands and plates to buckle and, if the weld is eccentric to the effective centroid of the resisting structure, will lead to an overall bowing of the fabrication. Although these longitudinal distortions are of considerable significance in general fabrication they are of only limited importance in connection design. As discussed in Chapter 7, only stocky plate and elements should be used in connections, and these can sustain the overall longitudinal shrinkage without any local buckling. The second type of longitudinal distortion is only of significance in long elements which rarely occur in connections.

Transverse shrinkage is likely to produce both angular and out-of-plane distortions, as shown in Figure 2.6(a). There are three possible approaches to the control of these distortions. First, prior to

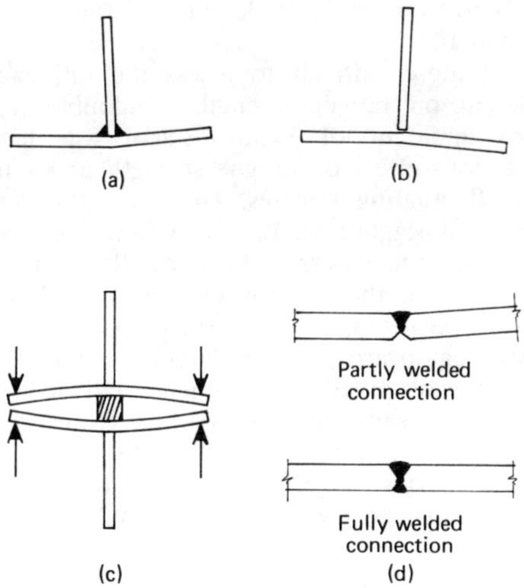

Figure 2.6 Welding distortion and its control. (a) Unrestrained distortion due to fillet welds; (b) presetting to eliminate distortion in completed structure; (c) restraint and flexure to eliminate distortion; (d) counteracting distortions

welding the elements may be preset so that after welding and its associated distortion the final geometry is correct. Second, the structure can be sufficiently restrained to keep the distortions at acceptable levels. Third, welding sequence can be modified so that compensating distortions of opposite sign cancel each other out. In practice, more than one method may be used at the same time for a particular situation.

Figure 2.6(b) shows the first kind of preventative action. The end plate is pre-dished, either by pressing or by spot heating, and the web plate is set a few degrees off the right angle to the end plate. Alternatively, the end-plate distortion could have been reduced by clamping it to a strong back to increase its flexural strength during welding. If this was combined with the elastic bending shown in Figure 2.6(c) it would be possible to eliminate the distortion completely. Figure 2.6(d) shows the third type of distortion control where distortions of opposite sign are offset against each other to achieve an acceptable final configuration.

2.6 Preheating

As discussed in greater detail in Section 2.7, rate of cooling is an important parameter in the control of hydrogen cracking of the heat-affected zone. It is possible to reduce the cooling rate and so reduce the susceptibility to cracking by preheating the elements before and during welding. In critical cases the 'preheat' is maintained for a considerable period after welding. Guidance on preheating is given in BS 5135; Figure 2.7 shows a typical graph of preheating temperature as a function of combined plate thickness, arc energy input and parent metal composition.

2.7 Weld defects

2.7.1 Undercut

Figure 2.8 shows typical examples of undercut or toe groove defects. The wide rounded profile shown in Figure 2.8(a) is generally formed when too much parent metal is washed into the weld pool and some disturbance of the pool prevents deposition at that point. For example, this disturbance might be due to too high a current, producing excessive turbulence in the pool. Probably the most common situation for

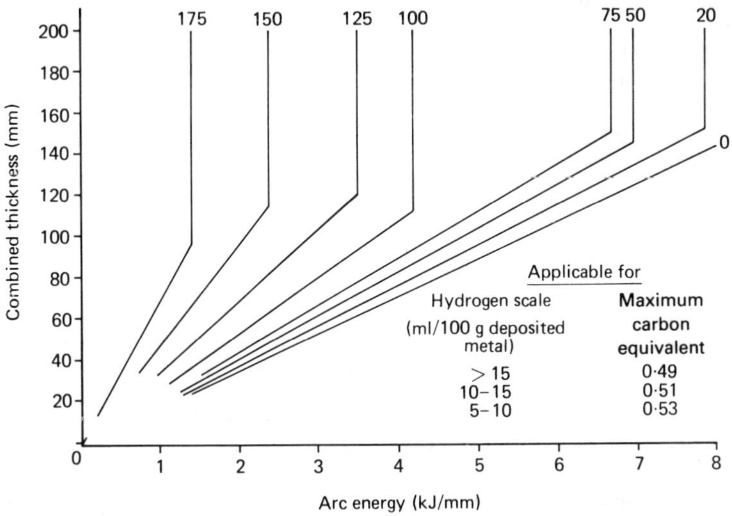

Figure 2.7 Typical graph of minimum preheat temperature (°C) from BS 5135

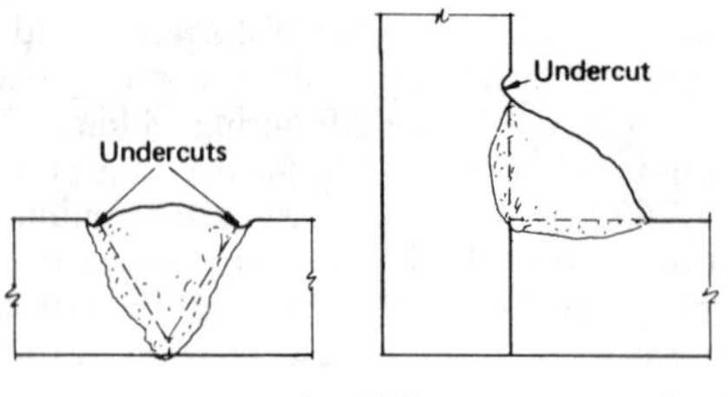

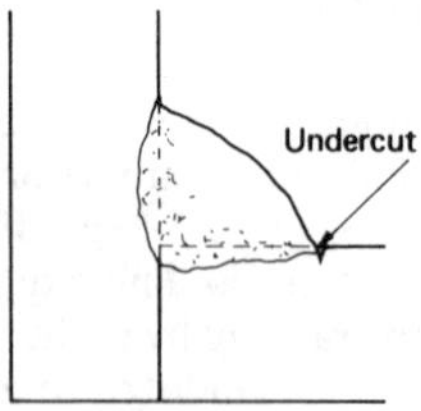

Figure 2.8 Typical examples of undercut defects. (a) Wide and curved; (b) narrow and cracklike

this type of undercut occurs with the deposition of large single runs of weld metal in the horizontal/vertical position to form a fillet weld. With the large weld pool the weld metal tends to sag away from the vertical plate.

The narrow and cracklike defects shown in Figure 2.8(b) are likely to occur if there is insufficient heating of the parent plate in the immediate vicinity of the weld pool and a consequent severe transition in thermal shrinkage at that point. This is most likely to occur if an incorrect weaving technique, with insufficient pauses at the weave ends, is used in vertical welding.

It is not easy to assess the significance of undercuts, and this fact is reflected in the wide variation that exists in specified defect tolerances in different specifications. In carrying out this assessment it should be appreciated that some microflaws, due to minute slag inclusions, are almost unavoidable in this region. These defects are almost impossible to detect by conventional means because they are unlikely to be more than 0.25 mm in depth, but they can have a significant influence on fatigue performance.

In a statically loaded structure, given adequate toughness, assessment can be based on straightforward considerations of loss of effective area. Thus if they are parallel to the applied stress these defects are unlikely to be significant. If they are transverse to it the loss in strength has to be considered in proportion to that of the element as a whole. Thus matching undercuts of 0.5 mm on each side of a 10 mm plate would lead to a loss of 10% in tensile capacity and 19% in flexural strength. Corresponding figures for similar undercuts in 20 mm plate are 5% and 10%.

It is more difficult to assess the influence of undercut on fatigue strength. Undoubtedly, the sharp undercut of Figure 2.8(b) will have a deleterious effect on fatigue strength under transverse fluctuating stresses. However, the smooth profile of Figure 2.8(a) is unlikely to be so damaging; it may even be beneficial if it reduces the stress field at the point where the microflaws are likely to exist. Minute variations in surface geometry in this region are probably largely responsible for the scatter observed in experimental fatigue lives. If remedial measures are proposed for critical fatigue locations it is almost certainly preferable to grind out the surface irregularity associated with the undercut than to reweld (see Section 2.8).

2.7.2 Slag inclusions

These are non-metallic particles, usually derived from the flux, that are trapped by the weld pool, as shown in Figure 2.9. They are likely to be of considerable length and can therefore influence strength if their cross-section is large compared to that of the weld pool. They can arise in multipass welds and, in that situation, are usually due to inadequate cleaning between passes, often in conjunction with unfavourable bead shape or incorrect bead sequence. Alternatively, they can occur in the weld root, usually as a result of too narrow a root gap. With the correct weld preparation, fit and procedure they are readily avoided.

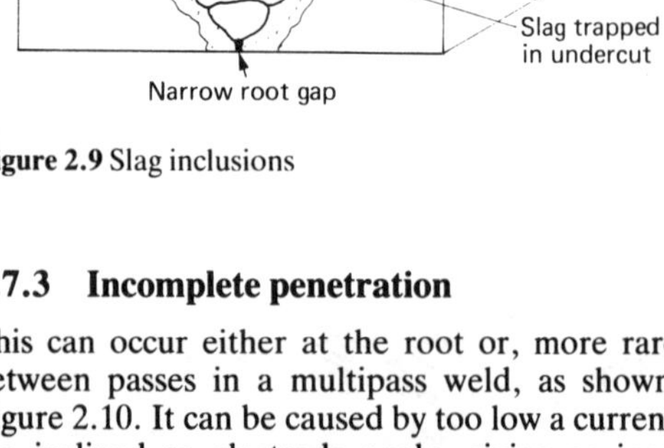

Figure 2.9 Slag inclusions

2.7.3 Incomplete penetration

This can occur either at the root or, more rarely, between passes in a multipass weld, as shown in Figure 2.10. It can be caused by too low a current or too inclined an electrode angle, giving an insufficient concentration of energy into the weld pool. Alternatively, it can arise with some geometric problem: the electrode may be too large for the

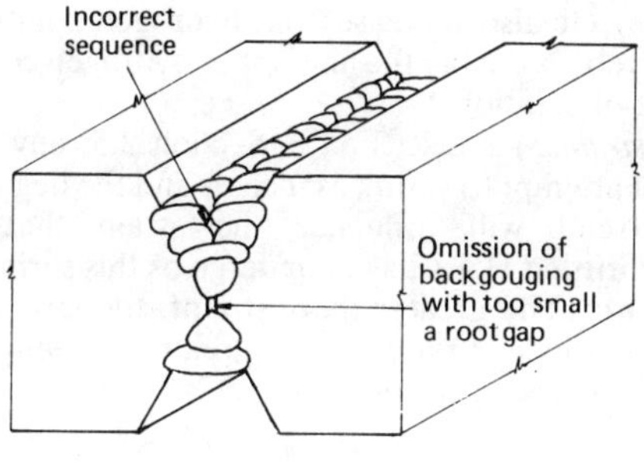

Figure 2.10 Lack of penetration

joint; the root gap may be too small; or the weld bead sequence may be incorrect. Within the body of the weld these problems can be readily corrected by a modification to the procedure. Incomplete penetration of the root gap can be more difficult to eradicate because of inevitable variations in gap along the joint. For this reason it is common, with manual welding, to backgouge the root runs and reweld from the reverse side. Lack of root penetration is less of a problem with submerged arc welding because of the higher welding currents; backgouging is therefore less common.

2.7.4 Lack of fusion

This defect is a less extreme form of lack of penetration. There are no voids left in the weld metal but the individual runs have not entirely fused or attached themselves to the previously laid weld metal or edges of the weld preparation or plate surface (see Figure 2.11). This can be caused by milder forms of the same shortcomings that lead to lack of penetration and may, in addition, be a result of contamination by rust or mill scale of the joint surface. It may be corrected by better cleanliness and/or a modification to the weld procedure.

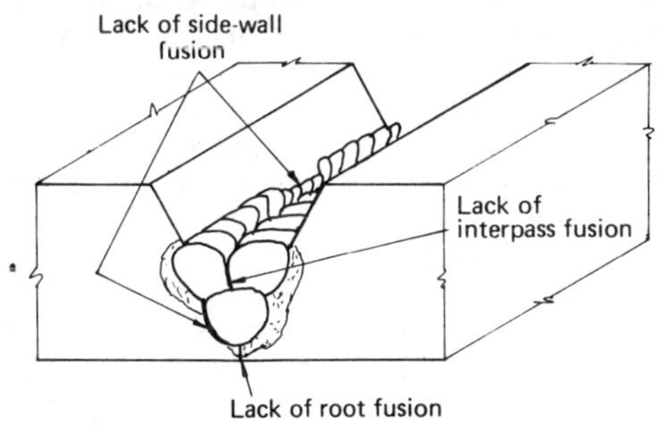

Figure 2.11 Lack of fusion

2.7.5 Porosity

Porosity is the formation of small cavities in the weld metal, and is caused by the trapping of gas in the molten weld pool. The cavities are generally spherical but can be elongated to give piping or worm holes. It arises because gas solubility in the weld pool diminishes as the temperature drops. The gas may arise from some contaminant on the plate or electrode; too turbulent a gas flow with gas-shielded processes; or too long an arc if basic fluxes are used. 'Start porosity' is a particular problem because, when the arc is first struck, the protective gas atmosphere is not immediately established and the start of the weld is exposed to air. This can be overcome either by restarting above and proud of the previous run so that the porous weld metal can be removed after welding by grinding, or by restarting in front of the end of the previous weld, moving the weld pool back to the previous weld and then coming back over the restart position, remelting the weld metal, and allowing the gases to escape.

2.7.6 Hydrogen, heat-affected zone or cold cracking

This type of cracking, shown in Figure 2.12(a), generally occurs in the heat-affected zone after welding. The cracks are most likely to form as this region cools below 300°C but may not occur until some time after welding; indeed they can develop considerably later during the service life of the structure. The mechanism of cracking depends on several interrelated factors. The cooling rate in the HAZ can be similar to that normally associated with quenching and, depending on the composition of the

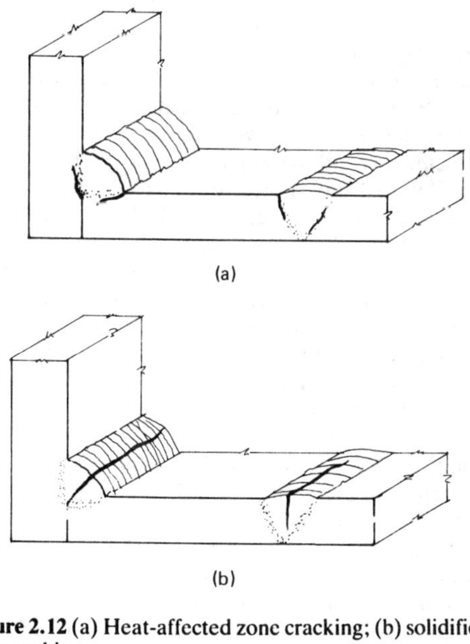

Figure 2.12 (a) Heat-affected zone cracking; (b) solidification cracking

steel, can cause considerable hardening and loss of ductility. Hydrogen may be introduced into the weld arc region by moisture in the flux or hydrocarbon contamination. At high temperatures both the weld metal and the HAZ dissolve significant quantities of any hydrogen that is available; as the temperature drops the solubility of the hydrogen diminishes and concentrations of gas build up in microscopic voids with very high pressure. The weld metal can accommodate this because of its ductility and its toughness but if the HAZ has become too brittle, cracks will develop.

The key parameters for eliminating this form of cracking are:

1. *Material composition*. For a given weld technique and geometry, susceptibility to cracking depends on hardenability of the parent material, which in turn depends on its composition. The higher the carbon and alloy content of the steel, the greater is its tendency to embrittlement. The following *Carbon Equivalent* formula is used as an empirical measure of this property:

$$\text{Carbon Equivalent (CE)} = \%C + \frac{\%Mn}{6} + \frac{\%Ni + \%Cu}{15} + \frac{\%Cr + \%Mo + \%V}{5}$$

   As a very general rule, once the carbon equivalent reaches 0.41% and/or the plate thickness reaches 30 mm there is a significant chance of hydrogen cracking unless special precautions are taken.
2. *Hydrogen level*. Irrespective of the composition of the parent material, electrodes must always be properly dried and the joint must be free of contamination. Above the carbon equivalent/thickness combination given above, basic (controlled hydrogen) electrodes should be used.
3. *Cooling rate*. This parameter influences susceptibility to cracking in two ways. The slower the rate of cooling, the less is the quenching effect and consequently the less is the degree of embrittlement. In addition, a slower cooling rate gives more time for the excess hydrogen to diffuse out of the weld and HAZ as its solubility diminishes.

   In the absence of any preheat, cooling rate primarily depends on the ratio between the heat input and the heat loss. The former is simply a function of arc energy and the latter depends on the size of the path for the dissipating heat. As the sum of the thicknesses of the plates down which heat can flow increases, so does the cooling rate. Preheat can be used to reduce the temperature difference between the HAZ and the surrounding steel and thus reduce the cooling rate. (It also increases the hydrogen solubility, thereby reducing the amount of hydrogen coming out of solution.)
4. *Restraint*. As discussed in Section 2.5, any weld will attempt to shrink as it cools and the degree of restraint will influence the strain that the embrittled HAZ has to undergo as this shrinkage occurs. The greater the restraint, the larger will be the local strain and the greater the tendency for a crack to develop.

2.7.7 Solidus, weld metal solidification or hot cracking

This form of defect is usually a longitudinal crack down the centre of the weld which forms shortly after the weld metal has solidified. Typical examples are shown in Figure 2.12(b). Because of heat flow patterns, this central portion of the weld pool is the last to solidify. Since most impurities have lower melting points than steel, they can collect in this region and form semi-continuous films of segregates along the grain boundaries. As the weld continues to cool after solidification it attempts to contract and may crack at these regions of weakness.

Note that the molten electrode material in the weld pool has been diluted by the drawing in of parent metal. Composition is the most important parameter in the control of this form of cracking. Thus impurities, of which sulphur and, to a lesser extent, phosphorus are the most important, have to be minimized in both the electrode and parent metal compositions. As with HAZ cracking, degree of restraint is also a significant parameter, as thermal shrinkage in the weld increases with degree of restraint.

As structural steels become cleaner, this form of cracking is becoming less significant. It is most likely to develop with deep-penetration submerged arc welding because of the high dilution. If it occurs in this situation, a change of weld procedure to more, smaller, weld passes, with lower arc energy and consequently less penetration, should alleviate the problem. When manual welding high-sulphur steels, basic electrodes with controlled hydrogen should be used because they have an inherently better ductility than other electrodes. Alternatively, high-manganese electrodes may be used because the manganese sulphide that is formed has a much higher melting point than ferrous sulphide and therefore cannot collect at the centre of the weld.

2.7.8 Lamellar tearing

This is caused by defects arising from the rolling process of the parent steel and the shrinkage strains after welding, and it has considerable practical significance in connection design.

As steel is rolled during manufacture, small

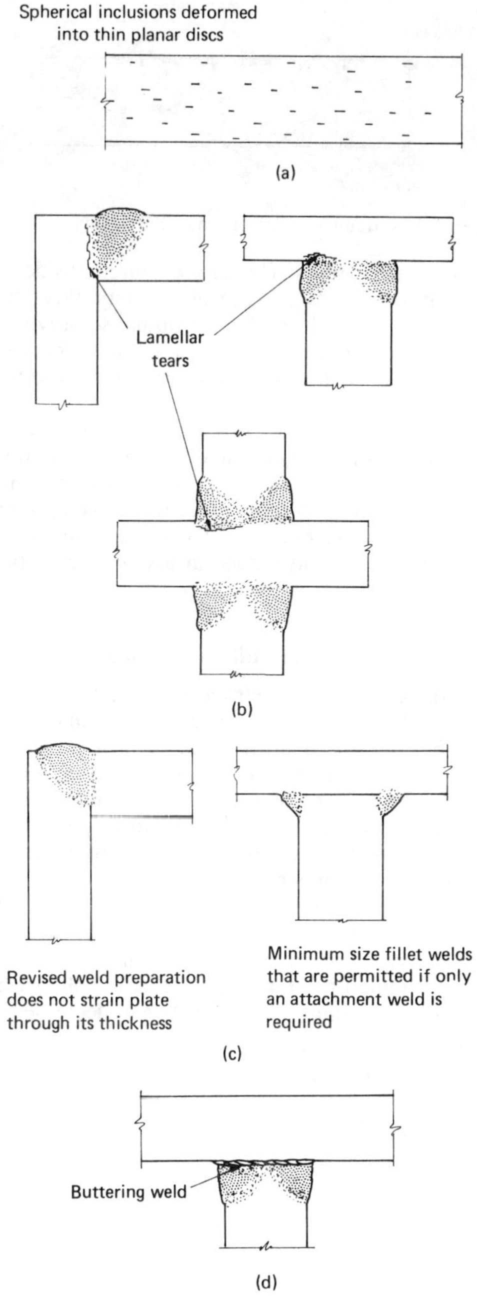

Figure 2.13 Causes and prevention of lamellar tearing

spheroidal particles of impurities become elongated into disc-shaped elements, as shown in Figure 2.13(a). These discs reduce the through-thickness strength and ductility of the steel compared to its properties in other directions.

If, as shown in Figure 2.13(b), such plates are incorporated into welded connections where they are subjected to significant through-thickness strains from welding shrinkage in the presence of restraint then lamellar tearing may occur. This defect is characterized by a stepped form of crack as the fracture progresses from one inclusion to the next at a different level. They are usually entirely sub-surface and therefore not detectable by visual inspection.

The incidence of lamellar tearing can be reduced by attention to detailing; Figure 2.13(c) shows revised versions of some details which are less susceptible to tearing. However, in some joints severe through-thickness strains are unavoidable, and in these situations two preventative measures are available. The traditional approach, as shown in Figure 2.13(d), is to butter the plate prior to welding with low-strength and high-ductility weld metal which can accommodate the strains because it is free from inclusions. The second approach is to specify steels that have guaranteed through-thickness properties. These exceptionally clean steels, developed to overcome the lamellar tearing problems of offshore construction. have differing levels of through-thickness ductility to suit the varying degrees of restraint and shrinkage. The ductility is measured by monitoring minimum percentage reduction in cross-sectional area when a tensile coupon, with its axis normal to the plate surface, is tested to failure. However, it should be noted that these steels are considerably more expensive than other material and their usage should be limited to situations with a significant likelihood of lamellar tearing and where structural integrity is essential for safety.

2.8 Fitness for purpose and the specification of weld repairs

The optimum welded joint is one that fulfils the requirements of the structure during its service life and which is produced at minimum cost. Concepts of fitness for purpose, taking account of the latest advances in fracture mechanics, have an important role in attempting to achieve this optimum. When a defect is found in a weld that is outside specified tolerances it is prudent to assess the significance of that defect in its particular location and only to proceed with a repair if it is clear that it is unacceptable. Repairs are always expensive, and can be very difficult to carry out properly on a completed structure because of difficulties of access and increased restraint. A poor repair is very likely to reduce the structural integrity due to the inevitably increased distortion and possibly increased cracking that occur in its execution.

In this context it is interesting to quote a recent study of welded ships' structures.[21] Non-destructive testing in ships involves spot checking and non-acceptable defects are automatically remedied. Because the inspection was less than 100% it was estimated that about 2000 major (i.e. planar) internal defects remained in each of the six ships studied after construction. A number of planar defects were therefore likely to be present in butt welds in parts of primary importance. The ships were surveyed after 4–6 years' service, and no reported damage was found to be related to an internal defect. On the other hand, some cracks found in service were related to repairs made during construction.

2.9 Weld inspection and non-destructive testing (NDT)

2.9.1 Visual inspection

The most fundamental form of weld checking is by visual inspection, and this can be used to check:

1. Quality of weld preparation and fit-up;
2. Root pass cracks;
3. Weld pass alignment in a multipass weld;
4. Cleanliness between passes;
5. Undercut, porosity and surface profile;
6. Final weld geometry.

2.9.2 Dye penetrants

The penetrant is a coloured liquid of very low viscosity that is drawn into any surface defect by capillary action. It is sprayed onto the weld and, after a brief interval for penetration, excess fluid is carefully removed. A white 'developer', which is either a dry powder or a powder suspended in a volatile solvent, is then sprayed onto the weld and surrounding material. This will draw the penetrant from any defects into which it has penetrated, clearly indicating their presence. Wide cracks produce a spread of penetrant while sharp ones often appear as a series of dots in a line. Rounded surface defects are also clearly visible.

This is a cheap and moderately quick method of inspection that does not require highly trained inspectors. Its principal disadvantage is that it can only detect surface defects.

2.9.3 Magnetic particle inspection

This technique monitors the leakage in magnetic flux that occurs in the presence of any surface or near-surface defects when the parent body is magnetized, as shown in Figure 2.14. The surfaces are cleaned and then coated with a white, flat, quick-drying paint. Magnetizable powder or ink is applied to the surface and a magnetic flux is induced in the parent body by magnets, magnetizing coils or

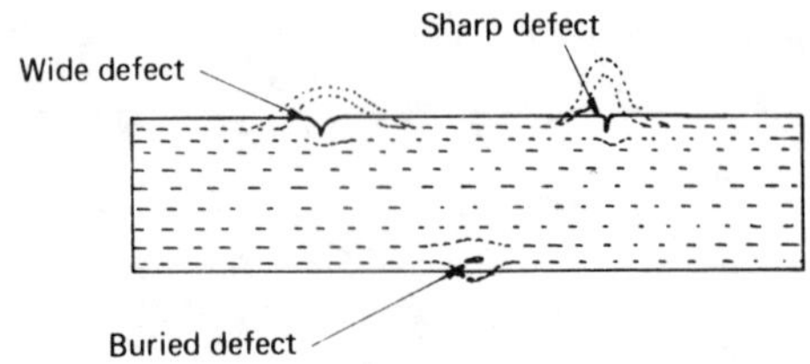

Figure 2.14 Influence of defects on magnetic flux

magnetizing currents. The dark-coloured powder or ink is drawn to any leakage of magnetic flux, thus highlighting the defects. For maximum sensitivity, a range of flux directions should be used because cracks will only show up if they cut across the flux lines.

This is a cheap and quick method of inspection that also does not require highly trained inspectors. The magnetizing system can be rather cumbersome for site use, and it is more amenable to component testing on an assembly line. Its principal disadvantage is that it can only detect surface or near-surface defects.

2.9.4 X-ray and gamma-ray radiography

Both these forms of electromagnetic radiation pass more readily through substances of lower than those of higher density. Thus if a weldment is subject to either form of radiation there will be a higher emission in regions where there is a defect for a significant proportion of the radiation path. This variation can be recorded by means of a photographic film, as shown in Figure 2.15.

It is necessary to vary the intensity and wavelength of the emissions in order to achieve optimum penetration of the weldment. With X-rays this is obtained by modifying the tube voltage; an increase will magnify the intensity and decrease the

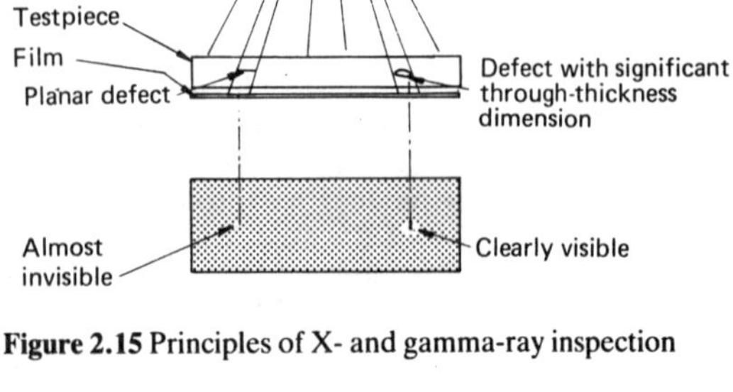

Figure 2.15 Principles of X- and gamma-ray inspection

wavelength. This gives a harder X-ray with greater penetration. Gamma-rays generally have shorter wavelengths than X-rays, giving greater penetration. Different sources have different energies and wavelengths; the lower the energy of the source, the better will be the film contrast. A popular source is iridium-192, which gives a very high-quality negative for plate thicknesses of 12–60 mm.

It is necessary to check the sensitivity of both systems during operation. This is carried out by placing an image quality indicator (IQI) on the surface of the weldment being inspected. A common type of IQI consists of wires of different diameters mounted side by side in a polythene tag. The sensitivity of the particular radiograph is defined as:

$$\frac{\text{Smallest diameter of wire that is visible}}{\text{Thickness of weldment being radiographed}} \times 100\%$$

Both systems are much more sophisticated than the techniques described earlier. They require fully trained operators, complex equipment and elaborate safety precautions to avoid health risks from the radiation. Their principal advantage is that they enable subsurface defects to be detected. However, their sensitivity to defects that do not have a significant dimension in the through-thickness direction is poor; the physical constraint of having to arrange the radiation source opposite the film can create difficulties with complex geometries.

X-rays have the advantages that the intensity of radiation can be varied, that it is more sensitive and that, since the source can be switched off, it is only necessary to take safety precautions during the exposure time rather than be concerned with continuous heavy shielding. Gamma-rays have the advantages that the equipment is smaller, cheaper and more mobile than the X-ray unit and that these rays have greater penetration than X-rays and can be used for thicker sections.

2.9.5 Ultrasonic inspection

This inspection technique is based on the reflectivity of ultrasonic waves within the weldment from boundaries in the steel. The quantities measured are the time taken for a pulse echo to return to the receiver and the strength of the echo. Output is measured on a cathode ray tube. The pulse generator produces a pulse which is fed to the transmitting crystal and which transmits a sound wave through the steel; at the same time the pulse transmits a signal on the tube. Some of the sound wave is reflected back by the defect or boundary and is transmitted back to the receiver probe, which transmits another signal to the tube. Because the sound waves take time to travel to and from the defect or boundary, when the resultant electrical signal reaches the tube it is displaced along the *X*-axis or time base by a distance proportional to the distance it has travelled in the steel. The time base can be varied and is preset to give different depth or distance ranges. The amplitude of the signal is proportional to that of the received sound.

Two modes of operation are employed. Where, as in Figure 2.16(a), the pulse direction is normal to the surfaces (compression wave probe) it is possible to locate the depth of any defect directly because the system is self-calibrating. That is, there will be separate echoes from the defect and the far boundary and simple proportion may be used to determine defect depth. Where the pulse is at an angle to the surface (shear wave probe) there will be

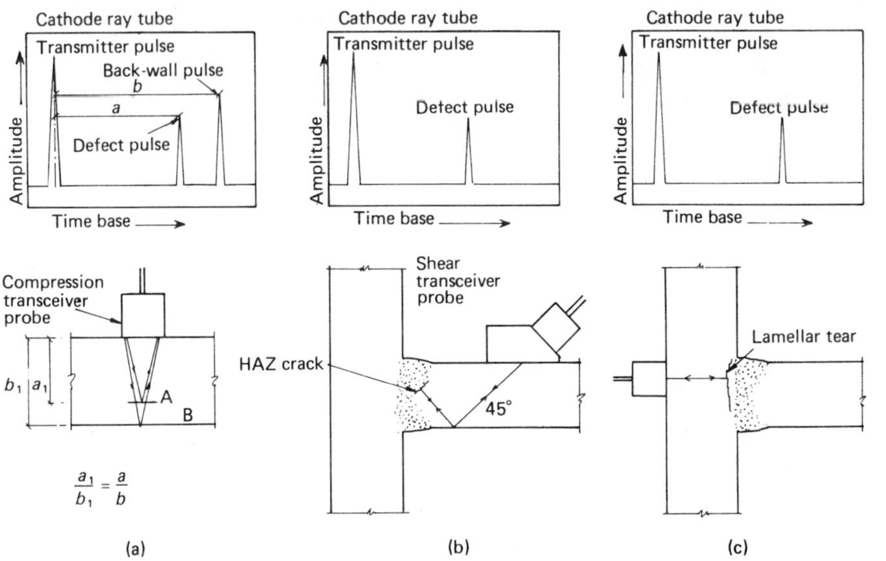

Figure 2.16 Examples of ultrasonic defect detection. (a) Planar defect in plain plate; (b) HAZ crack in Tee-butt weld; (c) lamellar tear in Tee-butt weld

no echo from the far face and any echo will therefore be evidence of a defect, as shown in Figure 2.16(b). The same effect can sometimes be achieved with a compression wave probe, as shown in Figure 2.16(c). In many cases different probe directions will have to be used to examine different parts of the same weld.

With ultrasonic inspection it is possible to search for internal defects in a wide variety of joints; the equipment is portable and safe to use and the method can detect all the most common defects found in welding, with adequate sensitivity. The main disadvantages are that the system relies completely on the skill and integrity of the operator; there are no permanent records; defect sizing may not be very accurate; and the system may be so sensitive that defects are found that are too small to be deleterious to the structure.

Bibliography and References

Welding processes in structural engineering

1. Pratt, J. L., *Introduction to the Welding of Structural Steelwork*, The Steel Construction Institute, Ascot, 1979.
2. Gray, T. G. F., Spence, J. and North, T. H., *Rational Welding Design*, Butterworths, London, 1975.
3. Houldcroft, P. T., *Welding Process Technology*, Cambridge University Press, Cambridge, 1977.
4. AWS, *Welding Handbook*, 8th edn, 1986: Volume 1, Welding Technology; Volume 2, Welding Processes; Volume 3, Materials and Applications (distributed by Macmillan, London).
5. Prior, H., *et al.*, 'Welding with self-shielded flux cored wire (Scottish branch sponsored meeting)', *Metal Construction*, August, **18**, No. 8 (1986).
6. Allum, C. J., 'MIG welding – time for a reassessment', *Metal Construction*, June, **15**, No. 6 (1983).
7. Kennedy, N., 'Narrow gap submerged arc welding of steel, Part 1', *Metal Construction*, November, **18**, No. 11 (1986).
8. Pumphrey, W. I. and McKeown, D., 'Arc welding processes and consumables in the UK 1985–90', *Metal Construction*, July, **17**, No. 7 (1985).

Welding fluxes and electrode classification

9. BS 639: 1986, Specification for Covered Carbon and Carbon Manganese Steel Electrodes for Manual Metal Arc Welding, British Standards Institution.
10. Davis, L., *An Introduction to Welding Fluxes for Mild and Low Alloy Steels*, The Welding Institute, Abington, Cambridge, 1981.

Weld preparations

11. BS 5135: 1984, Specification for Process of Arc Welding of Carbon and Carbon Manganese Steels, British Standards Institution.

Control of distortion

See also reference 2.

12. *Control of Distortion in Welded Fabrications*, 2nd edn, The Welding Institute, Abington, Cambridge, 1968.
13. Burdekin, F. M., 'To control distortion, keep it flat, keep it straight', *Metal Construction*, October, **13**, No. 10 (1981).

Preheating

See reference 11.

Weld defects (and acceptance criteria)

See also references 4 and 11.

14. PD 6493: 1980, Guidance on Some Methods for the Derivation of Acceptance Levels for Defects in Fusion Welded Joints, The British Standards Institution.
15. Bailey, N., *Faults in Fusion Welds in Constructional Steels*, The Welding Institute, Abington, Cambridge, 1986.
16. Jubb, J. E. M., 'Undercut or toe groove – the Cinderella defect', *Metal Construction*, February, **13**, No. 2 (1981).
17. Farrar, J. C. M. and Dolby, R. E., *Lamellar Tearing in Welded Steel Fabrication*, The Welding Institute, Abington, Cambridge, 1972.

Fitness for purpose and specification of weld repairs

18. Wells, A. A., 'Fitness for purpose and the concept of defect tolerance', *Metal Construction*, November, **13**, No. 11 (1981).
19. Burdekin, F. M., 'Some defects do – some defects don't', *Metal Construction*, February, **14**, No. 2 (1982).
20. 'The use (and abuse) of CTOD', *Metal Construction*, May, **14**, No. 5 (1982).
21. Johansson, B. G., Kjellander, S. L. and Leide, N. G., 'Is it good enough? Tolerances for welded ships' structures', *Metal Construction*, March, **14**, No. 3 (1982).

Weld inspection and non-destructive testing

22. Halmshaw, R., *Non-destructive Testing of Welded Joints*, The Welding Institute, Abington, Cambridge, 1988.
23. IIW, *Handbook on the Ultrasonic Examination of Welds*, The Welding Institute, Abington, Cambridge, 1977.
24. IIW, *Handbook on Radiographic Approaches and Techniques*, The Welding Institute, Abington, Cambridge, 1974.

3

Bolts and bolting, rivets and riveting

3.1 Scope

A full coverage of the topics covered in this chapter is inappropriate in a general book on connection design, and the presentation is therefore limited to those aspects that are of greatest importance to the designer. A bibliography is provided at the end of the chapter for the reader who wishes to pursue a topic in greater detail. A list of references is given for each of the main headings in the text. The order within each list indicates the relative standard of the references; more general and elementary texts are listed first, followed by more advanced ones.

3.2 Dowel bolts

3.2.1 General description

Dowel, ordinary or bearing bolts are the most economic and widely used type of structural fastener. They transfer shear loading directly by bearing between the bolt and the internal surfaces of the holes in the plates in conjunction with shearing on the bolt, as shown in Figure 3.1. They can also be used to resist static tensile loading.

Generally, they are used in clearance holes whose diameter is 2 mm greater than the nominal bolt diameter. Traditionally, they were used with a single washer under the nut, but many authorities will now permit this to be omitted. They are usually tightened by hand to a 'spanner-tight' condition.

3.2.2 Mechanical properties

These bolts, nuts and washers are now manufactured to common international standards, although national standards are generally used as implementation documents.

Bolts

The mechanical properties of the bolts are defined by the ISO strength grading which takes the form X·Y. The first figure (X) is one-tenth of the minimum ultimate tensile strength of the material in kg/mm^2; the second is one-tenth of the percentage of the ratio of the minimum yield stress to the minimum ultimate tensile strength. As indicated in Table 3.1, other significant mechanical properties are defined for a particular strength grade.

Nuts

As shown in Table 3.2, the mechanical properties of the nuts are less rigorously defined than those of the bolts. The ISO strength grading for nuts is simply related to the required proof stress.

Note that the proof stress referred to does not develop in the nut itself. It is the minimum stress on the tensile area of the associated bolt that the nut must be capable of developing during proof testing. Comparison of Tables 3.1 and 3.2 shows that the

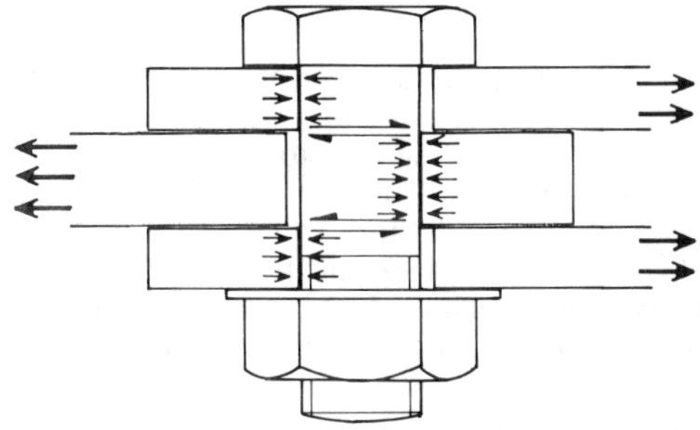

Figure 3.1 Bearing bolt in shear

Table 3.1 Mechanical properties of structural bearing bolts

Mechanical property			Strength grade designation 4.6	8.8	10.9	12.9
Tensile strength	(min.)	kgf/mm^2	40	80	100	120
		N/mm^2	392	785	981	1177
	(max.)	kgf/mm^2	55	100	120	140
		N/mm^2	539	981	1177	1373
Vickers hardness	(min.)		110	225	280	330
	(max.)		170	300	370	440
Yield stress	(min.)	kgf/mm^2	24	–	–	–
		N/mm^2	235	–	–	–
Stress at permanent set limit R_{02}	(min.)	kgf/mm^2	–	64	90	108
		N/mm^2	–	628	883	1059
Stress at proof load		kgf/mm^2	22.6	58.2	79.2	95.0
		N/mm^2	222	571	777	932
Elongation after fracture (min %)			25	12	9	8
Charpy impact strength	(min.)	kgfm/cm^2	–	6	4	3
		ft lbf	–	22	14	11

Table 3.2 Mechanical properties of nuts for structural bearing bolts

Strength grade designation		*4*	*8*	*12*
Proof load stress	kgf/mm^2	40	80	120
	N/mm^2	392	785	1177
Vickers hardness (max.)		310	310	370

required nut proof stress is equated with the minimum ultimate tensile strength of the bolt. Thus, in practice, if the bolt strength is greater than the minimum specified it is possible for either the nut or bolt to fail when the combination is tested to failure in tension.

Washers

Where they are used, washers will usually be of mild steel, even if high-tensile bolts are used. However, where such bolts are subject to high-tensile loads, mild steel washers will distort considerably under the high local compressions from the nut and bolt head. (The washers lack the triaxial containment and continuity which enables the equally soft plate material to resist these forces without significant distortion.) In such circumstances consideration should be given to either dispensing with the washers altogether or to specifying high-tensile washer material.

3.2.3 Bolt, nut and washer geometry

Overall geometry is also controlled by ISO recommendations. Bolt sizes are defined in terms of nominal diameter, length under the head and thread length. Washer and nut sizes are defined by nominal bolt diameter. Table 3.3 summarizes the most important dimensions for the common structural sizes and also illustrates the identification marks for both the bolts and nuts. Table 3.4 shows the preferred sizes for both common grades of bearing bolts.

3.2.4 Thread profile and tolerances

The basic ISO thread profile that is used for all structural nuts and bolts is shown in Figure 3.2(a). Clearance has to be provided between the bolt and nut threads. As shown in Figure 3.2(b), this is achieved by a combination of fundamental deviations and tolerances. The former define the closest that the nut and bolt may approach the nominal interface and the latter the additional clearances that may be permitted in order to accommodate practical manufacturing variations. Both sets of quantities are defined in terms of diameter. Thus the minimum total clearance between corresponding dimensions on the nut and bolt is the sum of their fundamental deviations. The maximum total clearance is obtained by adding both the nut and bolt diameter tolerances to the minimum total clearance.

As indicated in Table 3.5, different classes of thread tolerance are specified for the different types

Table 3.3 Bolt, nut and washer geometry and identifying marks

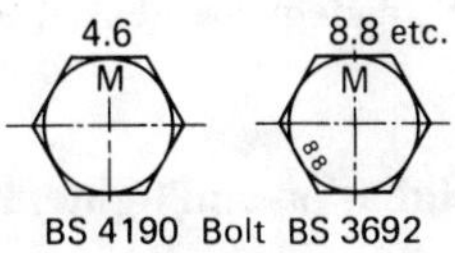

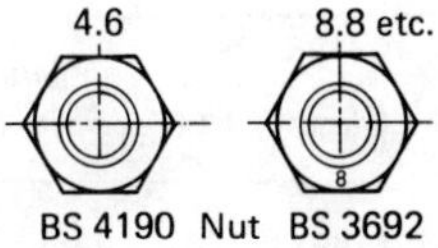

ISO Metric coarse threads	*(M12)*	*M16*	*M20*	*(M22)*	*M24*	*(M27)*	*M30*	*(M33)*	*M36*
Pitch (mm)	1.75	2.00	2.50	2.50	3.00	3.00	3.50	3.50	4.00
Tensile stress area (mm^2)	84.3	157	245	303	353	459	561	694	817
Basic effective diameter (Pitch diameter) (mm)	10.863	14.701	18.376	20.376	22.051	25.051	27.727	30.727	33.402
Length of threads									
BS 4190 Up to and inc. 125 mm	30	38	46	50	54	60	66	72	78
and Over 125 mm up to and inc. 200 mm	36	44	52	56	60	66	72	78	84
BS 3692 Over 200 mm	49	57	65	69	73	79	85	91	97
BS 4190 Up to and inc. 125 mm		24	30	33	36	40			
(Short thread length)	(Short thread lengths may also be available for BS 3692 bolts – to special order)								
Dimensions									
Max. width across flats	19.0	24.0	30.0	32.0	36.0	41.0	46.0	50.0	55.0
Max. width across corners	21.9	27.7	34.6	36.9	41.6	47.3	53.1	57.7	63.5
Nominal head depth of bolts	8.0	10.0	13.0	14.0	15.0	17.0	19.0	21.0	23.0
Nominal depth of nuts	10.0	13.0	16.0	18.0	19.0	22.0	24.0	26.0	29.0
Nominal washer thickness	2.5	3.0	3.0	3.0	4.0	4.0	4.0	5.0	5.0
Washer ext. diameter	24	30	37	39	44	50	56	60	66
Washer int. diameter	14	18	22	24	26	30	32	36	39

Table 3.4 Manufacturers' recommended and preferred range of bolt sizes

	Length (mm)																								
Dia.	*25*	*30*	*35*	*40*	*45*	*50*	*55*	*60*	*65*	*70*	*75*	*80*	*90*	*100*	*110*	*120*	*130*	*140*	*150*	*160*	*180*	*200*	*220*	*260*	*300*
M12	x	x	x	x p	x p	x p	x p	x p	x p	x p	x p	x p	x p	x p	x p	x p	p	x		x	x	x	x	x	x
M16		s	s	s	sp	xsp	xsp	xsp	xsp	xsp	xsp	xsp	xsp	xsp	x p	x p	x p	x p	p	x p	x	x	x	x	x
M20				s	s	s	sp	xsp	xsp	xsp	xsp	xsp	xsp	xsp	x p	x p	x p	x p	x p	x p	x	x	x	x	x
M24								s	p	xsp	p	xsp	xsp	xsp	x p	x p	p	x p	p	x p	x	x	x	x	x

x: Grade 4.6 standard thread length.
s: Grade 4.6 short thread length.
p: Grade 8.8.

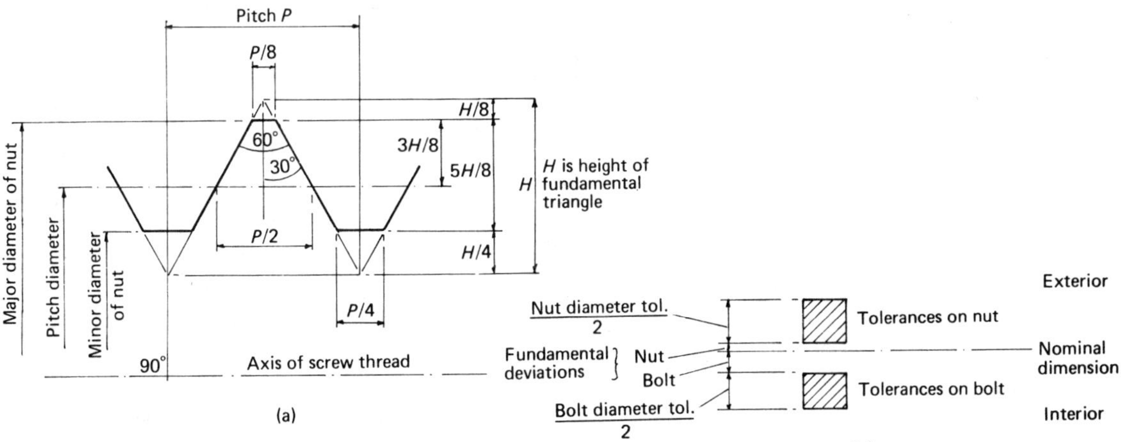

Figure 3.2 (a) Basic profile of ISO metric thread; (b) tolerance positions for nut and bolt threads

Table 3.5 Thread tolerances

Specification	*Class of fit*	*Tolerance class* *Bolts and screws*	*Nuts*
BS 4190	Free	8g	7H
BS 3692	Medium	6g	6H

Threads will be in accordance with BS 3643: Part 2.

of structural bolt. Table 3.6 lists the basic data for these classes and Table 3.7 presents the resulting maximum and minimum dimensions for the common structural bolts.

It should be appreciated that, in specifying these tolerances on thread profile, a careful compromise has been achieved between ease of manufacture and structural soundness. The overall geometry and tolerances have been developed after considerable research and have been carefully optimized. Fasteners which depart from permitted tolerances should not be used in structural connections because, as discussed in detail in Chapter 5, such a departure can cause undesirably abrupt modes of failure. If there is any doubt about thread geometry the practical methods of measurement outlined in Section 3.7 should be adopted.

A particular problem exists with coated threads that is most severe where galvanizing is used, because of the coating thickness. Only one component (usually the bolt) is coated to avoid the cold welding problems referred to in Section 3.3.2. In the UK the bolts are overcut to accommodate the coating. In some other countries, notably the USA, the nut is overcut. It is clearly essential to monitor supply to ensure that only one overcut component is used, otherwise inadequate tensile capacity will result.

3.2.5 Bolt installation and tightening

No special provision is necessary for bolt installation. For predominantly static loading the bolts need only be 'spanner-tight'. If subject to fluctuating tensile loads the bolts should be tightened to at least 70% of their proof load; this is to ensure that the bolt is 'protected' from significant fluctuations in load by its preload, as discussed in Section 6.2.3. Bearing bolts should not be used for shear loads that are subject to reversal.

3.3 HSFG bolts

3.3.1 General description

HSFG bolts are designed to be used in connections where shear loads are transmitted by friction between the plies, as shown in Figure 3.3. The bolts are tightened to their proof load in order to develop high bearing forces between the plies. Frequently the contact surfaces are specially prepared to improve the coefficient of friction. Even though these bolts are used in clearance holes, deformations under shear loading are very small until the slip load is reached.

Special provision is necessary to ensure that the bolts are tightened reliably to their proof load. Because of the high bolt tension, hardened washers should be used under the element which is rotated during tightening.

Table 3.6 Principles and basic data for ISO metric screw threads

		Tolerance class	*Pitch* *1.75*	*2.0*	*2.5*	*3.0*	*3.5*	*4.0*
	Fundamental deviations Nut	H	0	0	0	0	0	0
	Bolt	g	0.034	0.038	0.042	0.048	0.053	0.060
Crest	Nut minor diameter tolerance	6	0.335	0.375	0.450	0.500	0.560	0.600
Crest		7	0.425	0.475	0.560	0.630	0.710	0.750
Crest	Bolt major diameter tolerance	6	0.265	0.280	0.335	0.375	0.425	0.475
Crest		8	0.425	0.450	0.530	0.600	0.670	0.750
Pitch	Nut pitch diameter tolerance	6	0.200	0.212	0.224	0.265	0.280	0.300
Pitch		7	0.250	0.265	0.280	0.335	0.355	0.375
Pitch	Bolt pitch diameter tolerance	6	0.150	0.160	0.170	0.200	0.212	0.224
Pitch		8	0.236	0.250	0.265	0.315	0.335	0.355
	Root contours – May not transgress basic profile – Grade 8.8 bolts and above shall have non-reversing curvature; no part of which shall have a radius of less than		0.219	0.250	0.313	0.375	0.438	0.500
	Deviations on bolt minor dia. for stress calculations	6 & 8	−0.287	−0.327	−0.403	−0.481	−0.558	−0.637

All dimensions in millimetres.

Table 3.7 ISO metric screw threads: limits and tolerances for finished uncoated threads

Nom. dia.	Pitch	Tol. class	Fund. dev.	External threads – bolts, screws: Major diameter (Max.)	Major diameter (Tol.)	Major diameter (Min.)	Pitch diameter (Max.)	Pitch diameter (Tol.)	Pitch diameter (Min.)	Minor dia. (min.)	Tol. class	Fund dev.	Internal threads – nuts: Major dia. (min.)	Pitch diameter (Max.)	Pitch diameter (Tol.)	Pitch diameter (Min.)	Minor diameter (Max.)	Minor diameter (Tol.)	Minor diameter (Min.)
12	1.75	6g	0.034	11.966	0.265	11.701	10.829	0.150	10.679	9.602	6H	0	12.000	11.063	0.200	10.863	10.441	0.335	10.106
		8g	0.034	11.966	0.425	11.541	10.829	0.236	10.593	9.516	7H	0	12.000	11.113	0.250	10.863	10.531	0.425	10.106
16	2.00	6g	0.038	15.962	0.280	15.682	14.663	0.160	14.503	13.271	6H	0	16.000	14.913	0.212	14.701	14.210	0.375	13.835
		8g	0.038	15.962	0.450	15.512	14.663	0.250	14.413	13.181	7H	0	16.000	14.966	0.265	14.701	14.310	0.475	13.835
20	2.5	6g	0.042	19.958	0.335	19.623	18.334	0.170	18.164	16.624	6H	0	20.000	18.600	0.224	18.376	17.744	0.450	17.294
		8g	0.042	19.958	0.530	19.428	18.334	0.265	18.069	16.529	7H	0	20.000	18.656	0.280	18.376	17.854	0.560	17.294
(22)	2.5	6g	0.042	21.958	0.335	21.623	20.334	0.170	20.164	18.624	6H	0	22.000	20.600	0.224	20.376	19.744	0.450	19.294
		8g	0.042	21.958	0.530	21.428	20.334	0.265	20.069	18.529	7H	0	22.000	20.656	0.280	20.376	19.854	0.560	19.294
24	3.0	6g	0.048	23.952	0.375	23.577	22.003	0.200	21.803	19.955	6H	0	24.000	22.316	0.265	22.051	21.252	0.500	20.752
		8g	0.048	23.952	0.600	23.352	22.003	0.315	21.688	19.840	7H	0	24.000	22.386	0.335	22.051	21.382	0.630	20.752
(27)	3.0	6g	0.048	26.952	0.375	26.577	25.003	0.200	24.803	22.955	6H	0	27.000	25.316	0.265	25.051	24.252	0.500	23.752
		8g	0.048	26.952	0.600	26.352	25.003	0.315	24.688	22.840	7H	0	27.000	25.386	0.335	25.051	24.382	0.630	23.752
30	3.5	6g	0.053	29.947	0.425	29.522	27.674	0.212	27.462	25.305	6H	0	30.000	28.007	0.280	27.727	26.771	0.560	26.211
		8g	0.053	29.947	0.670	29.277	27.674	0.335	27.339	25.182	7H	0	30.000	28.082	0.355	27.727	26.921	0.710	26.211
(33)	3.5	6g	0.053	32.947	0.425	32.522	30.674	0.212	30.462	28.305	6H	0	33.000	31.007	0.280	30.727	29.771	0.560	29.211
		8g	0.053	32.947	0.670	32.277	30.674	0.335	30.339	28.182	7H	0	33.000	31.082	0.355	30.727	29.921	0.710	29.211
36	4	6g	0.060	35.940	0.475	35.465	33.342	0.224	33.118	30.654	6H	0	36.000	33.702	0.300	33.402	32.270	0.600	31.670
		8g	0.060	35.940	0.750	35.190	33.342	0.355	32.987	30.523	7H	0	36.000	33.777	0.375	33.402	32.420	0.750	31.670

All dimensions are in millimetres.

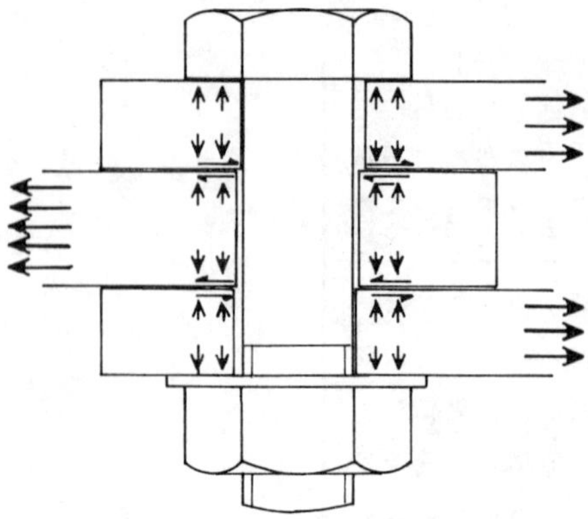

Figure 3.3 HSFG bolted connection in shear

These bolts may also be used in connections where they are subject to externally applied tension. Because of their high preload and rigidity under shear loading, they are very suitable for use in shear and/or tension under fluctuating loads and fatigue conditions.

3.3.2 Mechanical properties

The mechanical properties for bolts, nuts and washers are summarized in Table 3.8.

Bolts

Two classes of bolt material may be specified: general grade material which has mechanical properties similar to that of Grade 8.8 bearing bolts, and higher-grade material, which corresponds to Grade 10.9 with the ISO strength grading.

Nuts

Once again, nut strength is specified simply in terms of the proof stress which it must be capable of developing in the tensile section of the associated bolt. However, it is worth noting a significant variation between bearing and frictional nuts. With the former the nut proof stress was set equal to the minimum u.t.s. of the bolt material; with the latter it is set equal to the maximum likely u.t.s. Thus, unlike bearing bolts, nut failure is effectively precluded with HSFG bolts. This is in recognition of the possible abrupt nature of nut-related failure.

More care is also given to the relative hardness of the nut and bolt. Although not entirely achieved, it is notable that there is a practical attempt to ensure that nuts have a lower hardness than bolts. This separation of hardness is to reduce the tendency of the threads to lock-up by cold welding. If two surfaces of similar hardness are rubbed together at high pressure both will be cleaned by the abrasion and the surfaces may fuse together. With dissimilar hardness all the wear takes place on the softer surface and there is less tendency for cold welding.

Washers

The only property that is controlled is the hardness. Once again, in order to minimize the risk of cold welding, the washers are made appreciably harder than the nuts against which they bear.

3.3.3 Bolt, nut and washer geometry

Table 3.9 summarizes the geometry of the various types of bolts, nuts and washers.

Table 3.8 Mechanical properties for HSFG bolts, nuts and washers

Mechanical property		*General grade* $\ngtr$ *M24*	*General grade* > *M24*	*Higher grade*
Bolts				
Tensile strength (N/mm^2)		827	725	981
Vickers	min.	260	225	280
hardness	max.	330	292	380/370
Stress at permanent set limit $R_{0.2}$ (N/mm^2)		635	558	882
Stress under proof load (N/mm^2)		587	512	776
Elongation after fracture (min. %)		12	12	9
Nuts				
Proof load stress (N/mm^2)		1000	1000	1176
Vickers	min.	175	175	258
hardness	max.	310	310	370
Washers				
Vickers	min.	362	362	362
hardness	max.	440	440	440

Table 3.9 HSFG bolt, nut and washer geometry and identifying marks (Note that M12 are not recommended)

General grade to BS 4395: Part 1 — Higher grade to BS 4395: Parts 2 and 3

Hexagon bolt — Countersunk bolt — Nut — Hexagon bolt — Countersunk bolt (Part 2 only) — Nut

	Bolt diameter								
	M12	*M16*	*M20*	*M22*	*M24*	*M27*	*M30*	*M33*	*M36*
Thread lengths									
Bolt length up to 125 (Part 1)	30	38	46	50	54	60	66	72	78
Parts 2 and 3	36	44	52	56	60	66	72	78	84
125–200 mm (Part 1)	36	44	52	56	60	66	72	78	84
Parts 2 and 3	44	50	58	62	66	72	78	84	90
over 200 mm (Part 1)	49	57	65	69	73	79	85	91	97
Parts 2 and 3	54	62	70	74	78	84	90	96	102
Dimensions									
Width across flats	22	27	32	36	41	46	50	55	60
Width across corners	25.4	31.2	36.9	41.6	47.3	53.1	57.7	63.5	69.3
Nominal head depth of bolts	8	10	13	14	15	17	19	21	23
Nominal depth of nuts	11	15	18	19	22	24	26	29	31
Diameter of countersunk head	24	32	40	44	48	54	60	66	72
Diameter of waisted shank (Part 3 only)	–	12.0	15.2	17.1	18.3	21.1	23.3	26.2	–
Nominal washer thickness	2.6	3.2	3.5	4.0	4.0	4.0	4.0	4.4	4.4
Washer external diameter[a]	30	37	44	50	56	60	66	75	85
Washer internal diameter	13.6	17.6	21.3	23.2	26.2	29.2	32.6	35.6	38.6

[a]When required, washers may be clipped on one side.
Note that hardened square tapered washers are also available for use with tapered sections.

Bolts

Two principal types of bolt geometry may be used, either with parallel or with waisted shanks. The former are available with both material grades, the latter only with the higher-strength material. Countersunk bolts may also be used if a flush exterior surface is required. The hexagonal bolt head is the same depth as that for the corresponding ISO bolt but the across-flats dimension has been increased (wherever possible, to that of the next size up ISO bolt) in order to match the dimensions of the nuts.

Nuts

The nuts have both a greater depth and a greater across-flats dimension than those of the corresponding ISO bolts. These increases are necessary to achieve the greater strength required by the specifications for mechanical properties, in the presence of lower material hardness.

Washers

Standard washers are shown in Table 3.9. These have a considerably larger external diameter than the dimension across the corners of the nuts. Where the full washer cannot be accommodated (perhaps because of the proximity of a corner radius in a rolled section) it is permissible to clip the washer as indicated.

3.3.4 Thread profile and tolerances

The thread form and tolerances for all HSFG bolts are the same as those for ISO metric precision bolts, i.e. standard ISO coarse pitch threads with medium fit (6g for bolts and 6H for nuts). Details are given in Section 3.2.4.

Note that a long thread length is used for ductility. A sensible thread length should be within the grip (say, three threads for general grade and five for higher grade, parallel shank bolts). The influence of thread length on tensile behaviour is discussed in Section 5.3.

3.3.5 Bolt installation and tightening

At the time of assembly the contact surfaces in the connection must be free from any contamination which will reduce the coefficient of friction or would prevent solid seating of the parts. A washer must be placed under the bolt head or nut, whichever is rotated during tightening. The rotated bolt head or nut should always be tightened against a surface that is normal to the bolt axis – tapered washers may be necessary to achieve this with some rolled sections. These washers should also be used under the non-rotated surface if that surface is 3° or more from being normal to the bolt axis.

Each bolt in the bolt group has to be tightened to a minimum tension of its proof load for general grade bolts and 85% of its proof load for parallel shank, higher grade ones. The methods of tightening which are most widely used are as follows.

Part-turn of nut method

In essence, this method imposes an extension on the bolt by applying a prescribed turn of the nut. To be effective the nut must react against a solid base. It is therefore essential that, before applying any prescribed turn to the nut, all plies of the connection at all bolt positions are pulled into close contact, the 'snug-tight' condition. This is achieved by pretightening all the bolts in a group, in sequence, to refusal at the 'spanner-tight' condition. After making suitable marks so that the relative rotation between the bolt and nut may be observed, the nut is then subjected to an appropriate rotation. This is probably the most common method of tightening, and it is shown graphically in Figure 3.4. This method is not permitted by some Codes for higher grade bolts with parallel shanks because of concern over their limited ductility.

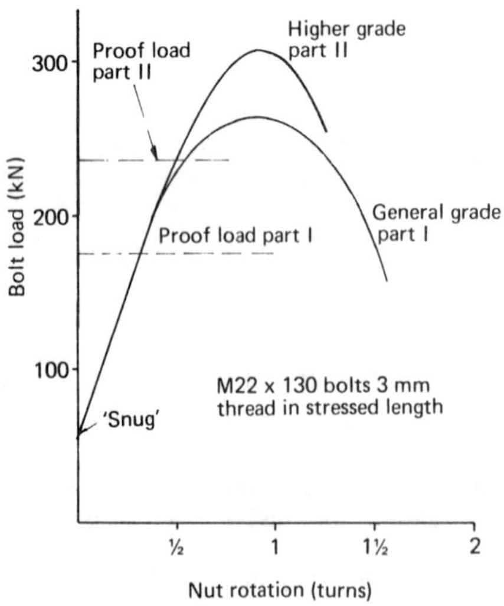

Figure 3.4 Induced tension/nut rotation relationships for parallel shank HSFG bolts

Part-torque, part-turn method

This is a variation of the part-turn method where the pulling up to the snug-tight condition is quantified by applying a specified torque to all bolts in the group in sequence to the point of refusal. It overcomes the problems that can arise with small-diameter bolts and overenthusiastic erectors, when it is quite possible with the turn-of-nut method for the bolts to be yielded during the snug-tightening sequence, and to be taken to the point of failure under the subsequent turn of the nut. This bedding torque is usually set at one-third or one-quarter of the estimated final torque. The turn of nut is also prescribed as a half-turn for M16 to M22 bolts for grips of up to 115 mm and M24 to M36 for grips of up to 160 mm, otherwise a three-quarter turn for grips up to 90 × thread pitch. This method is not permitted by some Codes for higher grade bolts with parallel shanks, because of concern over their limited ductility.

Torque control

Here each bolt in the group is subjected to a prescribed torque in sequence to the point of refusal. To overcome possible interaction between bolts if the plies do not readily draw up, the sequence should be repeated until all bolts refuse to turn further.

It should be appreciated that most of the torque effort (80–90%) goes to overcome friction in the threads. It follows that small variations in friction will lead to substantial variations in bolt load: great care is therefore necessary to achieve consistent results. The torque control device must be calibrated at least once per shift. Using bolts of the same diameter and batch, of similar length and with similar thread lubrication, the torque required to develop 1.1 × proof tension is determined. This setting is then used to torque the bolts. Where bolts of different diameters, batches or significantly different grip lengths are used, separate calibration is necessary for each class of bolt. Lubrication can be used both to reduce friction and improve consistency. The most consistent results have been obtained with bolts that have been carefully cleaned and then dipped in molten tallow. Even with careful execution, the coefficient of variation of bolt tension in a group is likely to be at least 15%.

This method is not popular, either with contractors or inspectors. Its use is likely to be limited to higher grade bolts with parallel shanks, where turn of nut is not permitted, and to exceptional situations where interaction between neighbouring bolts exists in practice because of some springiness in the system and where repeated 'elastic' tightening of the bolts is required. (A practical example of such an exceptional situation is a repair clamp on the tubular leg of an offshore jacket, where the clamp is designed not to close onto itself at full tension, to overcome variations in leg diameter.) Its principal advantage is that it does not cause yielding of either the bolt or any load-indicating device, thus permitting the repeated tightening referred to above.

3.3.6 Proprietary load-indicating devices

Various load-indicating washers and bolts are manufactured, and three proprietary makes are illustrated in Figure 3.5.

The first two of these devices indicate the load by producing a measurable change in some gap in the presence of local yielding. The third is tightened by turning the nut against the protruding nib until the latter shears off. (This has the advantage of not inducing any torsion in the main body of the bolt and is very simple to use. The nib shearing is only a measure of the torque applied to the nut and is therefore prone to the same variations in bolt tension described in the section on torque control above. However, given good quality of threads and good control of site cleanliness, these bolts should behave satisfactorily.) Both the gap changes and the nib shearing are, of course, irreversible and therefore these systems cannot indicate current bolt load. All they can show is that the particular bolt was once tensioned to its proof load or above. They should therefore not be used in situations where there is an interaction between neighbouring bolts. In any event, the connection should always be pulled up to its snug-tight condition before proceeding with final tightening, in order to minimize interaction between neighbouring bolt tensions. This method is becoming increasingly popular, largely because there is a simple permanent record of the tightening. However, for the reasons outlined above, adequate inspection is still required during tightening if reliable bolt tensions are to be achieved.

3.4 Rivets

3.4.1 General description

Rivets have become uneconomic and are now rarely used in structural steelwork. However, a brief description is included here, partly for completeness and partly to assist in any assessment of existing riveted structures.

Structural rivets are hot driven, being inserted in holes which are 2 mm (1/16 in) clearance on diameter. They give a satisfactory and rigid connection in shear. The hot forming of the head causes the shank of the rivet to expand, taking up the clearance on the hole diameter. As the rivet cools it contracts, producing a clamping force on the plies which further increases the connection rigidity. Rivets are considered to be less reliable in tension and designers have traditionally avoided subjecting them to such loading where possible.

3.4.2 Mechanical properties

Structural rivets are manufactured by hot or cold forging from mild or high-yield bar stock. The properties of undriven rivets therefore closely correspond to Grade 43 and Grade 50 steel sections.

3.4.3 Geometry

Figure 3.6 summarizes the principal types of rivet head. Snap head rivets are used where there is no restriction on clearance and universal or flat head rivets if clearance is limited. Countersunk rivets are used where a flush surface is required.

3.4.4 Rivet installation

Figure 3.7 illustrates the installation and head formation of rivets. The backing bar is required to provide a reaction block while the head formation is carried out, normally by pneumatic hammer.

3.5 Holding-down and foundation bolts

3.5.1 General description

Figure 3.8 shows the traditional types of holding-down bolt. The J-bolt and washer plate bolt are suitable for light construction and static loads. The J-bolt in particular becomes unwieldy to fix in lengths above 600 mm and diameters above 24 mm because of its lack of symmetry. These bolts are generally used in the spanner-tight condition. For heavy construction, tubed bolts should be used. Under static loading they may be used in the spanner-tight condition but they should be stressed to at least 70% of their proof load for fatigue conditions.

Low-alloy steel bolts are suitable where corrosion is unlikely to be significant and where the bolts are not subject to fatigue loading. In a corrosive environment appropriate corrosion resistant material should be used. In fatigue conditions, where the

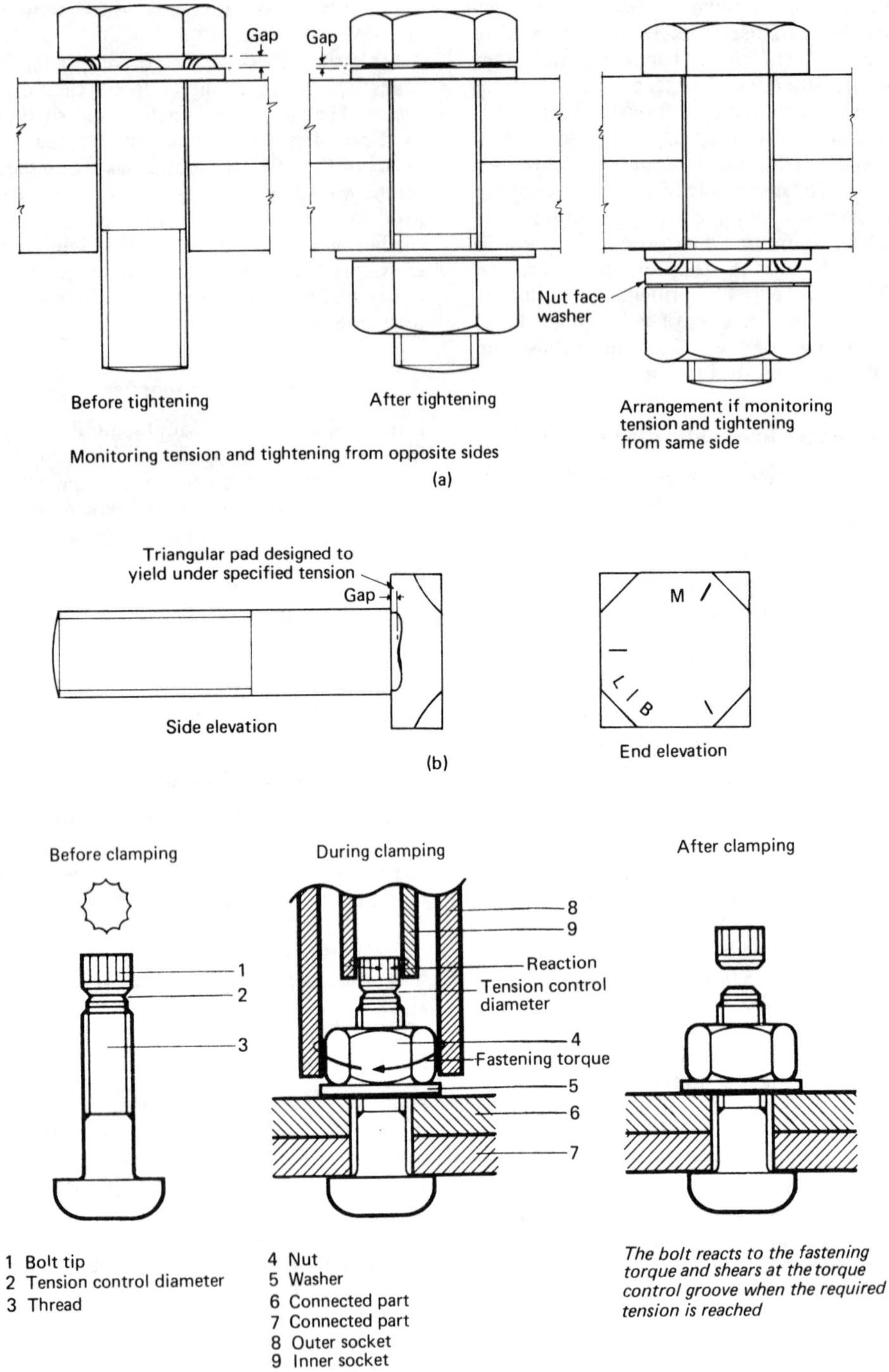

Figure 3.5 Proprietary load-indicating bolt systems. (a) Coronet load-indicating washer; (b) load-indicating bolt; (c) tension control bolt

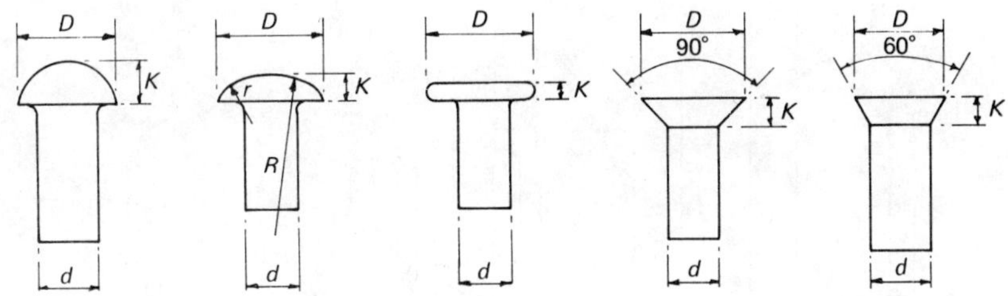

Head diameter					
Hot forged	$D = 1.6d$	$D = 2d$			$D = 1.5d$
Cold forged	$D = 1.75d$	$D = 2d$	$D = 2d$	$D = 2d$	
Head depth					
Hot forged	$K = 0.65d$	$K = 0.4d$			$K = 0.43d$
Cold forged	$K = 0.6d$	$K = 0.4d$	$K = 0.25d$	$K = 0.5d$	
Radii		$r = 0.6d$ $R = 3d$			
	Snap head rivets	Universal rivets	Flat head rivets	Countersunk rivets	Countersunk rivets

Figure 3.6 Standard rivet geometry

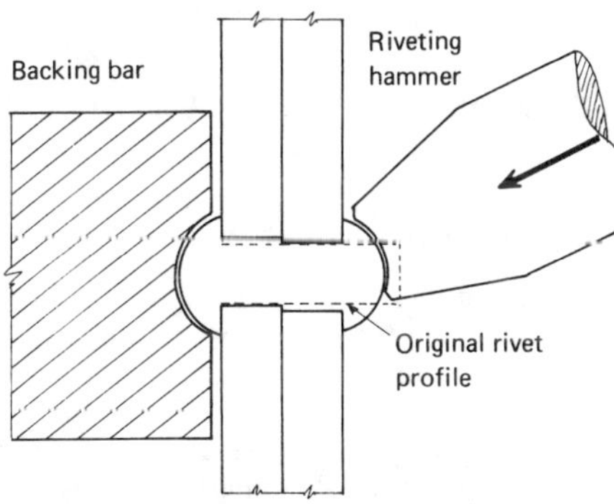

Figure 3.7 Schematic view of hot riveting

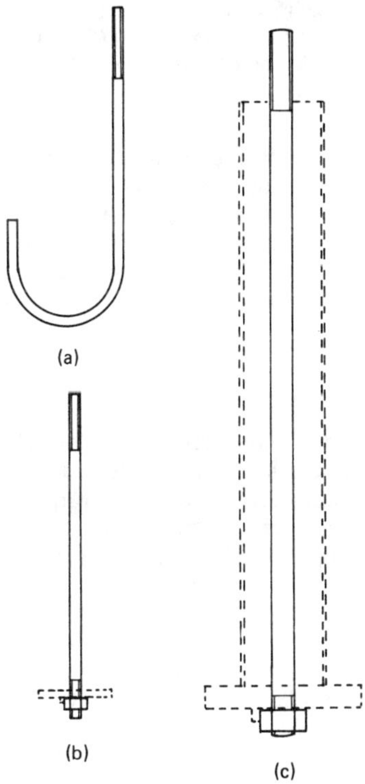

Figure 3.8 Cast-in-place holding-down bolts. (a) J-bolt; (b) washer plate bolt for light construction; (c) tubed bolt for heavy construction

bolt will be debonded from the concrete in some way for a considerable length, it would seem prudent always to specify an appropriate corrosion-resistant high-tensile steel.

Figure 3.9 shows examples of various types of foundation bolt. Types (a) to (d) are usually grouted into holes cast or drilled in the concrete, although they may be set in place prior to concreting. The grout may consist of either epoxy resin or cementitious mortar. Type (e) is an example of an expanding anchor bolt which is also set into a drilled hole. Tightening the bolt causes a wedging action to develop in the expanding outer barrel, gripping the concrete surface of the hole. Types (f) and (g) are examples of larger foundation bolts that are set in drilled holes. The block-ended bolt may be used where no horizontal tolerance in position is required and where shear forces may have to be resisted. The tube bolt provides for horizontal adjustment but cannot resist shear forces.

With the exception of type G, all these foundation bolts can resist both shear and tension. However, caution should be exercised in using such fasteners

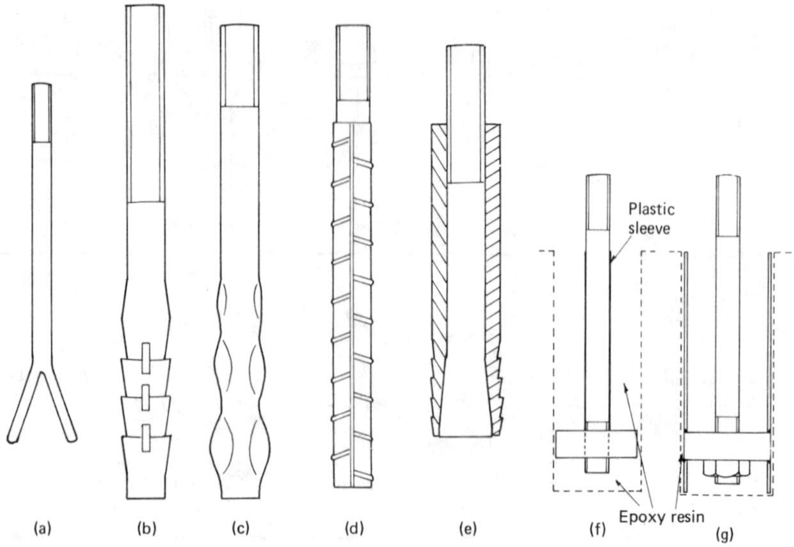

Figure 3.9 Examples of foundation bolts. (a) Split-end bolt (Cooper and Turner); (b) rag bolt (GKN); (c) indented bolt (GKN); (d) rebar bolt; (e) expanding bolt (Rawlplug); (f) block-ended bolt (Holst Wales); (g) tube bolt (Holst Wales)

to resist significant tensile forces, particularly under long-term loading. Most of these fasteners have inherently less tensile capacity than traditional holding-down bolts and are more likely to lose capacity in the presence of any local cracking in the concrete. This is discussed in greater detail in Section 11.8.

3.5.2 Mechanical properties

Low-alloy holding-down bolts are now generally specified by means of the ISO system referred to in Section 3.2.2. Corrosion-resistant bolts are generally supplied to manufacturer's specification.

The mechanical properties of proprietary systems can only be obtained from the supplier. Discretion must be used in translating laboratory results into design values which have to recognize the variations in workmanship that will exist in practice between laboratory and site conditions. This is particularly important for properties that depend on degree of compaction, surface texture of hole and cleanliness (for example, bond stresses between epoxy resin and concrete).

3.5.3 Geometry

The individual designer has considerable freedom to determine the geometry of holding-down and foundation bolt systems. Reinforced concrete design criteria should be used to determine required dimensions such as hook bolt radii, anchor plate sizes, etc.; thread geometry may be obtained from Section 3.2.3.

3.5.4 Installation and tightening

Considerable difficulty is experienced in practice with the positioning and concreting of traditional holding-down bolts. Almost all this difficulty can be traced back to an underestimate of the forces that a wet, vibrated, concrete mix can exert on the bolts and any supporting frame. Detailed guidance is beyond the scope of this book, but there are useful publications which discuss the matter in some depth. These books also give guidance on appropriate procedures for other types of anchor and foundation bolt.

Holding-down bolts usually pass through holes in the steelwork with 2 mm clearance on diameter. Consideration should be given to the specification of larger clearances in appropriate circumstances.

Where the holding-down bolts are not subject to fluctuating uplift forces they only need to be spanner-tight. However, in the presence of fatigue loading it is essential that bolts are sufficiently preloaded to ensure that they are protected from fluctuations in tensile stress, as discussed in Chapter 6. Loss in prestress due to concrete creep will be minimized by the use of high-tensile bars of sufficient length. In situations of severe fatigue loading it would be appropriate to specify a further tensioning after, say, 12 months, when most of the creep losses have occurred.

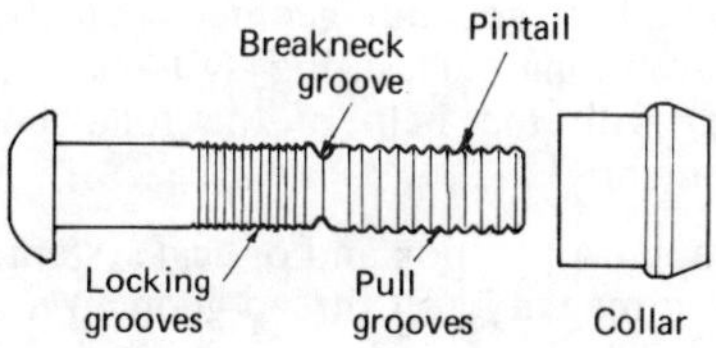

Figure 3.10 Huck high-tensile fastener system

3.6 Special fasteners

3.6.1 Huck bolts

The Huck bolt or lockbolt shown in Figure 3.10 is a special type of high-tensile fastener, with the special merit that it is considerably faster to install than a conventional torqued bolt. They are generally installed with a single-operation tightening system which cannot be repeated. Such a system is only suitable for use as a replacement to a conventional HSFG bolt in situations where the initial 'snug-tightening' can be dispensed with. If this condition is not satisfied it is possible, with care, to use a two-stage tightening procedure which will pull the plies together. Alternatively, bolts may be installed in some of the holes in the bolt group and pulled up to achieve the snug-tight condition for the bolt group. Huck bolts can then be installed in the remaining holes and the initial set of bolts finally replaced by Huck bolts. Unfortunately, this three-stage process diminishes the time saving which is the system's chief advantage. The tightening sequence is shown diagramatically in Figure 3.11.

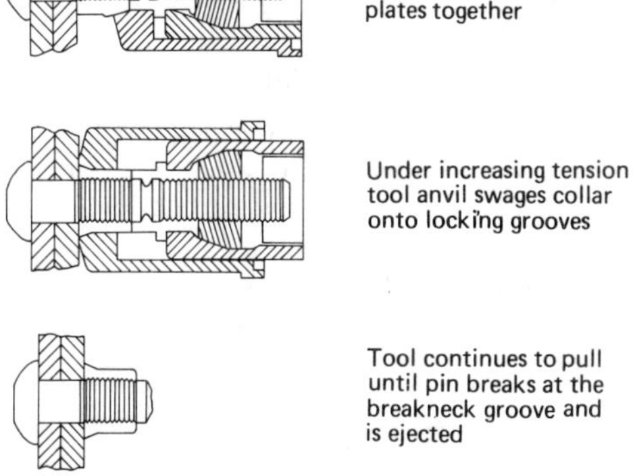

Figure 3.11 Tightening sequence for Huck fastener

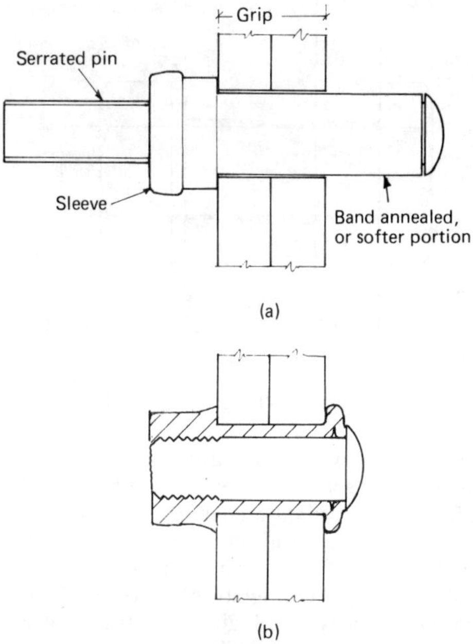

Figure 3.12 Huck BOM fastener. (a) Prepared for tightening; (b) installed fastener

3.6.2 Huck blind fastener

Figure 3.12 shows the Huck BOM (blind, oversized, mechanically locked) fastener which may be used in situations where it is only possible to gain access to one side of a connection. It is used with the same tightening tool as the Huck bolts. The principal difference is that the collar is now attached to a sleeve which has the same diameter as the pinhead. The end portion of the sleeve is softened so that the first effect of pulling on the end of the pin is to distort the end of the sleeve to form an effective head. Subsequent pulling produces a sequence of results similar to those shown in Figure 3.11.

Unfortunately, these fasteners are only available in sizes up to M12, which considerably limits their applicability to constructional steelwork.

3.6.3 Resin-packed bolts

Where a rigid connection under shear loading is required and friction-grip connections cannot be used, the injection bolt system shown in Figure 3.13 may offer an economic alternative to fitted bolts. The system is particularly effective at resisting transient overloads where no account need be taken of creep in the resin.

The bolt is assembled and tightened with the ventilation tube on the washer positioned as shown. Where possible, the vent should be on the upper side of the bolt. Taking appropriate care for the

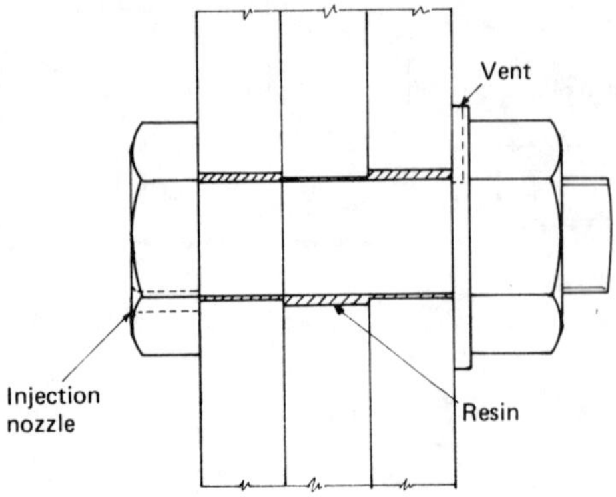

Figure 3.13 Resin-packed bolt in clearance hole

removal of rainwater, a resin is continuously injected through the nozzle on the bolt head until it issues from the ventilation channel. Once the resin has hardened, the connection will resist shear deformations until the resin starts to deform plastically. Design bearing strengths of 300 N/mm^2 and 200 N/mm^2 may be obtained for transient and long-term loading, respectively.

3.7 Bolt inspection and testing

The following tests are generally specified for the mechanical properties of bolts:

1. Tensile testing and proof load testing of full-sized bolts;
2. Tensile testing of machined test pieces to determine tensile strength, yield point stress, stress at permanent set limit of 0.2%, percentage elongation after fracture;
3. Hardness tests;
4. Wedge testing on full-size bolts (to ensure that the bolt can accommodate non-uniform bearing under the head);
5. Head soundness test (ISO bolts only);
6. Impact testing (ISO bolts only Grade 8.8 and above).

Nuts are subject to the following tests of mechanical properties:

1. Proof load;
2. Hardness.

Most aspects of bolt and nut geometry can be checked by conventional measuring systems and thread geometry with the help of the following additional equipment:

1. A shadow graph (a lightbox and optical system) may be used to magnify bolt thread geometry.
2. A thread gauge (a hardened template of ideal thread profile) may be used to check thread profile, accuracy of pitch, etc.
3. Go-gauges may be used for nut threads.

In practice, much can be achieved with an 'informal' checking of thread geometry by means of micrometer, thread gauge and standard taps and dies.

3.8 Bolt and rivet holes

3.8.1 Normal clearance holes

Almost all bolts and rivets are used in clearance holes to alleviate problems of fit during erection. Bearing bolts of all diameters are inserted in holes which are 2 mm clearance on nominal bolt diameter. Holes for HSFG bolts are similar for bolts of up to and including 24 mm diameter but may have 3 mm clearance for larger bolts.

3.8.2 Oversize and slotted holes

Some codes permit the use of oversize and slotted holes for the inner plies of HSFG bolted joints provided that an appropriate reduction is made in design capacity. Table 3.10 lists the standard sizes for the various categories of hole.

3.8.3 Fitted bolts

Fitted bolts are still occasionally used in situations where a rigid connection under shear loading is required and it is not practicable to use a friction connection. Such holes must have a clearance on the nominal bolt diameter of not more than 0.15 mm. They are used in conjunction with precision-bearing bolts.

3.8.4 Method of preparation

Generally, holes will be drilled, although some codes permit punching to be used for statically

Table 3.10 Maximum hole sizes for HSFG bolts to BS 5950: Part 1, 1985, Table 35

d *(mm)*	*Standard clearance hole*	*Oversize hole*	*Short-slotted hole*	*Long-slotted hole*
≤22	$d + 2$	$d + 5$	$(d + 2) \times (d + 6)$	$(d + 2) \times 2.5d$
24	27	30	27 × 32	27 × 60
≥27	$d + 3$	$d + 8$	$(d + 3) \times (d + 10)$	$(d + 3) \times 2.5d$

loaded joints of limited plate thicknesses. The use of punched holes is restricted because of concern over the embrittlement of the work-hardened material that immediately surrounds the hole. If used outside permitted limits, there is the possibility that punching could lead to the initiation of brittle fracture of the connection in the presence of significant tensile strains on material of low toughness. Holes for fitted bolts will generally be underdrilled and reamed after a trial assembly.

3.9 Bolt layout within the connection

3.9.1 Pitch and edge distance criteria

Almost all codes of practice stipulate pitch and edge distance criteria, and minimum criteria are specified for the following reasons:

1. To provide adequate clearance for tightening;
2. To limit any adverse interaction between high bearing stresses on neighbouring bolts. This is discussed in greater detail in Section 5.2.6;
3. To eliminate any tendency to bursting or in-plane deformation during drilling or punching; this reason particularly relates to minimum edge distance criteria;
4. To ensure adequate resistance to tear-out of the bolts.

Maximum pitch and edge distance criteria are required:

1. To eliminate local buckling of outer plies;
2. To ensure, in corrosive conditions, that an integral paint film is maintained across the plate interfaces, thus preventing any corrosion between the surfaces in contact.

Note that there is an adverse interaction between these two considerations. Because corrosion products occupy a greater volume than the sacrificial metal, any corrosion between the surfaces will cause them to separate, increasing any tendency to buckling. For this reason, maximum pitch criteria are more severe than other geometric criteria that control local buckling.

3.9.2 Clearance for bolt tightening

Figure 3.14 presents the necessary clearances for the various methods of tightening bolts. Bearing bolts in shear or static tension will usually be tightened by spanner and under fluctuating tension have to be tensioned to at least 70% of their proof load and all general grade HSFG bolts must be tensioned to at least their proof load. Up to and including M20 diameter such tightening can usually be most efficiently carried out by hand. Above that, clearance should be provided for a power wrench or torque multiplier.

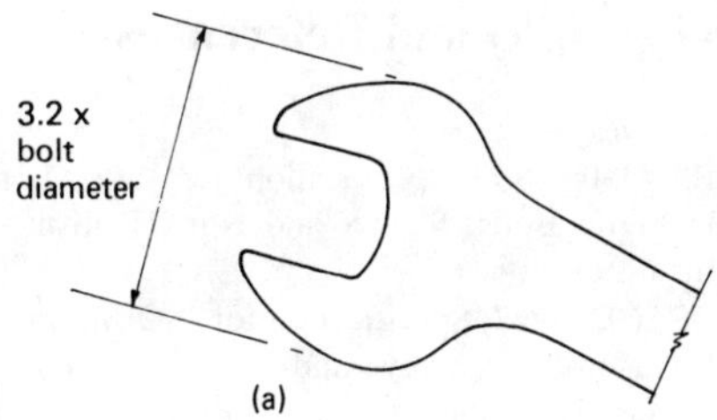

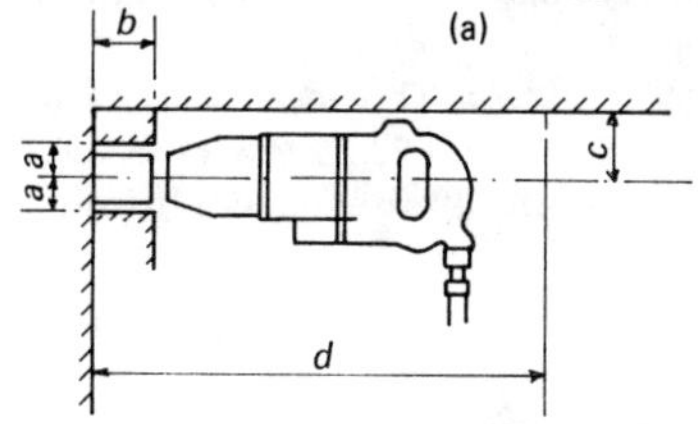

Size of bolt	*a*	*b*	*c*	*d*-power
M12	23	27	30	500
M16	30	46	60	500
M20	30	46	60	600
M24	36	65	60	600
M30	49	78	70	700
M36	49	97	100	700

(b)

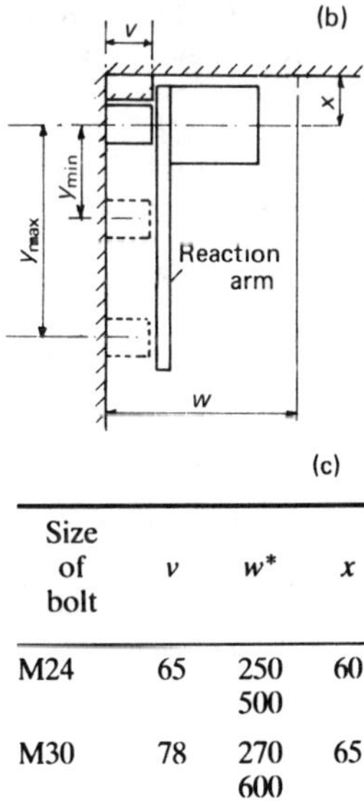

Size of bolt	*v*	*w**	*x*	min. *y* to max. y_1
M24	65	250 500	60	82 210
M30	78	270 600	65	89 260
M36	97	300 600	65	89 260

Upper figure is for hand wrench: lower figure is for pneumatic wrench.

Figure 3.14 Clearances for bolt tightening. (a) Hand spanner for ordinary bolts; (b) impact wrench for HSFG bolts; (c) torque multiplier for HSFG bolts

Note that the clearances given in Figure 3.14 are the minimum values for convenient working. Lesser values than these can be used where necessary, after consultation with the equipment manufacturers.

Bibliography and References

Dowel bolts

1. BS 4190: 1967, Specification for ISO Metric Black Hexagon Bolts, Screws and Nuts, British Standards Institution.
2. BS 3692: 1967, Specification for ISO Metric Precision Hexagon Bolts, Screws and Nuts. Metric Units, British Standards Institution.
3. BS 3643, ISO Metric Screw Threads. Part 1, 1981: Principles and Basic Data; Part 2, 1981: Specifications for Selected Limits of Size, British Standards Institution.

HSFG bolts

4. BS 4395, Specification for High Strength Friction Grip Bolts and Associated Nuts and Washers for Structural Engineering. Part 1, 1969: General Grade; Part 2, 1969: Higher Grade Bolts and Nuts and General Grade Washers; Part 3, 1973: Higher Grade (Waisted Shank) Nuts and General Grade Washers, British Standards Institution.
5. BS 4604, Specification for the Use of High Strength Friction Grip Bolts in Structural Steelwork. Metric Series. Part 1, 1970: General Grade; Part 2, 1970: Higher Grade (Paralleled Shank); Part 3, 1973: Higher Grade (Waisted Shank), British Standards Institution.
6. *Coronet Load Indicators*, Cooper and Turner Ltd, Sheffield.
7. Struik, J. H. A., Oyeledun, A. O. and Fisher, J. W., 'Bolt tension control with a direct tension indicator', *Engineering Journal*, AISC (1973).
8. Cheal, B. D., *Design Guidance Notes for Friction Grip Bolted Connections*, CIRIA TN98, London, 1980.

Rivets

9. BS 4620: 1970, Specification for Rivets for General Engineering Purposes, British Standards Institution.

Holding-down and foundation bolts

10. Blake, A. *et al.*, *Holding Down Systems for Steel Stanchions*, The Steel Construction Institute, Ascot, 1980.
11. *Threaded Bars*, McCalls Special Products, Sheffield, 1981.

Special fasteners

12. Hyler, W. S., Humphrey, K. D. and Croth, N. S., 'An evaluation of the high tensile Huck-Bolt fastener for structural applications', *Report 72*, Huck Manufacturing Co., Detroit, USA.
13. *The BOM Fastener*, Huck Manufacturing Co., Detroit, USA (available from Huck UK Ltd, Telford, Staffs).
14. 'Tests on connections with bolts whereby the clearance between the bolt shank and the hole has been filled with a resin according to patent no. 6804977', *Report 6-69-2*, Slevin Laboratory, Delft, the Netherlands.

Bolt inspection and testing

See the relevant sections in references 1, 2 and 4.

4

Static behaviour and design of welds

4.1 Butt welds

4.1.1 Behaviour

The most critical form of loading on a butt weld is transverse tension. Figure 4.1 shows the deformed shape of a parallel-sided tensile coupon, after testing, which included a full-penetration butt weld transverse to the applied load. It is clear that this properly executed butt weld had a greater static strength than the parent metal. In addition, it can be seen from its reduced transverse contraction that the yield stress of the weld metal and HAZ parent metal was considerably higher than that of the parent metal which had not been affected by welding heat.

Table 4.1 presents typical values for the mechanical properties of weld metal[1] and contrasts these with the values for the unaffected parent metal. The elevation of the yield stress of the HAZ metal is primarily due to the quenching effect which occurs with the very high cooling rates in this region immediately after the weld has been deposited. The increased yield stress of the weld metal arises partly from this effect and partly from additional alloying constituents in the electrodes. (These alloying elements will also have migrated into the heat-affected zone, where they have a secondary effect on mechanical properties.)

The increases in ultimate strength shown in Table 4.1 are due to similar effects. In the welded coupon shown in Figure 4.1 there is an additional effect arising from the differences in yield stress. Because of this difference and the associated variation in area reduction between the weld and other cross-sections in the coupon the latter are subjected to a higher true stress than the former. This is a further encouragement for the specimen to fail away from the weld.

Note that these increases in strength usually occur at the expense of both ductility and toughness. These adverse countereffects have been minimized by developments in electrode and flux chemistry, but it is still unlikely that values as high as those of the parent metal can be achieved. For many practical situations an appropriate choice of electrode is all that is necessary to ensure satisfactory

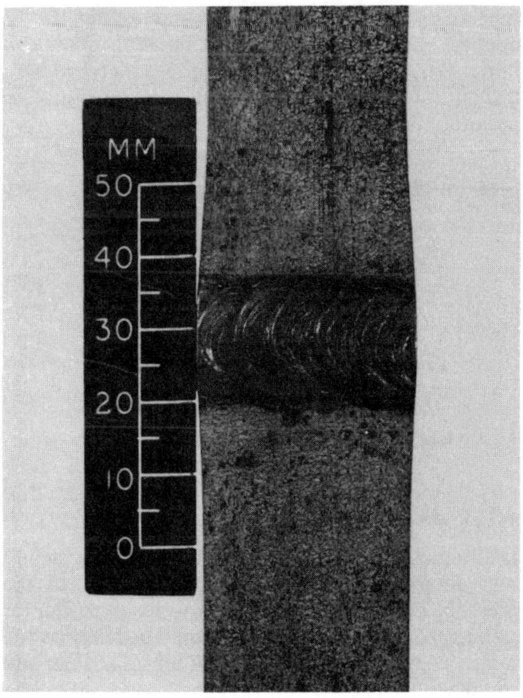

Figure 4.1 Parallel-sided tensile coupon containing full-strength transverse butt weld, after testing

Table 4.1 Tensile properties of base metal and butt welds

Material		*Direction*	*0.2%Y.S.* (*N/mm²*)	*T.S.* (*N/mm²*)	*El. in 50 mm* (*%*)	*R.A.* (*%*)	*Fractured point*
			Tensile properties[a]				
Specified value			≥310	≥490	–	≥30	–
Base metal		Longitudinal	360	524	34.6	75.8	–
		Transverse	363	523	35.5	75.2	–
		Through-thickness	360	526	31.0	70.6	3/8 t
Weld joint	SMAW	Transverse	371	520	21.4	73.0	Base metal
	SAW		352	516	25.7	72.5	Base metal
Weld metal	SMAW	Longitudinal	483	592	25.9	74.9	–
	SAW		453	549	27.2	73.3	–

[a]Average of two test specimens.

behaviour. However, post-weld heat treatment[2] should be considered in critical cases.

The above discussion on strength only relates to full-penetration butt welds. Partial-penetration welds differ from the former in two important respects. As shown in Figure 4.2, their use inevitably leads to a local reduction in cross-section. Thus any tensile overload will lead to severe local plastic straining of the weld metal and HAZ and, as discussed above, these materials already suffer from reduced ductility. The other problem with such welds is that it is very difficult to be certain about weld root quality. None of the repair techniques for full-penetration butt welds, such as backgouging and sealing, can be applied. Post-weld inspection of root quality is impossible. Variations in root gap, which could have a significant effect on weld root quality, cannot be monitored after welding.

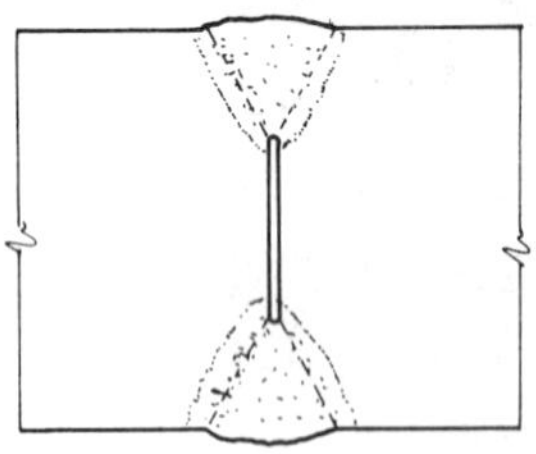

Figure 4.2 Partial-penetration butt weld

4.1.2 Design

It follows from the first part of the previous section that no particular account need be taken of the presence of full-penetration butt welds in design for static strength. Design strength may be taken as the same as parent metal strength for tension, shear and compression.

Partial-penetration welds require greater care in design. Because of concern over loss of ductility

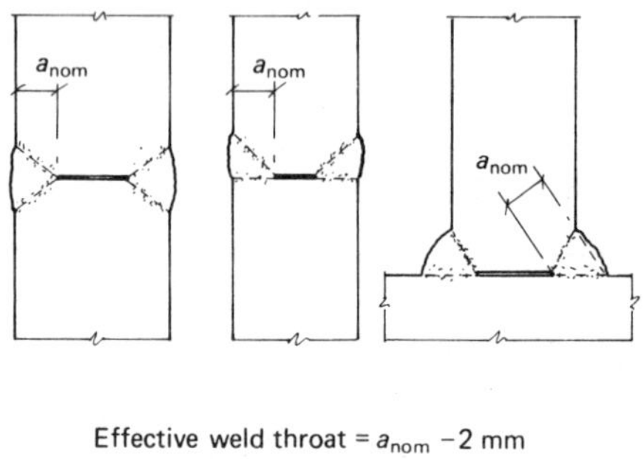

Figure 4.3 Definitions of effective weld throat for partial-penetration welds

some codes preclude their use in tension. Others generally require that they should be treated in the same way as fillet welds. In this latter case it is necessary to define the effective throat. Figure 4.3 summarizes the definitions permitted by draft Eurocode 3.[3] It can be seen that no account can be taken of weld reinforcement, unless a superimposed fillet weld is specified.

As discussed in greater detail in Section 4.2, simplified average stress concepts for welds are only applicable if it is possible to satisfy equilibrium with uniform stresses across the weld throat. Single-sided partial-penetration welds loaded in tension are clearly unable to satisfy this criterion, and their use to transmit tension should be avoided if possible. If not, the effects of eccentricity should be allowed for in design.

Because of the difference in design strength between full- and partial-penetration butt welds it is necessary to define the changeover between the two configurations. Some codes require, by implication, that any lack of penetration should cause the weld to be treated as a partial-penetration weld. However, draft Eurocode 3 permits a degree of lack of penetration provided that this is compensated by suitable reinforcing fillet welds. Figure 4.4 gives the

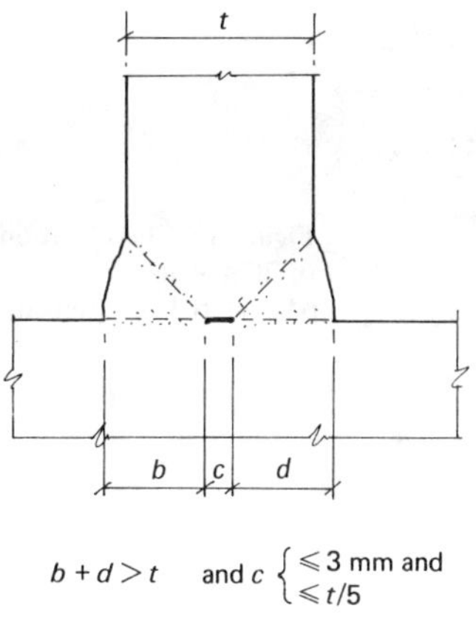

Figure 4.4 Criteria for a partial-penetration weld to be considered as a full-penetration weld for static loading

details. Note that this relaxation only applies to static strength; the fatigue performance of the weld shown in Figure 4.4 would be markedly worse than that of a true full-penetration butt weld.

4.2 Fillet welds

4.2.1 Behaviour

Figure 4.5 shows the variations in fillet weld behaviour with relative direction of the load vector to the weld axis.[4] When $\theta = 0°$ the weld axis is normal to the load vector, the so-called end-fillet situation, and the weld develops a high strength. Average throat stress at failure is similar to the weld metal tensile strength. However, ductility is very limited, with deformation at rupture being less than 1 mm. At the other extreme, when $\theta = 90°$, the weld axis is parallel to the load vector, the side-fillet situation, and the weld shear strength is limited to little more than half the weld metal tensile strength. However, a side fillet exhibits considerably more ductility; rupture occurs at over 2 mm deformation. Intermediate orientations show intermediate values of both strength and ductility. Figure 4.6 shows practical examples of side- and end-fillet weld configurations.

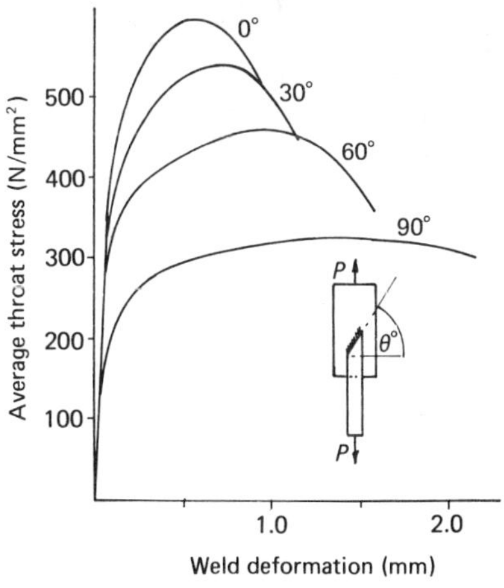

Figure 4.5 Load/deformation curves for an 8 mm fillet weld inclined at Θ°. Weld length 50 mm; plate thickness 19 mm. Ultimate strength of plate = 511 N/mm^2; ultimate strength of weld metal = 565 N/mm^2

The true distribution of stresses in a fillet weld is complex. Figure 4.7 shows the total distribution of stress resultants on a fillet weld under longitudinal shear (P_s), transverse 'shear' (P_t) and transverse tension (P_n). In addition to the applied loading there will be:

1. Moments M_1 and M_2 acting on the faces of the weld. These may be necessary for overall equilibrium;
2. Interface reactions Q arising from weld shrinkage during cooling;
3. Residual stresses. R, the longitudinal residual stress, is shown but there will also be self-equilibrating residual stresses on the faces of the weld;
4. Stresses which are part of the overall stress system on the welded element. An example would be the longitudinal bending stresses on the flange/web welds in a plate girder under a bending moment. (These stresses can generally be ignored because the contribution of the weld area to the overall resistance of the section is negligibly small and these stresses can, in any case, redistribute to the surrounding plates.)

Attempts at analysing weld behaviour rigorously have not been very successful, with only limited agreement with experimental data. Residual stresses present particular difficulty, in part because of their variability. Good agreement is unlikely to be achieved until analysis can take proper account of:

1. The multiaxial stress state;
2. Variations in yield stress with cooling rate;
3. Residual stresses and their influence on initiation of yield;
4. The flow rules which govern strain rates once yielding has commenced;
5. Strain-hardening effects.

In many circumstances it is possible to limit the analysis to a simplified set of average stresses on the weld throat. These are shown in Figure 4.8, the notation used being the same as that of the IIW test series discussed below. This simplification can only be justified if the ductility of steel is recognized and, most importantly, if it is possible to achieve equilibrium with the applied loads with such a set of average stresses. For example, it would not be acceptable to base any analysis on average stresses

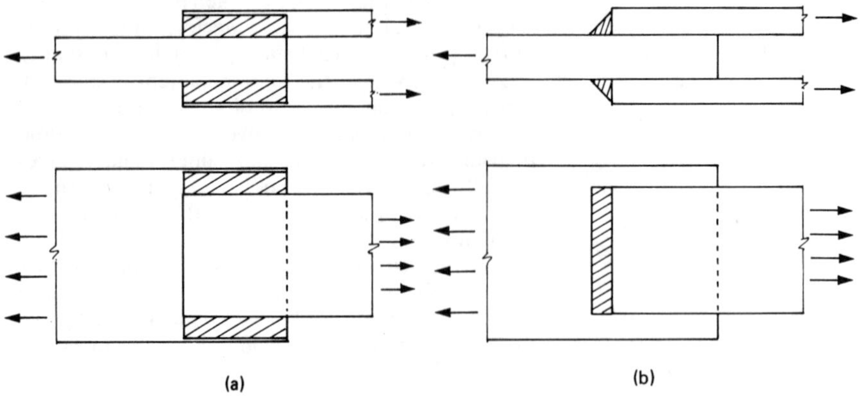

Figure 4.6 Classification of fillet welds. (a) Side (shear); (b) end (tension)

where there was any primary moment applied to the weld throat. In this case equilibrium could only be achieved with some variation in $\sigma_\perp$ across the throat.

The most successful simplified analytical approach is that by Kamtekar.[5,6] He analysed the fillet weld under a simplified stress system, only including the other reactive stresses that were necessary for equilibrium. He further postulated that longitudinal residual stresses would take a value which maximized strength: this was justified because he was using a lower-bound approach to the analysis. This gave good agreement with some experimental results, particularly those for tension fillet welds. However, agreement in other cases was less satisfactory. One of the most important conclusions that he came to was to point out that fillet weld shape is very significant for end-fillet welds. If the leg lengths become unequal, making the critical throat more nearly parallel to the direction of the applied tension, this will reduce strength considerably. The optimum weld shape is to make the 'shear' leg $\sqrt{3}$ times the 'tension' leg. By contrast, small variations in side-fillet profile have little influence on strength.

Because of the difficulty of carrying out satisfactory analysis, considerable effort has been devoted to obtaining experimental data, and one of the most effective early reviews of experimental data[7] led to the form of presentation of results that is shown in Figure 4.9.

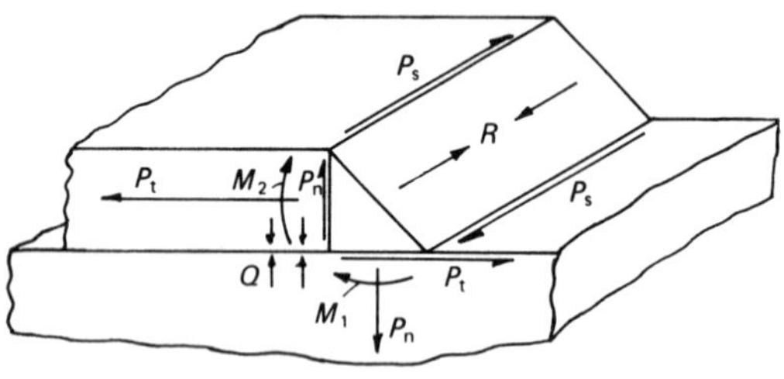

Figure 4.7 Force systems on fillet welds

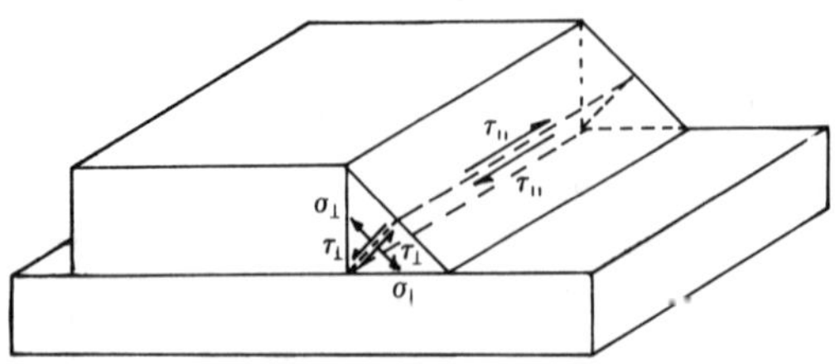

Figure 4.8 Simplified average stress systems on fillet weld throat

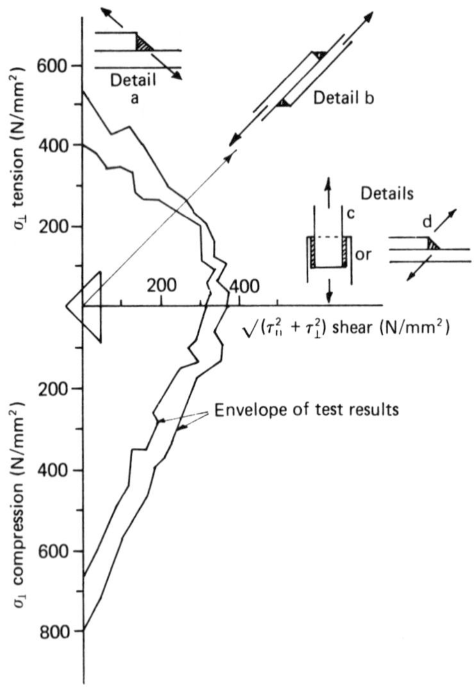

Figure 4.9 Fillet weld strength under combined stresses. Classification of details: a – pure tension weld; b – nominal 'tension' weld; c – longitudinal shear weld; d – transverse shear weld

Subsequently, an international test series was instigated by the International Institute of Welding in 1960. In this collaborative test programme experiments were carried out in ten countries over a period of eight years. These tests had the particular objective of obtaining a better understanding of the partition of load between end- and side-fillet welds in practical joints.[8]

The study was limited to those situations where average stresses on the weld throat could justifiably be used; all the results are presented in terms of those stresses and are summarized in Figure 4.9. The vertical axis is the normal stress at failure on the weld throat $\sigma_{\perp}$. Originally a three-dimensional presentation was used with separate, orthogonal horizontal axes for $\tau_{\perp}$ and $\tau_{\parallel}$. However, it was found that the interaction surface so created was, within the limits of experimental variation, a surface of revolution about the $\sigma_{\perp}$ axis. The two-dimensional presentation shown was therefore adopted, with a horizontal axis of the vector sum of the two shear stresses.

It can be seen that the weld is somewhat stronger in nominal tension (i.e. with equal magnitudes of tension and transverse shear on the weld throat) than in pure shear. It can also be seen, as might be expected, that fillet welds are stronger in compression than in tension.

4.2.2 Design

Background

The simplest approach to design is to ignore the variation in fillet weld strength with load direction and to limit the throat stress to a certain value. The applied loads per unit length are summed vectorially and then divided by the weld throat size to determine the average throat stress. This throat stress is sometimes, erroneously, called the weld shear stress; in practice it could contain elements of tension or compression. This design procedure is effectively that of working to some inscribed circle within the envelope of weld strength shown in Figure 4.9.

There is some international disagreement about the value of design strength to be used with this design approach.[9] Table 4.2 summarizes codified values. Variations in design strength are probably more a reflection of differing levels of concern over workmanship rather than fundamental differences in understanding of weld metal strength.

Alternatively, design strength can be varied with direction of applied load vector. Most methods are variations of the basic International Institute of Welding strength formula that is given below. As the test programme proceeded it became clear that, provided equilibrium could be achieved in the presence of average throat stresses, strength was governed by the effective stress (σ_w) given by:

$$\sigma_w = \beta\sqrt{[\sigma_{\perp}^2 + \gamma(\tau_{\perp}^2 + \tau_{\parallel}^2)]}$$

where $\sigma_{\perp}$ is the direct stress on the weld throat,
$\tau_{\perp}$ is the shear stress on the weld throat normal to the weld axis,
$\tau_{\parallel}$ is the shear stress on the weld throat parallel to the weld axis.

It was found that γ could take varying values between about 1.8 and 3.0. It was finally decided to make $\gamma = 3.0$, at least in part because of the consequent similarity with the von Mises yield criterion. In any case, the higher value is more conservative because it make the shear stress more damaging. However, it should be noted that the above formula relates to interactive resistance to rupture, not onset of yielding. β was a function of parent metal and weld metal strength. If the design strength σ_w is related to parent metal strength then the lower the parent metal strength, the lower could be the value of β. This is because the ratio of weld metal strength to parent metal strength increases with decreasing parent metal strength. Initially (and indeed in some current codes) β was made a step function of parent metal strength. Typical values are:

$\beta = 0.7$ for $240 \nless \sigma_y$
$\beta = 0.8$ for $240 < \sigma_y < 280$
$\beta = 0.85$ for $280 < \sigma_y < 340$
$\beta = 1.0$ for $340 < \sigma_y < 400$

However, as shown in Figure 4.10, this produces undesirable steps in a graph of weld metal strength against parent metal strength. Other codes have

Table 4.2 Codified values of design strength of fillet welds – simple (direction-independent) method (N/mm²)

	BS 449[a] *× 1.6*	*BS 5950*	*BS 5400: Part 3*	*ECCS*	*Draft Eurocode 3*	*AISC*[a] *× 1.6*	*Ontario Bridge*
Grade 43	184	215	187	177	198	199	207
Grade 50[b]	256	255	193	241	205	232	247

[a]Allowable stress code, stated load factor applied to code value.
[b]Assuming E51 electrode is used.

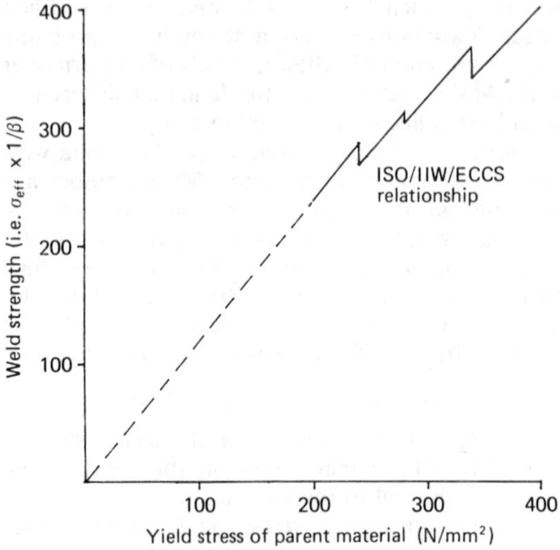

Figure 4.10 Relationship between weld 'strength' and material strength

managed to overcome this by making β a linear function of σ_y. Table 4.3 summarizes the different formulae used by the more common codes for direction-dependent fillet weld design.

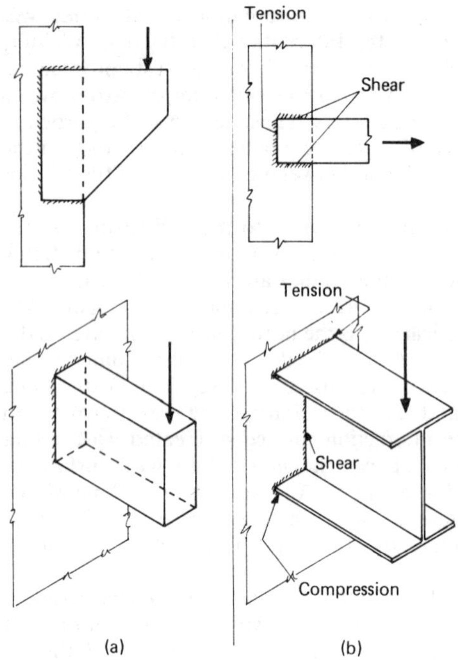

Figure 4.11 Connections where (a) simple weld design and (b) direction-dependent weld design methods are likely to be used

Choice of strength assessment method

Many modern codes give the designer a choice between the simple and direction-dependent methods of strength assessment. However, there is little guidance on which method to use. In practice, the choice is usually made on practical grounds, as illustrated in Figure 4.11. It is simply not feasible to use the direction-dependent method for situations where the load vector direction varies around the weld group. Thus this method is limited in application to cases of pure shear, tension or compression. Where the simple method is used, the stress is taken as the vector sum of the force components acting on the critical portion of the weld, divided by the throat area.

Design of 'full-strength' fillet welds

It is frequently necessary to design fillet welds to develop the full strength of the attached plate in shear, tension or compression. Table 4.4 summarizes the weld sizes necessary to achieve this for the common codes and different methods of strength assessment. The advantage of using the direction-dependent method for tensile and compressive loading can clearly be seen. It is interesting to compare the results obtained with the traditional 'rule of thumb' for such welds in mild steel, which was simply to ensure that the sum of the weld throats (Σa) at least equalled the plate thickness (t). For Grade 50 steel the corresponding rule is that the sum of the throats (Σa) should be at least 1.2 times the plate thickness (t).

Table 4.3 Codified design approaches for fillet welds, taking account of variation in strength with direction of loading

BS 449	nil
BS 5950	nil
BS 5400: Part 3[a]	$\sqrt{[\sigma_\perp^2 + 3(\tau_\perp^2 + \tau_\parallel^2)]} \ngtr \dfrac{(\sigma_y + 455)}{2\gamma_m\gamma_{f3}}$
ECCS	$\sigma_{eff} = \beta\sqrt{[\sigma_\perp^2 + 3(\tau_\perp^2 + \tau_\parallel^2)]} \ngtr \sigma_y$ and $\sigma_\perp \ngtr \sigma_y$
Eurocode 3	Similar to ECCS
AISC	nil
Ontario Bridge code	nil

[a]Discounting k in equation 14.6.3.11.2 (code error).

Table 4.4 Ratios of minimum throat size to attached plate thickness (a/t) to develop full strength of attached plate, with symmetric fillet welds

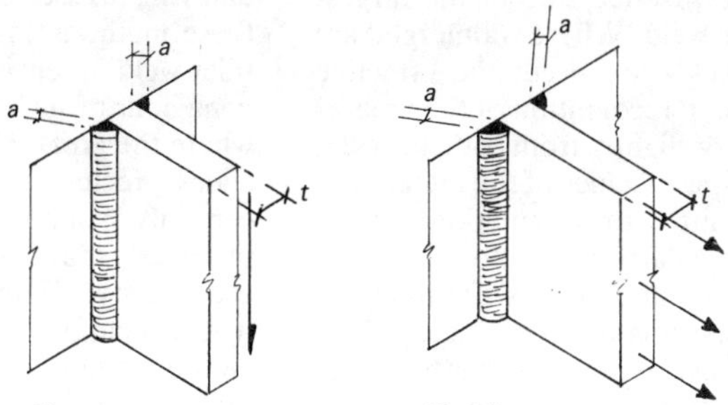

Shear Tension

	Simple		*Direction dependent*		*Simple*		*Direction dependent*	
Grade	43	50	43	50	43	50	43	50
Code								
BS 449	0.50	0.50	–	–	0.67	0.67	–	–
BS 5950[d]	0.38	0.42	–	–	0.50	0.50	–	–
BS 5400	0.43	0.51	0.39[c]	0.46[c]	0.49[a] 0.68[b]	0.57[a] 0.79[b]	0.56[c]	0.65[c]
ECCS / Eurocode 3	0.40	0.50	0.40	0.50	0.70	0.87	0.57	0.71

[a]Value for end fillets in end connections.
[b]Value for other tension connections.
[c]Discounting k in equation 14.6.3.11.2 (code error).
[d]Based on design strengths of 275 N/mm^2 for Grade 43 and 355 N/mm^2 and E51 electrodes for Grade 50.

Design of welds of minimum size for ductility of the connection

It is not uncommon to have to design welds of a minimum size for the connection to behave in a ductile manner under tension or shear (i.e. so that the parent metal yields rather than the weld). This usually arises because of some uncertainty over detailed load paths through the connection. In such circumstances full-strength fillet (or butt) welds should generally be specified. In circumstances where considerable ductility may be required it would seem appropriate to ensure that:

For Grade 43 $\Sigma a \nless 1.2t$
For Grade 50 $\Sigma a \nless 1.4t$ (i.e. leg length = t)

The reason for this additional conservatism is to cover the situation where the parent plate has mechanical properties that are significantly greater than guaranteed minimum values.

4.2.3 Geometry

In addition to considerations of strength, any method of strength assessment has to consider details of weld geometry. Fillet welds are usually defined by their leg length, but calculations relate to the weld throat.

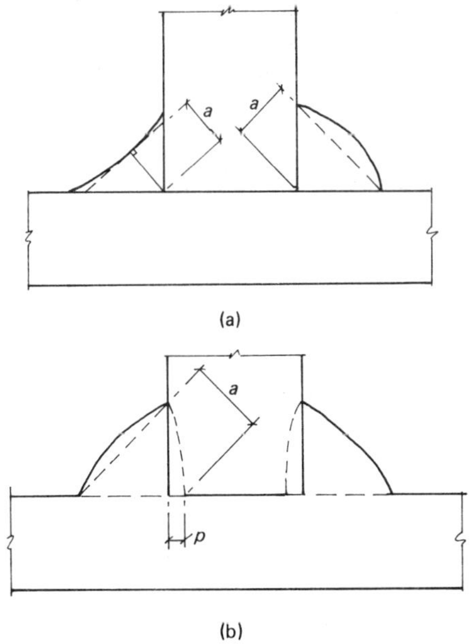

Figure 4.12 Definitions of effective throat for (a) normal welding and (b) deep-penetration and submerged arc welding. p is penetration of deep-penetration weld, demonstrated by procedure trials. For submerged arc welding, a may be increased by the lesser of $0.2a$ and 2 mm

As indicated in Figure 4.12(a), the effective weld throat is usually defined as the length of the perpendicular from the inclined side of the largest inscribed triangle of the weld. Where submerged arc or deep penetration welds are used, the effective throat may be increased in recognition of the greater penetration. Details vary slightly from code to code; Figure 4.12(b) shows typical values of the increase in throat dimensions permitted for calculation.

Fillet welds are particularly prone to defects at start and finish positions, and craters are likely to form as the weld pool cools and contracts. More rigorous codes of practice require strength assessment to be based on effective weld length, that is, weld length minus a leg length for each start and stop position. For continuous welds the difference is not very great but it can become significant for intermittent ones.

4.3 Secondary considerations in fillet weld design

4.3.1 Single-sided welds in tension

Figure 4.13 compares symmetric and non-symmetric fillet welds in tension. It is not possible to postulate any uniform distribution of stress on the fillet weld throats of the non-symmetric welds in Figure 4.13(a) which can be in equilibrium with the applied load. The eccentricity between the line of action of the load and the throat centroid inevitably creates a moment on the weld throat. This is in contrast to the symmetric arrangements of Figure 4.13(b). Here, although there may initially be some variation in stress across the weld throats, only modest ductility will be required for this variation to be redistributed. By symmetry, uniform stress fields on the weld throats can be in equilibrium with the applied load.

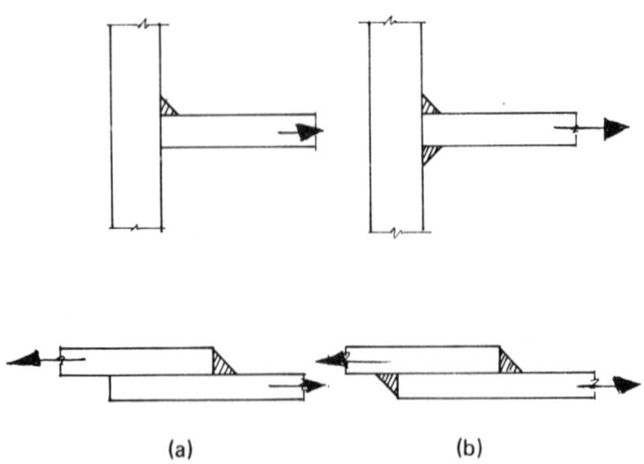

Figure 4.13 Examples of (a) unsatisfactory and (b) satisfactory welds for tension connections

All the methods of strength assessment discussed in Section 4.2 are based on a consideration of uniform stresses on the weld throat. It follows that these methods cannot be used for any single-sided fillet weld in tension. In practice, it is found that the behaviour of such welds is very variable, particularly where the greater tension is on the weld root. Many codes preclude the use of such welds, and this is certainly sound practice. Others permit them, provided that proper account is taken of the eccentricity. Where such welds are permitted it would seem prudent to assume, conservatively, that all the moment is transmitted to the weld throat (in practice much of it may be carried by some other restraint) and then to ensure that an elastic summation of the weld throat stresses does not exceed the basic weld design strength.

4.3.2 Short, widely spaced side-fillet welds

For most practical purposes it is possible to ignore the influence of the deformation of the connected parts on the distribution of forces in the weld and hence on the connection strength. However, a particular problem arises with short, widely spaced

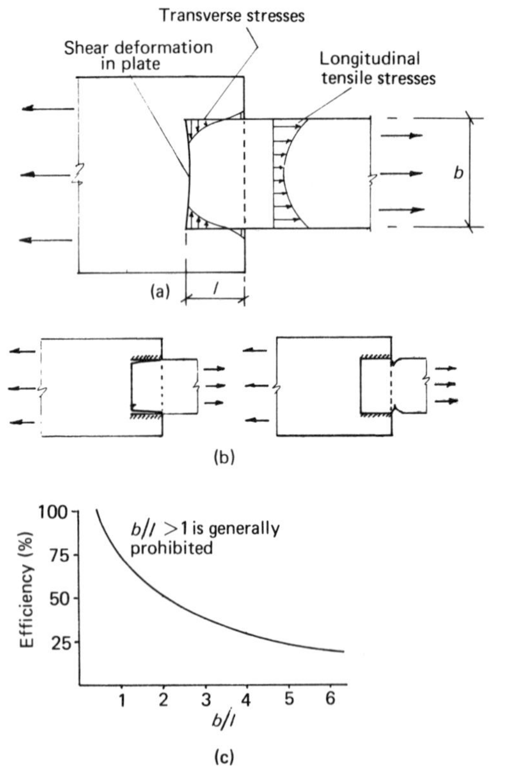

Figure 4.14 Behaviour of short, widely spaced fillet welds. (a) Stress distribution; (b) modes of failure; (c) loss of efficiency

side-fillet welds.[10] The shear deformation that occurs in the attached plate causes two effects. As shown in Figure 4.14(a), stress concentration develops at the end of the fillet weld that is superimposed on the stress concentration that would exist in any case due to extensional strains. Second, transverse contraction and in-plane bending occur, which produce additional loads on the welds.

Figure 4.14(b) shows the alternative modes of failure that can develop. If weld strength is critical, transverse tearing is likely to occur. If plate strength is critical, tearing may develop from the embrittled HAZ. In either case there can be a significant loss in efficiency. This is summarized in Figure 4.14(c). Most codes of practice preclude the use of connections with side fillets only where b, the spacing between the welds, is greater than l, the individual weld length. It can be seen that if this limit is adhered to, efficiency will not drop below 75%. Where b/l ratios of more than 1.0 have to be used, end fillets must be added to the connection.

4.3.3 Long weld effects

Figure 4.15 shows another situation where the deformation of the connected parts influences the distribution of forces in the weld and weld strength.

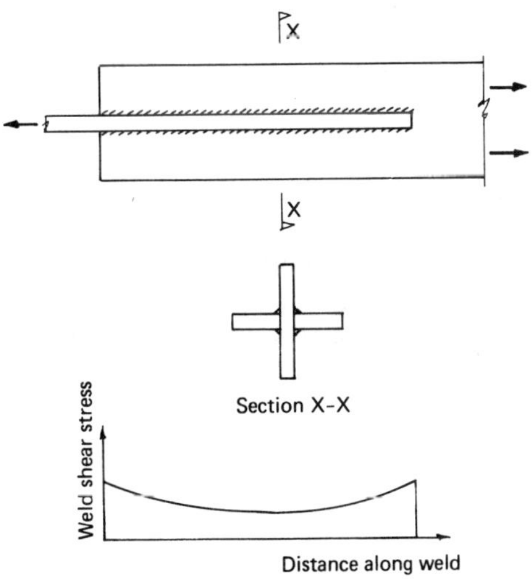

Figure 4.15 Behaviour of long fillet welds: effect of plate extensions on distribution of weld shear

In this case it is the variation in extension of each element along the length of the connection that causes the variation in weld shear along the length. There is a similar, but more severe, effect in a long-bolted splice, and a detailed discussion of the phenomena is presented in Section 5.2.6. In the context of welded connections the effect is of limited practical significance, because of the ductility of fillet welds in shear. The effect is covered in most codes but will only lead to a reduction in capacity for welds over 1.5 m in length.

It should be emphasized that the effect only arises because of the incompatibility of longitudinal strain between the two elements. Reductions need not be made for situations where no such incompatibility exists (for example, the web/flange welds in a plate girder).

References

1. Clark, A., George, M. J. and Kirkwood, P. R., 'Development of automatic welding procedures to meet toughness requirements for offshore structures', *Int. Conf. on Welding in Offshore Construction*, The Welding Institute, 1974.
2. *Heat Treatment of Welded Steels*, The Welding Institute, Abington, Cambridge (to be published).
3. Eurocode 3, *Common Unified Rules for Steel Structures*, First draft 1984, Commission of the European Communities, Brussels.
4. Clarke, A., MSc thesis, University of London, September 1970.
5. Kamtekar, A. G., 'A new analysis of the strength of some simple fillet welded connections', *JCSR*, **2**, No. 2 (1982).
6. Kamar, A. G., 'The strength of planar fillet weld groups subjected to an eccentric shearing force in the planes', *JCSR*, **7**, No. 3 (1982).
7. Vreedenburgh, G. G. J., 'New principles for the calculation of welded joints', *Welding Journal*, August, **53** (1954).
8. Ligtenburg, F. K., 'International test series, final report', *Int. Inst. of Welding, Doc XV-242-68* (1968).
9. Clark, P. J., 'Basis of design for fillet welded joints under static loading', *Proc. Conf. on Improving Welding Product Design*, The Welding Institute (1976).
10. Freeman, F. R., 'The strength of arc-welded joints', *Proc. ICE*, **23** (1931). Discussion by Dr C. H. Jobban.
11. Grinter, L. E., *Design of Modern Steel Structures*, Macmillan, New York, 1960.

Static behaviour and design of bolts and bolted connections

5.1 Introduction

With many connection configurations there is considerable interaction between behaviour of the bolts and that of the connected parts. It therefore seems most logical to discuss both sets of topics within the same chapter.

Under each general heading the behaviour of simple connections (i.e. single-bolt connections in shear, double-bolt connections in tension) is discussed first, because this is the clearest way of presenting the principal aspects of behaviour. Subsequently the discussion is extended to include the significant features of multibolt connections.

5.2 Dowel bolt connections in shear

5.2.1 General: implications of clearance holes

The behaviour of typical single-bolt connections in shear is shown in Figure 5.1.[1] If the connection is not assembled in bearing it will slip as soon as the applied load has overcome the modest friction at the interface. With normal clearance holes this slip may be as much as twice the bolt clearance in extreme circumstances, i.e. up to 4 or 6 mm, depending on the bolt size. In practice, in multibolt connections mismatch of the holes is likely to reduce this movement considerably.

Once in bearing, the connection will behave linearly until yielding takes place at one or more of the following positions:

1. At the net section of the plate(s) under combined tension and flexure;
2. On the bolt shear plane(s);
3. In bearing between the bolt and the side of the hole. (Near a plate edge 'bearing' is taken to include local shearing and other deformations in the plate that are caused by bearing action.)

The response of the connection becomes progressively more non-linear as plasticity spreads in the

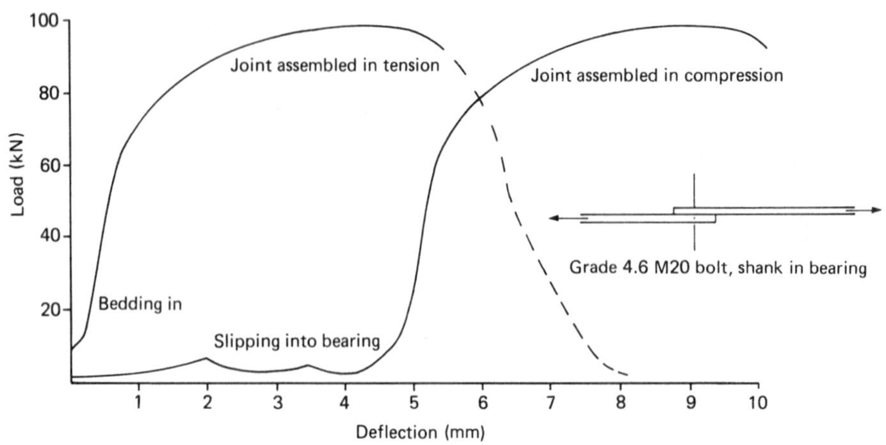

Figure 5.1 Behaviour of a single bearing bolt connection in shear

presence of strain hardening until failure occurs at one of the critical locations listed above.

The point of initiation of yielding and the mode of failure of the connection will depend on its proportions and the relative material strengths of the components. Each aspect of behaviour is discussed separately below.

5.2.2 Single-bolt connections – tension on the net section

If the connection is narrow, with edge distances close to the permitted minimum, it is likely that, at least for thin plates, the critical element will be the net section in tension.[2] If the speciment is symmetric, with the bolt in double shear, there will be no bending on the net sections. In such circumstances it will be found that the failure load will correspond to approximately 1.05 × net section area × material tensile strength. The slight increase in strength above that predicted by simple theory is due to the restraint offered to the net section by the less highly stressed, neighbouring gross section.

If the bolt is in single shear there will be co-existent out-of-plane bending stresses on the net section during loading. However, as such a speciment approaches failure, the bolt will rotate so that it is partly in tension and partly in shear. The effect of this will be to reduce the moment on the net section; this reduction will be most significant for thin plates, those most likely to be critical in tension. The effect of this reduced bending will still be to reduce the net section capacity below that given by simple theory. However, this capacity is unlikely to drop below 0.9 × net section area × material tensile strength.

Traditionally, the design strength of the net section was related to the product of the net section area and the basic tensile design stress. However, some codes of practice now permit a local increase in stress on the net section. It is argued that a local exceedance of yield over a small proportion of the member length at the collapse limit state is unlikely to lead either to unserviceability of the structure at the appropriate, lower, load factors or to cause any major redistribution of forces at the collapse limit state.

5.2.3 Bolt shear

Figure 5.2 shows the behaviour of both high-strength, Grade 8.8, and black, Grade 4.6,[3] bolts in shear. The most notable feature is that behaviour is influenced considerably by the position of the threaded portion of the bolt relative to the shear planes. The presence of threads in the shear planes reduces both strength and deformation capacity

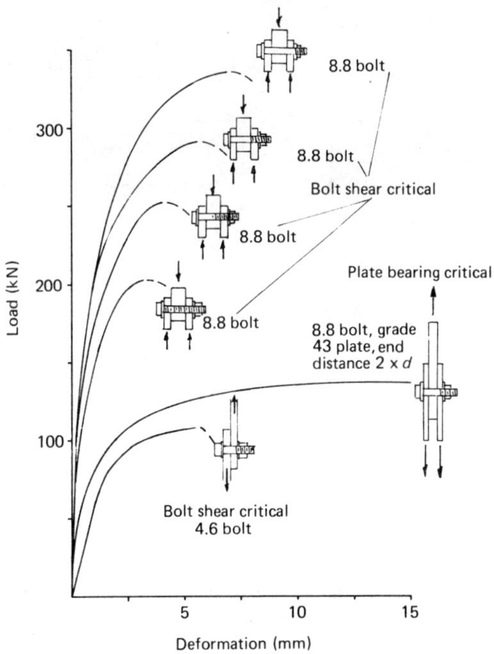

Figure 5.2 Typical load deformation graphs for 20 mm diameter bearing bolts in shear

markedly. Note that with modern detailing it is likely that threads will extend into one or both shear planes in many practical connections unless specific action is taken for their exclusion. Traditionally, design was based on shank area of the bolt and a reduced value of shear strength to account for the possible presence of threads. Current codes generally require the use of the true cross-sectional area and a more realistic design shear strength.

Figure 5.2 also draws attention to the considerable variation in shear strength between Grades 4.6 and 8.8 bolts. It would appear that they have similar deformation capacity, but this is rather misleading. If an 8.8 bolt is examined after failure in shear it will have been found to have deformed by approximately one-sixth of its diameter. The considerably greater total deformation shown in the figure includes the deformations to the edges of the holes, in bearing. A 4.6 bolt will typically have deformed by one-quarter of its diameter after failure. However, the total deformation of the assembly will not be markedly different from the less ductile (higher-strength) bolt. The greater bolt deformation will have been compensated by the lesser bearing deformation because of the lower bearing stresses.

It is customary to compare shear strength with tensile strength. For rivets, where there is little variation in cross-section along the length of the fastener, an experimentally determined ratio for

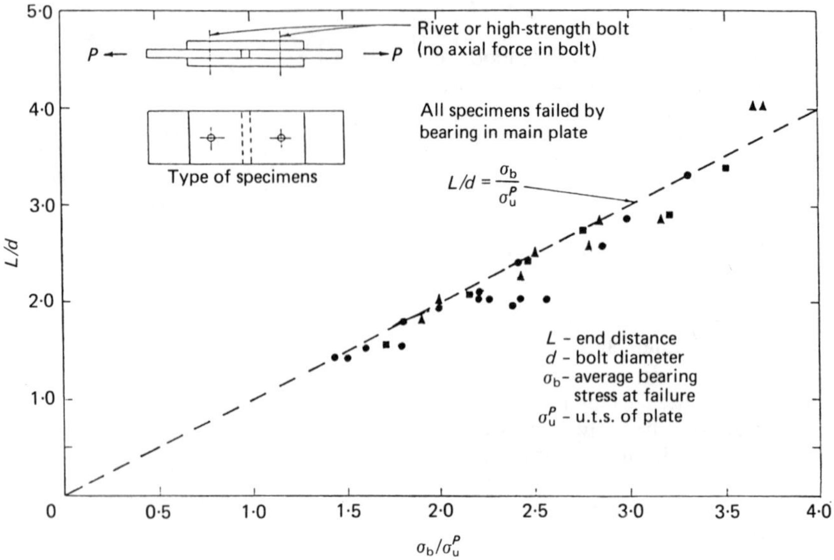

Figure 5.3 Relationship between end distance and bearing strength for double-shear specimens

single shear would typically be:

$$\frac{\text{Ultimate shear strength}}{\text{Ultimate tensile strength}} = 0.7\text{–}0.75$$

For bolts with the shank in the shear plane a corresponding ratio is 0.8. Here the increase is due to the different cross-sections involved; shear is acting on the gross section while tension is acting on the reduced section through the threads. For bolts with threads in the shear plane the ratio is likely to be 0.63, notably less than the traditional value for rivets. This reduction would seem to be related in some way to stress concentration effects associated with the threads.

5.2.4 Single-bolt connections – bearing on the plate

Where a large-diameter bolt passes through relatively thin plate the critical mode of behaviour will be the bearing of the bolt against the plate. (A superficially similar set of bearing stresses is created in the bolt. The reason that these will not generally govern design is discussed in the following section.) This mode of behaviour is generally monitored by regulating the nominal bearing stress, that is, the load transferred between the bolt and plate, divided by the projected nominal contact area (bolt diameter × thickness).

Figure 5.2 also shows the behaviour of a connection where bearing is critical.[2] The extreme ductility of this mode of behaviour can be seen. Deformations of over half the bolt diameter will usually occur prior to failure. However, only limited account can be taken of the latter part of the response curve, because of the need to limit deformations at working load.

The region of highest bearing stress in the plate is triaxially restrained. Transverse restraint is provided by the surrounding plate; through-thickness restraint develops from pressure under the bolt head and nut. Because of this triaxial restraint very high bearing stresses can be sustained, and these are discussed in detail below.

Because of this triaxial action, bearing behaviour is influenced by the proximity of plate boundaries. Away from a boundary or neighbouring holes, significant hole elongation commences at a nominal stress of about 2 × uniaxial plate yield stress but failure will not occur until about 3 × ultimate plate tensile strength. Figure 5.3 shows the effect of a free boundary close to the bolt and normal to the direction of loading.[4,5] Once the *end* distance is less than 3 × bolt diameter, the free boundary reduces the in-plane containment and modifies behaviour. If the specimen is wide, it will fail by shearing from the edges of the hole towards the plate edge. It is found that bearing strength is approximately a linear function of end distance. Thus ultimate bearing strength will be approximately 2.0 × plate u.t.s. for an end distance of 2*d*.

Edge distance will also influence behaviour.[2] The mode of failure described above (and illustrated in Figure 5.4(a)) will only occur if the specimen is sufficiently wide for the net section not to yield in tension prior to failure in bearing. If the specimen is

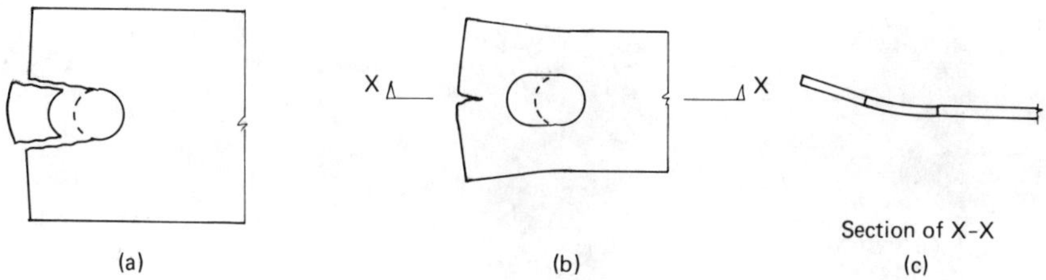

Figure 5.4 Failure modes of plates in bearing

narrow, so that yielding of the net section in tension precedes bearing failure, it will fail, as shown in Figure 5.4(b). The tensile yielding of the net section reduces the in-plane containment to the regions of high bearing compression. The end of the specimen splays out as transverse tension near the free edge develops to try to contain the bursting action. Finally, the specimen will fail in transverse tension at the free boundary, at an ultimate bearing stress of approximately 1.9 × plate u.t.s. for an end distance of $2d$. If the bolt is in single shear, so that it is free for its axis to rotate from its original orientation, out-of-plane effects modify behaviour further, as shown in Figure 5.4(c). Net section yielding will occur earlier due to the out-of-plane bending. The bolt head or nut will dig into the plate at the net section, tending to crop it. Because of these effects, the ultimate bearing stress will reduce to 1.75 × plate u.t.s. for a narrow specimen.

The presence of threads in the bearing zone will increase the flexibility of bearing behaviour.[2,6] At the lower deformations, which are of concern for serviceability, threads will almost double the deformation at a given bearing stress. However, threads in the bearing zone do not reduce bearing strength. Indeed, once the threads have dug into the plate they provide additional through-thickness restraint – in some instances this will increase bearing strength.

5.2.5 Bolt bearing

Surface-bearing stresses similar to those in the plate are induced in the bolt. However, the bolt is subject to even more effective triaxial containment than the plate. The shape of the bolt and the nature of the bearing contact provides circumferential restraint. Through inner plies, where bearing stresses are usually greatest, bolt flexure induces longitudinal compression, which provides restraint in the third direction. The result of this restraint is that in a bolted connection with similar materials (say, Grade 4.6 bolts and Grade 43 plate) only very modest indentation and marking of the bolt will occur even as the plate fails in bearing. If bolts of higher grade are used there will only be surface polishing on the bolt.

It is possible to induce bearing failure in bolts but only by using them in connections with very high-grade plates – a situation that does not occur in practice. Figure 5.5 shows three Grade 4.6 bolts that were tested in connections made from material with a yield stress of $700\,N/mm^2$.[7] It illustrates the interaction that will occur between bolt shear and bolt bearing in these extreme circumstances. In the left-hand example the bolt has primarily failed in double shear, despite being subjected to bearing stresses of over $1000\,N/mm^2$. In the central example the failure mode is an interaction of shear and bearing modes. It is only in the right-hand example that a true bearing failure has occurred. There is no significant deformation to the bolt face opposite the region of high bearing stress. Superficially, it appears that the bolt material has suffered a volumetric change; in practice, the material has flowed longitudinally, increasing the bolt length. This bolt material sustained bearing stresses in excess of twice ultimate tensile strength. This very high figure indicates why it is not generally necessary to consider bolt bearing in design. The excessive ductility of plate bearing is also absent, so the full strength can be mobilized. However, note that, in BS 5950, Part 1, the bolt capacity does govern the bearing design for Grade 4.6 bolts. The authors consider that this is simply an anachronism: there are no similar clauses in other codes.

5.2.6 Multibolt connections

When the preceding basic understanding of single-bolt connections is applied to practical multibolt ones some additional aspects of behaviour must be considered.

Critical net sections

In multibolt connections selection of the critical section may no longer be obvious. Where the connection is subject to torsion and shear, or where members are only partially connected, the critical section may not just be subject to tension. Guidance on net section considerations is given in Section 7.3.

Bolts in shear in mismatched holes

Figure 5.2 illustrates the limited deformation capacity of bolts in shear. Where threads are in the shear plane the bolt is likely to have failed before it

Figure 5.5 M20 Grade 4.6 bolts after failure under combined shear and bearing

has sustained a deformation of 4 mm. Hole mismatch with normal clearance holes can equal this value. It is clear that distribution of shear between bolts in the presence of a mismatch will be substantially different to that given by any theoretical analysis which ignores this misalignment. Some bolts in a group may not be in bearing contact, and therefore cannot be contributing to connection strength when others rupture. The effect of this was illustrated in Figure 1.3.[7] The 40% reduction in strength observed there might be exceeded in other situations. Because of this uncertainty, a conservative approach to connections where bolt shear is critical would seem prudent. Methods of analysis which assume or imply plastic redistribution of shear between bolts, other than that necessary to accommodate hole mismatch, should be avoided.

Bearing in multibolt connections

Because high bearing pressures only occur in the immediate locality of each bolt it is reasonable to discuss bearing behaviour in a multibolt connection in terms of the sum of the behaviours of the individual bolts, subject to the interactions discussed below. Figure 5.2 also illustrates the substantial ductility of individual bolted connections where plate bearing governs. Because of this ductility it is possible to be more sanguine about possible redistributions of load between bolts – provided that no redistribution of bearing stresses could lead to individual bolts being seriously overloaded in shear.

Because of the complex set of stresses set up in the plate by bolt bearing there is an interaction between neighbouring bolts if they are spaced at or near minimum pitch. Figure 5.6(a) illustrates one extreme case where the line of bolts is transverse to the applied load.[2] Here there is a beneficial interaction, because the in-plane bursting actions of neighbouring bolts cancel each other out. Thus it is only necessary for 'external' in-plane transverse restraint to be provided at the ends of the bolt group, as indicated by the transverse arrows, this restraint being provided by in-plane bending of the plate from the net section and lateral tension near the free end. Figure 5.6(b) shows the other extreme case where the bolts are in a line parallel to the direction of the applied load.[2] Here the bursting action of the neighbouring bolts is cumulative, as all the bursting actions are separately trying to spread the plate. For example, at a pitch of $2.5d$ there is insufficient material between the bolts to provide the same in-plane transverse restraint as a single bolt at an end distance of $2d$ (the single-bolt case with the comparable amount of transverse cross-section per bolt). With these dimensions it is found that there is a 10% reduction in ultimate bearing stress.[8] This is not sufficiently serious to warrant special treatment in design. However, if the material between the bolts was reduced further (for example, by some

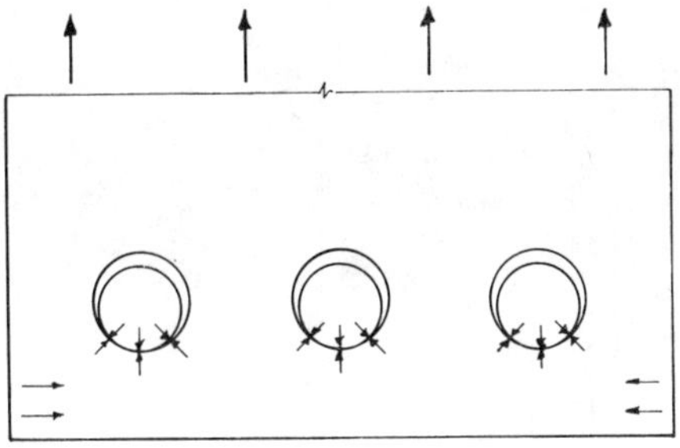

(a)

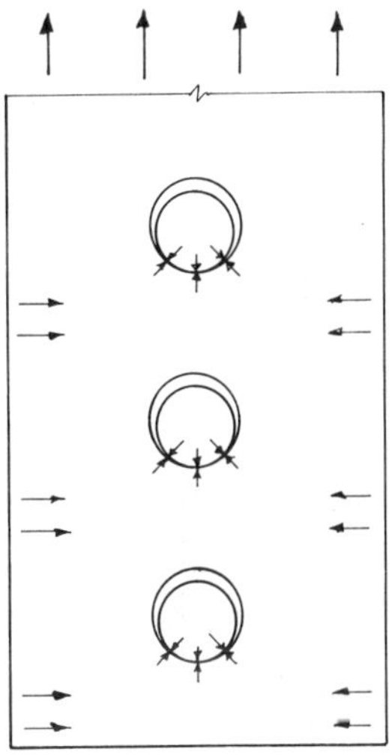

(b)

Figure 5.6 Bearing interaction between neighbouring bolts at minimum pitch. (a) Line of bolts transverse to applied load; (b) line of bolts parallel to applied load

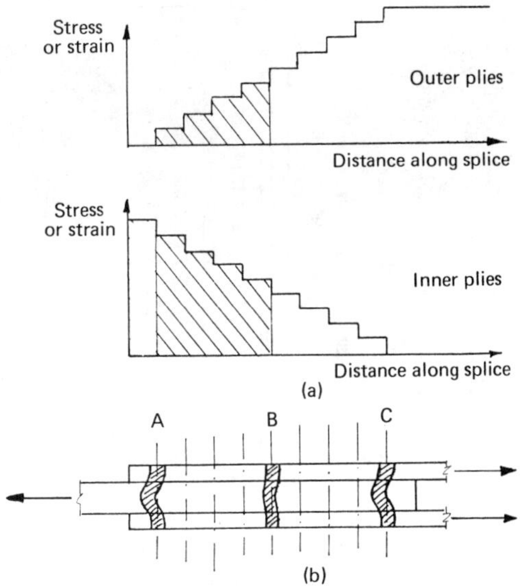

Figure 5.7 Effect of splice length on distribution of bolt forces in a long connection

modification to the hole sizes) the adverse interaction would become severe. In particular, if oversize or slotted holes are used for any reason it is important to ensure that the material cross-section between bolts is not reduced below that which would exist with normal clearance holes at minimum pitch.

Large connection effects

In general, methods of analysis for bolted connections ignore the deformation of the connected parts. In reality, incompatibilities will arise because of variations in strain between neighbouring elements. For any given magnitude of strain difference the incompatibility will be a linear function of size. With connectors of limited deformation capacity there will be a size above which the bolt deformation capacity will be insufficient to accommodate this incompatibility.

This scale effect is most marked in tension splices, and considerable research has been carried out to quantify the effects.[9] Figure 5.7(a) shows schematically the distribution of average stress, and hence to a different scale average strain, in the plies for a long splice. The shaded areas of the two graphs are therefore measures of the elongation of the plies between the mid-point and the left-hand end of the connection. Since the central ply is subjected to greater deformation than the outer over this portion of the splice it follows that, as shown in Figure 5.7(b), there will inevitably be greater deformation in bolt A than in bolt B. In the other half of the connection the relative magnitudes of the deformations are reversed, again leading to a greater shear deformation in bolt C than in bolt B.

Figure 5.8(a) shows the effect of this on the distribution of bolt forces in a long joint and the resulting loss of efficiency. The principal parameter influencing this loss in efficiency is overall joint length, and their relationship can be seen in Figure 5.8(b). In this context it should be noted that the relevant length is the length over which the load transfer takes place.

The above discussion relates to tension splices. Similar effects would exist in other large connections if similar incompatibilities could develop. In such circumstances it would seem prudent to apply the same reduction factor on connection strength.

5.2.7 Design

Tables 5.1 and 5.2 summarize the design strengths of the more common bearing bolts in accordance

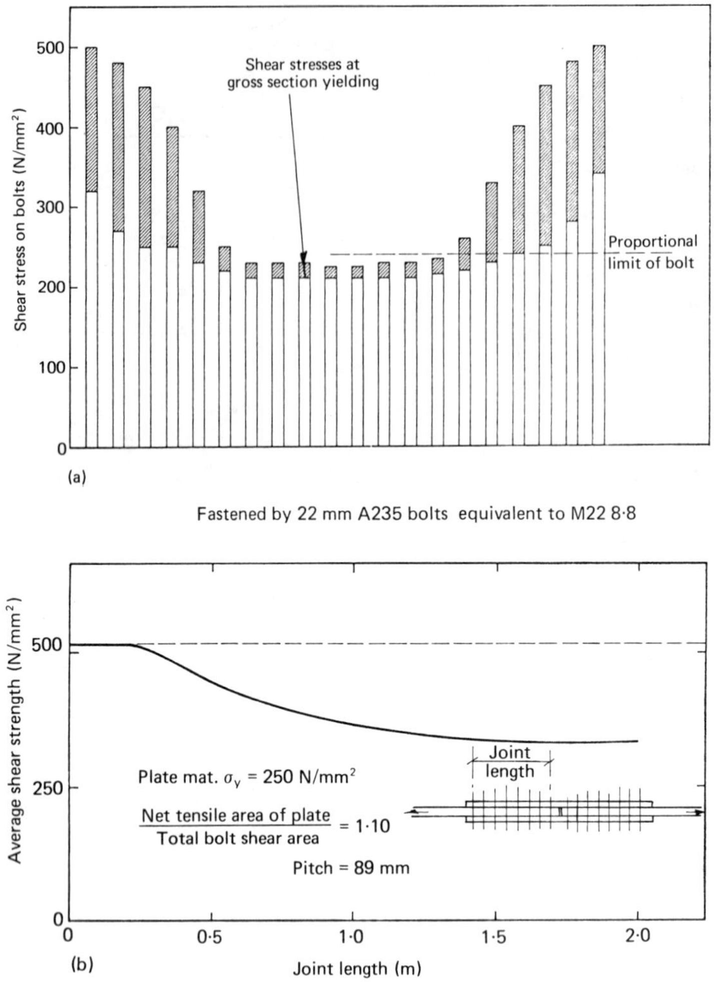

Figure 5.8 Incompatibility of axial deformations in long-bolted joints. (a) Load partition in long joint; (b) effect of joint length on joint strength

with current codes of practice. When determining design strength in shear the designer should, of course, take the lower of the shear and bearing strengths.

Appropriate calculations show that 'factors of safety' are higher for bolts than for most other structural elements. For shear strength this would appear to be because of concern over lack of ductility and the problems of unequal load distribution between bolts, as discussed in Section 5.2.6. For bearing strength the primary concern is to limit deformations at working load levels to acceptable values.

In addition to guidance on design strength, most codes of practice make reference to several points discussed in this section.

5.3 Tension connections

5.3.1 Bolts in tension

Figure 5.9 shows the load/deformation relationship of three typical bolts.[10,11] Tensile strength is governed by the threaded portion of the bolt, within the stressed length. The effective area of the threaded portion (the tensile stress area) is based on the core diameter plus the area of the single thread that cuts the failure section. For the ISO thread profile that is commonly used it has the value

$$\frac{\pi(d - 0.9382p)^2}{4}$$

where d is the shank or nominal diameter (mm),
p is the pitch (mm).

Table 5.1 Design data for ordinary bolts to BS 5950: Part 1
Design strengths

Bolt grade	4.6		8.8	
Steel grade	43	50	43	50
Tension p_t	195	195	450	450
Shear p_s	160	160	375	375
Bearing on connected ply p_{bs}	(460)	(550)	460	550
Bearing on bolt p_{bd}	435	435	(970)	(970)

Bolt capacities

Nom. dia. (mm)	*Stress area (mm²)*	*Tension (kN)*	*Shear*[a] *Single (kN)*	*Double (kN)*	*Bearing*[b] *(Grade 43/Grade 50)* *6 mm (kN)*	*8 mm (kN)*	*10 mm (kN)*	*12 mm (kN)*	*15 mm (kN)*	*20 mm (kN)*	*25 mm (kN)*
4.6 bolts											
12	84.3	16.4	13.5	27	31.3/31.3						
16	157	30.6	25.1	50.2	41.8/41.8	55.7/55.7					
20	245	47.8	39.2	78.4	52.2/52.2	69.6/69.6	87/87				
24	353	68.8	56.5	113.0	62.6/62.6	83.5/83.5	104.4/104.4	125.3/125.3			
30	561	109.4	89.8	179.5	78.3/78.3	104.4/104.4	130.5/130.5	156.6/156.6	195.7/195.7		
8.8 bolts											
12	84.3	37.9	31.6	63.2	33.1/39.6	44.2/52.3	55.2/66	66.2/—			
16	157	70.7	58.9	117.7	44.2/52.8	58.9/70.4	73.6/83	88.3/105.6	110.4/132	147.2/—	
20	245	110.3	91.9	183.7	55.2/66	73.6/88	92/110	110.4/132	133/165	184/220	
24	353	158.9	132.4	264.7	66.2/79.2	88.3/105.6	110.4/132	132.5/158.4	166/198	220.3/264	276/—
30	561	252	210.4	420.7	82.8/99	110.4/132	133/165	166/198	207/247.5	276/330	345/412.5

[a]Assumes threads in both shear planes.
[b]Assumes end distance not less than $2d$.

Table 5.2 Design strenghts of bearing bolts according to BS 5400: Part 3 (ultimate limit state)

Bolt grade		*4.6*	*8.8*
Tension	$\frac{\sigma_t}{\gamma_m \gamma_{f3}}$	$\frac{235}{1.2 \times 1.1} = 178$	$\frac{549.5}{1.2 \times 1.1} = 416$
Shear	$\frac{\sigma_q}{\gamma_m \gamma_{f3} \sqrt{2}}$	$\frac{0.85 \times 235}{1.1 \times 1.1 \times \sqrt{2}} = 116.7$	$\frac{0.85 \times 549.5}{1.1 \times 1.1 \times \sqrt{2}} = 273$
Bearing[b] (Grade 43, $L = 2d$, $T \leq 16$ mm)	$\frac{k_1 k_2 k_3 k_4 \sigma_y}{\gamma_m \gamma_{f3}}$	$\frac{0.85 \times 1.97 \times 1.2 \times 1.0 \times 235}{1.05 \times 1.1} = 409\ \text{N/mm}^2$	$\frac{0.85 \times 1.97 \times 1.2 \times 1.0 \times 275}{1.05 \times 1.1} = 478\ \text{N/mm}^2$

Bolt capacities

Nom. dia. (mm)	*Stress area (mm²)*	*Tension (kN)*	*Shear[a] Single (kN)*	*Shear[a] Double (kN)*	*Bearing[b] (Grade 43/Grade 50 for 8.8 bolts) 6 mm (kN)*	*8 mm (kN)*	*10 mm (kN)*	*12 mm (kN)*	*15 mm (kN)*	*20 mm (kN)*	*25 mm (kN)*
4.6 bolts											
12	84.3	15.0	9.8	19.7	29.4						
16	157	27.9	18.3	36.6	39.2						
20	245	44.6	28.6	57.2	49.0	65.3					
24	353	62.8	41.2	82.4	58.8	78.3	97.9				
30	561	99.9	65.5	130.9	73.4	98.0	122	147			
8.8 bolts											
12	84.3	35.1	23	46.0	34.4/44.4	45.9/—					
16	157	65.3	42.8	85.3	45.9/59.2	61.1/79.0	76.4/98.8	91.8/—			
20	245	102	66.9	133	57.4/74.0	76.4/98.8	95.2/123	114/148	143/—		
24	353	147	96	193	68.8/88.8	91.7/118	114/148	138/178	172/222	221/—	
30	561	233	153	306	86.0/111	115/148	143/185	172/222	214/277	277/360	346/—

[a] Assumes threads in both shear planes.
[b] Enclosed bearing values with an end distance of 2 × d

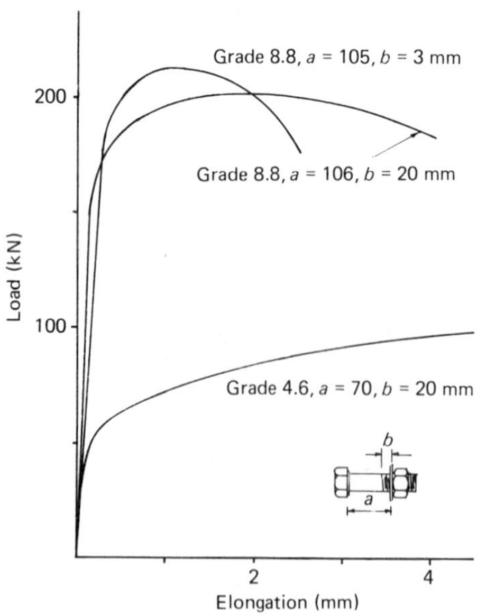

Figure 5.9 Tensile behaviour of dowel bolts

The deformation capacity of the bolt is a function of material properties, the ratio of the shank to the tensile stress area and the length of thread in the stressed length. Generally:

$$\frac{\text{Tensile area}}{\text{Shank area}} \doteqdot 0.8$$

Bolts of Grade 8.8 and above have a ratio of:

$$\frac{\text{Proof stress}}{\text{Ultimate stress}} = 0.8 \text{ or more}$$

Thus such bolts may have failed in tension on the threaded portion before they have yielded on the shank. As shown in Figure 5.9, the ductility of such bolts is influenced considerably by the length of the threaded portion in the stressed length.

Grade 4.6 bolts should have a ratio of yield stress/ultimate strength that will allow substantial yielding of the shank before rupture. However, it is not uncommon in such bolts for this ratio to rise above 0.8, in which case they will behave in a similar way to heat-treated bolts. In order to ensure reasonable deformation capacity it is therefore important to check that there are at least three full threads, plus the thread run-out, in the stressed length of Grade 4.6 and 8.8 bolts. Higher-grade bolts have lower minimum elongation requirements and should have correspondingly longer threaded portions in the stressed length. The Grade 4.6 bolt response shown in Figure 5.9 is that of a typical 'soft' bolt. The greater deformation capacity of such a bolt compared to a Grade 8.8 can clearly be seen.

If the nut is of the same strength grade as the bolt there is a considerable chance that fastener failure will be by thread stripping. This is particularly true of Grade 4.6 bolts because of the free fit permitted for the threads of such fasteners. However, provided that both thread forms are within specified geometric tolerances, such a mode of failure should always have been preceded by yielding of the bolt in tension and adequate ductility will be achieved.

However, if either the nut material is weaker than that of the bolt or the thread interlock is less than it should be because of some deviation from permitted tolerances then stripping of the threads may occur prior to yielding of the bolt.[12] Figure 5.10 illustrates the most likely mode of failure. Initially, yielding of the threads increases the bursting action on the nut (1). The lower portion of the nut is then likely to yield under the combined action of local shear, compression and flexure on the threads, hoop tension, cylindrical shear and bearing compression (2). The nut will expand during this yielding, reducing thread interlock and causing further bursting action. The associated nut taper may be visible (3). Also cylindrical shearing deformation of the top of the nut is usually visible prior to failure (4).

This mode of failure can be very abrupt,[13] particularly with long bolts which can store sufficient strain energy to increase the plastic damage to the nut once an individual nut/bolt combination has passed its maximum load-carrying capacity. There will usually be little or no warning of incipient failure other than slight evidence of nut deformation. If, in any tensile connection, nuts are observed

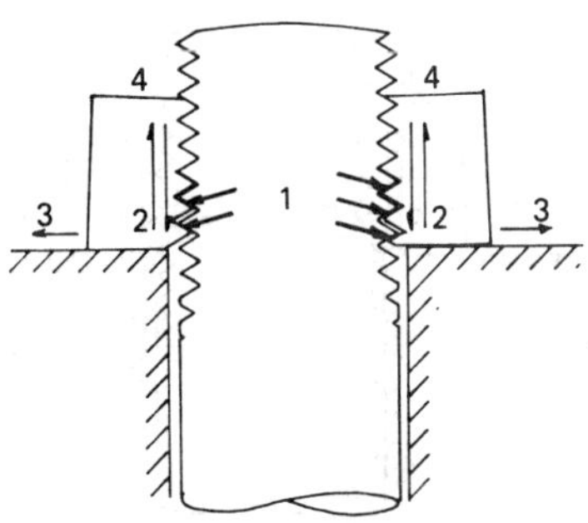

Figure 5.10 Thread-stripping mode of failure (nut weaker than bolt)

to have expanded or show signs of shear deformation on the unloaded face they should be replaced and the cause of the damage ascertained. They may well be very close to an abrupt, possibly explosive, failure.

In the early days of bolt development head failure would sometimes occur. This is another undesirably sudden mode of failure. Concern for it led to the introduction of both the head-soundness and the wedge-loading tests in current bolt specifications. These are important parts of the quality control of bolts. One author can recall a case where they were dispensed with through haste with a batch of Grade 12.9 bolts that had been manufactured to order. Some 24 hours after these bolts had been inserted and torqued to proof load most of the heads came off, in the absence of any external tension!

5.3.2 Bolted connections in tension: prying action

In practice it is not possible to separate the discussion of bolts in tension from that of the surrounding elements. Flexure of the connected parts may lead to a significant increase in bolt load due to prying action. Figure 5.11 shows the variations in behaviour that can occur in a simple, two-bolt, connection.[9]

Where, as in Figure 5.11(a), the end plate is relatively rigid it does not deflect significantly and it is possible to ignore its flexural action. For applied loads that are less than the sum of the bolt preloads there is no significant separation of the connection components and only modest changes in the bolt preload (such changes as do occur are due to through-thickness effects in the vicinity of the bolt holes). Once the applied load exceeds the sum of the bolt preloads, the end plate separates entirely from the base. From this point onwards to rupture the sum of the bolt loads equals the applied load.

However, if a flexible end plate is used behaviour is more complex. Each portion of the end plate bends into double curvature; the restraining moments at the bolt centreline develop from forces Q at or near the tips of the end plate. Overall equilibrium is now given by $2B = 2F + 2Q$. The effect of this amplification of the bolt forces is twofold: there is an earlier separation of the connection elements with a reduction in connection stiffness once separation has occurred; and the ultimate capacity is reduced (in the example shown from 260 kN to 190 kN).

Analysis of prying action has proved to be difficult, and there are the conventional problems of:

1. The need for elastic/plastic analysis;
2. The significance of imperfections and fit on the distribution of (prying) forces.

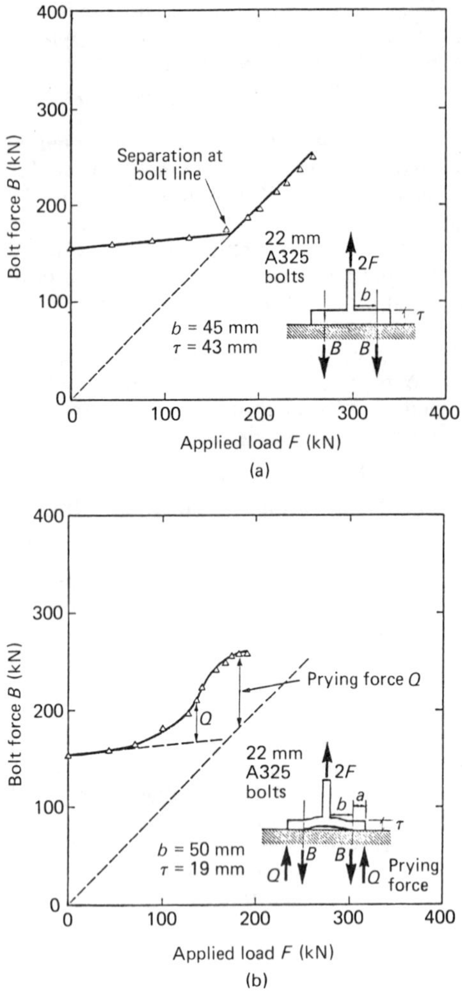

Figure 5.11 Influence of end-plate flexure on tension connections. (a) Thick plate; (b) thin plate

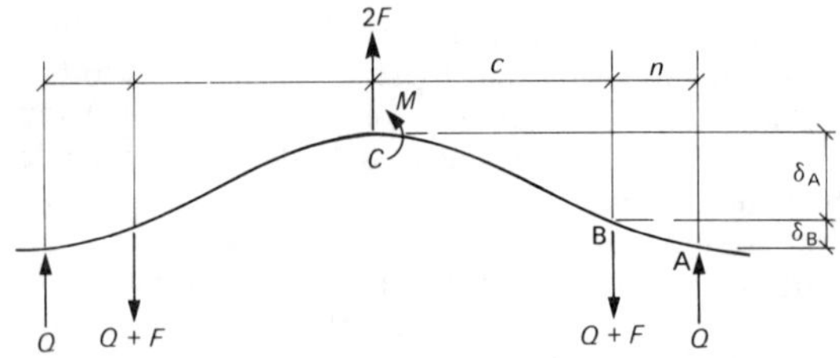

Figure 5.12 Forces and geometry of end plate for prying analysis

To these have to be added the particular difficulties of:

3. The assessment of true bolt stiffness;
4. Uncertainty of distribution of bearing between bolt head and end plate;
5. Uncertainty of line of action of prying forces due to initial and induced curvatures in end plate;
6. Local through-thickness effects in the vicinity of the bolt holes.

The following analysis is considered to combine simplicity with adequate representation of behaviour, and it leads to straightforward design formulae.

Figure 5.12 shows a schematic representation of the end plate in flexure. Most terms are defined in the figure, and additional definitions are:

T = flange thickness,
w = effective width of flange per pair of bolts,
b = distance from the bolt centreline to the toe of the fillet weld or weld reinforcement (for a welded Tee) or to half the root radius for a rolled section,
p_0 = minimum proof stress in bolt.

Consider elastic deformation of the Tee section. Taking origin at C, then by McCauley's method:

At C, $x = 0$; $\dfrac{dy}{dx} = 0$ and $y = 0$

$$EI\frac{d^2y}{dx^2} = M - Fx + (Q + F)[x - c] \tag{5.1}$$

$$EI\frac{dy}{dx} = Mx - \frac{F}{2}x^2 + \frac{(Q + F)}{2}[x - c]^2 \tag{5.2}$$

$$EIy = \frac{M}{2}x^2 - \frac{F}{6}x^3 + \frac{(Q + F)}{6}[x - c]^3 \tag{5.3}$$

At B, $x = c$; $y = \delta_A$

$$EI\delta_A = \frac{M}{2}c^2 - \frac{F}{6}c^3 \tag{5.4}$$

At A, $x = n + c \quad y = \delta_A + \delta_B$

$$EI(\delta_A + \delta_B) = \frac{M}{2}(n + c)^2 - \frac{F}{6}(n + c)^3 + \frac{(Q + F)}{6}n^3 \tag{5.5}$$

From equations (5.4) and (5.5),

$$EI\delta_B = \frac{M}{2}(n^2 + 2nc) - \frac{F}{6}(n^3 + 3n^2 - 3nc^2) + \frac{(Q + F)}{6}n^3 \tag{5.6}$$

Taking moments at C for ABC:

$$M = (Q + F)\,c - Q(n + c)$$
$$M = Fc - Qn \tag{5.7}$$

From equations (5.6) and (5.7),

$$6EI\delta_B = 3Fnc^2 - Q(2n^3 + 6n^2c)$$

$$Q = \frac{F - \dfrac{2EI\delta_B}{nc^2}}{\dfrac{2}{3}\left(\dfrac{n}{c}\right)^2 + 2\left(\dfrac{n}{c}\right)} \qquad (5.8)$$

In the calculations for the bending strength of the Tee connection the maximum bending moment (plastic hinge) adjacent to the supporting web is assumed to occur at a distance b from the bolt line, defined above.

Due to the local stiffening of the flange by the web of the Tee and the root fillets or welds, the effective value of c is between b and the distance from the centreline of the bolt to the centreline of the web.

It is more convenient to carry out the calculation of the minimum prying force in terms of b rather than c. This can be achieved by rewriting formula (5.8) using the following assumptions:

(1) $$\left[\frac{1}{\dfrac{2}{3}\left(\dfrac{n}{c}\right)^2 + 2\left(\dfrac{n}{c}\right)}\right] = \frac{b}{2n}$$

If the limitation on the value of n is adopted (see later) it is estimated that the maximum error due to this simplifying assumption would be about 4%.

(2) $c^2 = 1.5b^2$ (i.e. $b = 0.82c$) in the $\dfrac{2EI\delta_B}{nc^2}$ term

Hence formula (5.8) becomes:

$$Q = \frac{b}{2n}\left[F - \frac{4EI\delta_B}{3nb^2}\right] \qquad (5.9)$$

Where preloaded bolts are used, the aim in developing the prying force formula is to limit the loss of preload after the application of an external load. It is necessary to limit the loss of preload when the connection is subject to dynamic/fatigue loading or where the bolts are also used to carry shear as friction-grip bolts. Excessive loss of preload may also be undesirable from the point of view of loss of stiffness in the connection. Where non-preloaded bolts are used, the aim is to limit the loss of the nominal preload so that the bolts do not become loose after loading.

Preloaded bolts

When an external load is applied to a tension connection with preloaded bolts, in the initial stages there is a gradual increase in the load in the bolts (Figure 5.11(b)). After the load in the bolts reaches the yield point, it increases at a faster rate. If the applied load is removed after the load in the bolts has passed the yield point there is a loss of preload. The amount of this loss depends upon the extent of the plastic straining of the bolts. For a fully threaded bolt the elongation at proof load = $p_0/E \times$ grip length, where p_0 = minimum proof stress of bolt. The elongation per flange plate = $\delta_p \simeq p_0T/E$.

Consideration of typical load/elongation curves for general grade HSFG bolts indicates that if the plate deformation at the bolt line (δ_B) (which is also nominally the additional elongation of the bolt beyond its preload elongation) is limited to $\delta_p/3$ there would be a limited loss of preload (up to 10%) if the connection were overloaded to 1.67 × working load. After loading to the working load the loss of preload would be negligible.

Substituting in equation (5.9)

$$\delta_B = \frac{\delta_p}{3} = \frac{p_0T}{3E} \text{ and } I = \frac{wT^3}{12}$$

$$Q = \frac{b}{2n}\left[F - \frac{p_0wT^4}{27nb^2}\right] \qquad (5.10)$$

Non-preloaded bolts

In this case the preload is a nominal tightening and may be relatively small. Without preload the early increase in bolt load as the external load is applied is more rapid.

Consideration of typical load/elongation curves indicates that if δ_B is limited to $2\delta_p/3$ the loss of preload would be limited to less than 10% of the proof load if the connection were overloaded to 1.67 × working load, and would be negligible after loading to working load.

Substituting

$$\delta_B = \frac{2}{3}\delta_p = \frac{2p_0T}{3E} \text{ and } I = \frac{wT^3}{12}$$

in equation (5.9):

$$Q = \frac{b}{2n}\left[F - \frac{2p_0wT^4}{27nb^2}\right] \qquad (5.11)$$

A design formula for the minimum prying force can then be written as:

$$Q = \frac{b}{2n}\left[F - \frac{\beta\gamma p_0wT^4}{27nb^2}\right] \qquad (5.12)$$

where $\beta = 1$ for preloaded bolts
$= 2$ for non-preloaded bolts
$\gamma = 1.0$ for working load design
$= 1.5$ for factored load (limit state) design.

(*Note*: Units of p_0 must be consistent with units of F (for example, kN/mm² and kN.)

The $b\gamma p_0 wT^4/27nb^2$ term is usually relatively small. If it is neglected, formula (5.12) becomes:

$$Q = \frac{Fb}{2n} \tag{5.13}$$

This is the formula that is obtained if plastic hinges are assumed at the bolt line and the root, i.e. when minimum flange thickness design is used.

The design of a Tee connection, end plate or column flange would normally be carried out by assuming plastic hinges at the bolt line and the root. However, if it is desired to use a smaller size of bolt this is achieved by reducing the prying force and designing the flange for the increased moment at the web root. If the bolts had unlimited ductility, and it did not matter if the bolts lost their preload and became loose, no further action would be required. However, in practice this is not the case, and a minimum prying force calculated using formula (5.12), which is based on limiting the extension of the bolt, should always be adopted. Where bolts are preloaded to increase the frame stiffness, but are not designed to carry shear by friction and are not subject to fatigue or dynamic loading, consideration may be given to using $\beta = 2$ in the formula if the loss of stiffness due to an overload in one joint would be acceptable.

Position of the prying force

When the distance from the bolt line to the edge of the flange is relatively large the prying force (Q) will act at some intermediate point rather than at the edge of the flange. To determine the distance n to this intermediate point, assume that the slope of the flange at A is zero:

$$\text{At A, } x = n + c; \; \frac{dy}{dx} = 0$$

$$0 = M(n + c) - \frac{F}{2}(n + c)^2 + \frac{(Q + F)}{2} n^2 \tag{5.14}$$

The formula for Q, without simplification, would be

$$Q = \frac{F - [(\beta\gamma p_0 wT^4)/(18nc^2)]}{\frac{2}{3}\left(\frac{n}{c}\right)^2 + 2\left(\frac{n}{c}\right)} \tag{5.15}$$

From equations (5.7), (5.14) and (5.15):

$$M = \frac{\beta\gamma p_0 wT^4}{6n^2}\left(1 + \frac{n}{c}\right)$$

Assume $M = Kp_y x \dfrac{wT^2}{4}$ and $\gamma = 1.5$

where K = factor relating M to the plastic moment at the root of the web,
p_y = yield stress (or design strength) of the flange

$$\text{Then } n = T\sqrt{\left[\frac{\beta p_0}{Kp_y}\left(1 + \frac{n}{c}\right)\right]} \tag{5.16}$$

In the above 'elastic' analysis the moment at the bolt line is less than that of the plastic hinge. In minimum flange thickness design, when plastic hinges are assumed at the root of the web and the bolt line,

$$\text{moment at bolt line} = M_B = Qn = \frac{p_y}{\gamma} \times \frac{wT^2}{4}$$

$$\text{and } \gamma = 1.5 \tag{5.17}$$

If the slope of the flange at A is zero,

$$\delta_B = \frac{Qn^3}{6EI} \tag{5.18}$$

$$\text{If } \delta_B = \frac{\beta p_0 T}{3E} \text{ as before} \tag{5.19}$$

From equations (5.17), (5.18), (5.19) and $I = \dfrac{wT^3}{12}$

$$n = T\sqrt{\frac{\beta p_0}{p_y}} \tag{5.20}$$

The simplified analyses ignore effects such as local compression of the plates and shear deflections. These effects would be beneficial in that they would allow a higher value of n if they were taken into account. In the absence of tests to give guidance it is proposed that the following formula for the maximum value of n is used in place of formulae (5.16) and (5.20), i.e. with both the formula for minimum prying force and for minimum flange thickness design:

$$n = 1.1T\sqrt{\frac{\beta p_0}{p_y}} \tag{5.21}$$

For practical design using HSFG or Grade 8.8 bolts this leads to the following values of n:

For preloaded bolts $n = 1.5T$
For non-preloaded bolts $n = 2.0T$

Comparison with tests

Nearly all the formulae that have been proposed for calculating the prying force are semi-empirical. That is, although they have a 'theoretical' basis they are modified to agree with the results of laboratory tests. A large number of tests have been carried out on Tee connections[14–16] but unfortunately the range of size and proportion covered by them is extremely limited. Calibration against tests has been carried out for the prying force formulae presented here and the conclusions from this are as follows:

1. Prying forces must be taken into account in the design of tension connections, if full bolt strengths are to be mobilized safely. (See Section 5.3.4 for a discussion of BS 5950.)
2. Calculating the prying force from the moment at the bolt line will not always be adequate. The calculation of a minimum prying force using formula (5.12) is also necessary (except, of course, for a minimum thickness design under normal loading).
3. There appears to be no need to modify formula (5.12) to fit the test results.

5.3.3 Multibolt connections

Variations in behaviour between the simple, two-bolt, connections discussed above and multibolt connections are primarily related to the more complex behaviour of the connected parts, and their behaviour is discussed in Section 7.7.

5.3.4 Design

Tables 5.1 and 5.2 summarize the design strengths of the more common dowel bolts in accordance with current codes of practice. Once again it can be observed that 'factors of safety' are higher for bolts than for most other structural elements. This may partly be ascribed to concern about the limited ductility of such elements, combined with the difficulty found in assessing true load in the bolts. The difficulty arises mainly from the potentially complex interaction of the stiffnesses of the various elements of the connection and to lack of fit. In addition, it seems likely to be related to the difficulties of allowing for prying action, as discussed in Section 5.3.2. The design strengths given in BS 5950 for ordinary bolts appear to make a 20% allowance for prying, which is not required to be calculated. Where, as is certainly recommended by the authors for critical connections or those of unusual proportions, prying forces are calculated directly it would seem reasonable to base bolt design strength on the lesser of stress under proof load and 0.7 × ultimate strength.

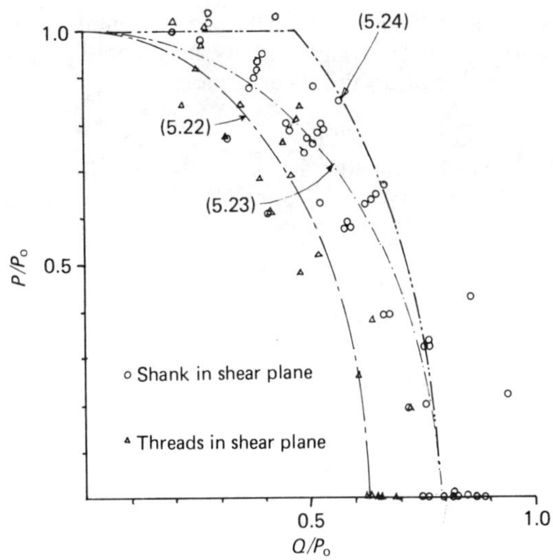

Figure 5.13 Interaction diagram for dowel bolts under combined shear and tension. P = tension in bolt at failure; Q = shear in bolt at failure; P_0 = tensile capacity of bolt

5.4 Bearing bolts under combined shear and tension

5.4.1 Behaviour

Figure 5.13 summarizes the experimental strength of bearing bolts subject to combined shear and tension.[17,18] It can be seen that bolts show considerable variation in strength under combined loading. The first point to note is the variations in the ratio of shear strength to tension strength. Where the shear plane cuts the threaded portion of the bolt the ratio varies from 0.63 to 0.68. Where it cuts the bolt shank it varies from 0.75 to 0.89.

This variation in relative strength must partly be explained by the scatter of results observed under separate components of loading. However, a further complexity arises because of possible uncertainty of failure plane position when the shear plane cuts the bolt shank. In such circumstances the bolt may fail in combined shear and tension on the shear plane. Alternatively, it may fail primarily in tension on the threaded portion. Indeed, if the threaded region is sufficiently far from the shear plane the former may be subject to negligible shear. A further complexity, reported in an earlier study of the subject,[17] is the influence of grip length on strength. Because more bending can develop in a long-grip bolt it will present an elliptical, non-orthogonal cross-section to the shear plane, thus increasing shear strength.

It is difficult to formulate an empirical interaction relationship in the presence of such variability and complexity. However, the elliptical relationships given below and shown plotted in Figure 5.13 would seem a reasonable fit to the experimental results, particularly if the high outliers are ignored.

For threads in the shear plane:

$$\left(\frac{\text{Applied tension}}{\text{Tensile strength}}\right)^2 + \left(\frac{\text{Applied shear}}{0.63 \times \text{tensile strength}}\right)^2 = 1.0 \qquad (5.22)$$

For shanks in the shear plane:

$$\left(\frac{\text{Applied tension}}{\text{Tensile strength}}\right)^2 + \left(\frac{\text{Applied shear}}{0.79 \times \text{tensile strength}}\right)^2 = 1.0 \qquad (5.23)$$

The coefficient 0.79 in the latter equation is simply 0.63/0.8, where 0.8 is the average ratio of tensile area to shank area. It recognizes that the shear is applied to the larger cross-sectional area.

An alternative approach for situations where the shear plane cuts the bolt shank would be to apply the basic interaction relationship (5.22) to the shank area and apply a separate check to the threaded portion in tension. When expressed in terms of bolt loads this leads to:

$$\left.\begin{aligned} &\left(\frac{0.8 \times \text{applied tension}}{\text{Tensile strength}}\right)^2 + \left(\frac{\text{Applied shear}}{0.79 \times \text{tensile strength}}\right)^2 = 1.0 \\ \text{and} \quad &\frac{\text{Applied tension}}{\text{Tensile strength}} \ngtr 1.0 \end{aligned}\right\} \qquad (5.24)$$

Although adopted as the basis for some codes it is apparent from Figure 5.13 that it is too optimistic. This presumably reflects the real uncertainty of failure plane which exists when the threads nearly extend to the shear plane. Nominally, the shank cross-section may govern; in practice, the nearby threaded portion may be more critical in the presence of significant shear and coincident bending.

5.4.2 Design

Because of the variability outlined above, it would seem prudent for strength factors to be conservative. As indicated in Sections 5.2.7 and 5.3.4, this has already been achieved for both shear and tensile loading separately. Thus any basis for strength assessment that uses these component strengths should have a satisfactory degree of conservatism.

Many codes of practice, including both BS 5400, Part 3, and draft Eurocode 3, are based on an elliptical interaction. In both these cases an effective coefficient of $1/\sqrt{2}$ rather than the 0.63 indicated in the previous section has been used.

BS 5950 takes another approach. Recognizing both the uncertainty of the strengths and the awkwardness of elliptical interaction rules for design office use, it proposes a trilinear interaction. Thus:

$$\frac{\text{Applied tension}}{\text{Tensile strength}} \ngtr 1.0$$

$$\frac{\text{Applied shear}}{\text{Shear strength}} \ngtr 1.0$$

$$\frac{\text{Applied tension}}{\text{Tensile strength}} + \frac{\text{Applied shear}}{\text{Shear strength}} \ngtr 1.4$$

In all cases the design should be based on the net tensile area where threads are in the shear plane. Where the shank is in the shear plane it is necessary to check both the shear plane under combined loading and the net tensile area under tension only.

5.5 HSFG bolted connections in shear

5.5.1 General behaviour

Figure 5.14 shows the behaviour of a single, 20 mm diameter, general grade HSFG bolted connection in double shear.[19] In this case the faying surfaces (i.e. those in contact) have been grit blasted; the holes are 2 mm clearance on bolt diameter. Initially the

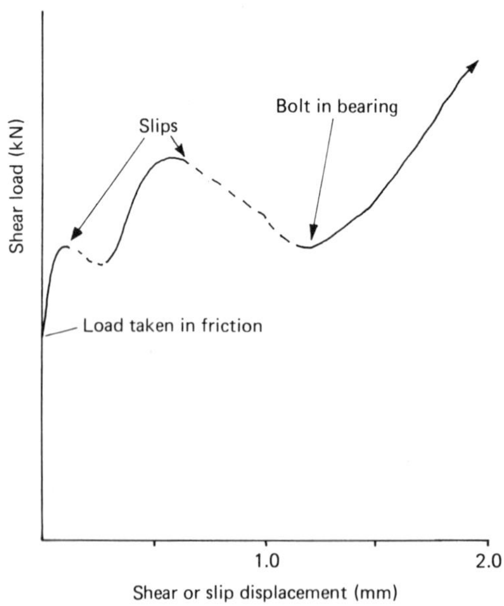

Figure 5.14 Load deformation response of a single, 20 mm diameter, general grade HSFG bolt in double shear

load is transferred by friction between the surfaces in contact, this friction being developed by the high interface pressures that arise because the bolt has been stressed to its proof load or higher. (A discussion of bolt types and methods of tightening is presented in Section 3.3.)

Initially the connection shows very little deformation because of the interlock between the interstices of the faying surfaces; this high stiffness is, of course, the most important characteristic of such connections. For bare metal surfaces this pre-slip movement is unlikely to exceed 0.1 mm (some metal coatings can cause pre-slip movements to be considerably greater, as discussed in Section 5.5.2). As the load increases the interstices start to yield significantly in shear and deformations increase.

Further loading will produce a slip, i.e. a sudden, usually audible, movement. If sufficient strain energy is available a single, large slip will take the bolt into bearing. In this instance there was insufficient energy available and the slip was arrested after a small movement. Further loading produced a second slip of greater magnitude, taking the bolt into bearing. For bolts in normal clearance holes the slip may total 4 mm. The variation in the two slip loads (in this case with the second being higher than the first, although in many cases the reverse will be true) and the variation in slip displacement is quite typical of this sort of test. It illustrates the general variability of most aspects of frictional behaviour.

Once the bolt is bearing on the sides of the holes it starts to behave partially as a dowel fastener. The load is thus being carried partly by bearing and partly by friction. As the bolt, already at the point of plastic deformation due to its pretension, is subject to additional, shear, forces, it will become plastic. This plastic deformation will lead to a reduction in preload (by the normality principle for plastic deformation) and a transition from frictional to bearing behaviour gradually takes place. However, in most cases a bolt/plate combination has a greater capacity in bearing and shear than in friction, and this change in mode of behaviour will be accompanied by an increase in resistance.

It can readily be demonstrated that a bolt will have a greater shear capacity than its slip capacity. Bolt proof load for a general grade HSFG bolt is 70% of ultimate tensile capacity (P_{ult}) and a coefficient of friction is unlikely to exceed 0.6. Therefore slip load per interface is unlikely to exceed $0.6 \times 0.7 \times P_{ult}$, i.e. $0.42\ P_{ult}$. However, as discussed in Section 5.2.3, bolt capacity in single shear is approximately $0.6\ P_{ult}$, a satisfactory reserve. Bearing capacity in the post-slip condition will be a function of plate thickness – for most practical connections it will not govern. Unlike the situation for bearing bolts, where the necessity to limit deformations at serviceability load levels reduces design strength in bearing at collapse, it is possible to mobilize the full bearing capacity in HSFG bolted connections, where design for the post-slip condition is only concerned with strength.

5.5.2 Friction

When HSFG bolted connections were first developed it was customary to specify bare steel surfaces, although tightly adhering mill scale was deemed to be acceptable, and design to a coefficient of friction (μ) of 0.45. It is now acknowledged that this is too simple and optimistic an approach for such a variable phenomenon. The variability of friction is well demonstrated by Figure 5.15, which summarizes a large number of tests for surfaces coated with mill scale. A comparison of this figure with the original design value for μ of 0.45 also shows the optimism of early approaches to design.

Design values for μ are dependent on choice of partial factors. Since different codes take different views of the variability of friction it can be misleading simply to look at design values. Table 5.3 therefore presents typical *average* values of coefficients of friction. However, note that these are based on measured bolt preloads which, in many cases, are significantly higher than nominal preloads. Any assessment of friction that is based on nominal bolt preload is likely to overestimate true friction.

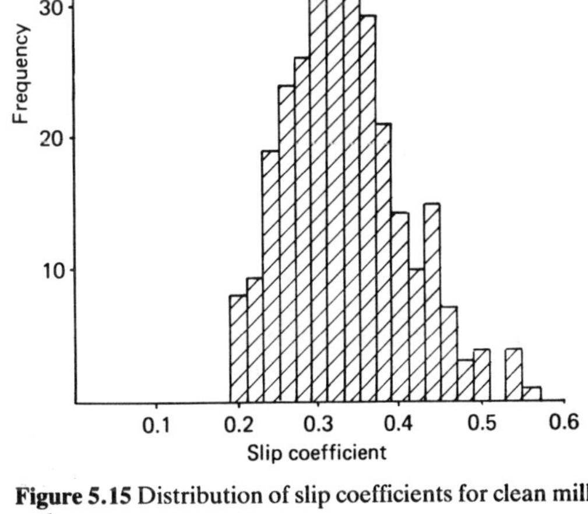

Figure 5.15 Distribution of slip coefficients for clean mill scale

Table 5.3 Typical average values for coefficients of friction

Clean mill scale	0.33
Grades 43 or 50, grit or shot blasted	0.48
Ditto, after light rusting	0.52
Very high tensile steel, grit blasted	0.33
Red lead paint	0.1 or less
Grit blasted and oiled	0.25
Galvanized	0.22
Galvanized, subsequently wire brushed or grit blasted	0.41
Metal sprayed with zinc	0.46
Metal sprayed with aluminium	0.51
Metal sprayed aluminium on zinc	0.49
Surfaces painted with alkali–zinc silicate coat (50–80 mm)	0.46

It is possible to identify some general characteristics of behaviour. First, any coating or foreign matter that can act as a lubricant must be excluded. Oil-based paints and drilling oil are particularly deleterious. In some circumstances mill scale can also act as a lubricant, and this partly explains the variability associated with this finish – it is generally not permitted in modern specifications, although its use is still allowed in some British Standards. *Loose* rust and an excessive build-up of some softer metal coating such as zinc from hot-dipped galvanizing can also reduce friction.

Surface roughness is an important criterion. Grit blasting generally gives a very satisfactory coefficient of friction because of the interlock between the interstices produced by the grit impacts. However, a markedly lower coefficient of friction is obtained with a very high-yield steel; a less rough surface is obtained on the tougher material with a less satisfactory interlock. *Light* rusting appears to improve the interlock provided that all loose rust is removed by wire brushing prior to assembly. However, heavy rusting and its associated pitting reduces friction capacity.

Metal coatings may have a significant influence on frictional behaviour. As mentioned above, if a thick coating of zinc is applied, as in galvanizing, it may act initially as a form of lubricant, giving very low frictional capacity. However, after some degree of slip, cold welding is likely to occur; this 'lock-up' phenomenon is discussed in greater detail below. If the excess zinc is removed and the surface is roughened by grit blasting a much more reliable and satisfactory coefficient of friction is achieved. Aluminium and zinc spray coatings give satisfactory coefficients of friction. The surface roughness is good and the controlled thickness prevents any possibility of lubricating action. Even with the double-thickness aluminium on zinc coating used for marine applications the friction is satisfactory. However, note that some of these surfaces tend to creep under sustained loading close to the limiting slip capacity. Both test procedures for slip coefficients and design values can be modified to take account of this phenomenon.

A standard test has been devised for the determination of slip coefficients for non-standard surfaces.[5] This is a practical test, rather than a refined study for research use. Figure 5.16 shows the general arrangement of the test pieces, which are detailed to ensure that there is no transfer of bolt preload between halves of the specimen from lack of fit. The greatest concession to practicality is that the bolts are tightened by standard procedure, without any direct monitoring of bolt preload. In calculation the bolts are assumed to be tightened only to their nominal preload. Their actual preload may well exceed this value by 20–30%, depending on the particular bolt material and method of tightening. In critical situations it is therefore important to have a consistent approach to bolt tightening between the test specimens and the practical application. Three tests are carried out with two separate half joints for each specimen. The lowest value is taken as the 'characteristic' slip value; undoubtedly a statistician would question the definition of the lowest of six values as the characteristic result.

The slip test is in effect a single sample (all the specimens are made in a batch); it is quite likely to have a higher μ than the production surfaces and have bolts tensioned above their nominal preload. For these reasons, conventional design of such joints against slip is more a serviceability check than a demonstration of adequate strength. Some modern codes specify a larger γ_m for such connections: some extra factor would certainly seem prudent for structures where slip could lead to collapse.

A discussion of friction in this form of connection cannot be entirely separated from one on the distribution of bearing pressures between the faying surfaces. Figure 5.17 shows the distribution of the region of high bearing pressure in a well-fitting connection. If the bearing pressure on the annular

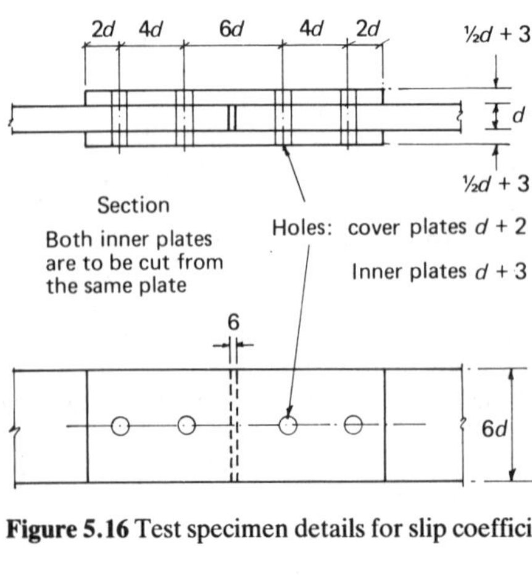

Figure 5.16 Test specimen details for slip coefficient tests

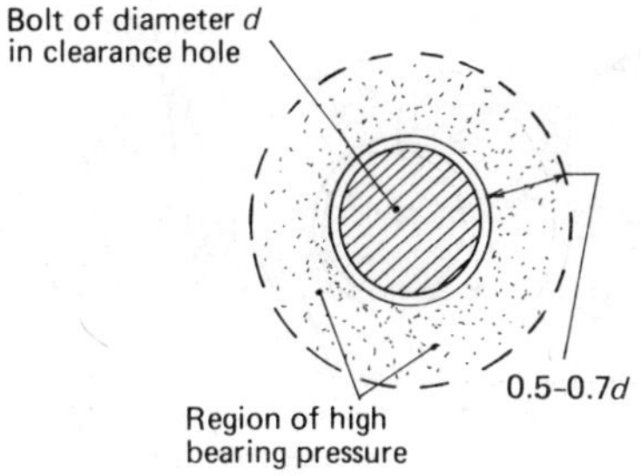

Figure 5.17 Distribution of bearing pressure at the faying surface in a friction connection

ring of contact is calculated it will be clear that the high spots on these roughened surfaces are likely to be yielding in compression. This conclusion is supported by the appearance of joints that have been separated after testing, where the high spots appear polished. It seems likely that the presence of this limited yielding contributes to the high values of friction obtained in these joints. (The coefficients of friction obtained are much higher than a mechanical engineer, say, would rely on for design in some other context.) There is certainly some evidence that, at these pressures, friction is not independent of bearing pressure. A reduction of effective bearing area (for example, with badly pitted surfaces or in a connection with oversize or slotted holes) will generally show a reduced coefficient of friction. The former should be precluded; the latter are subject to a reduction coefficient in design. Some codes also introduce a reduction coefficient for higher grade bolts, where the bearing pressure will be increased.

5.5.3 Bolt tensile behaviour

Figure 5.18 contrasts the behaviour of HSFG bolts under direct and torqued tensile loading.[9] Behaviour of similar, high-strength, bolts under direct tension has been discussed in Section 5.3.1. The variations in response under torqued tension arise because of the additional stresses in the stressed length that are caused by the torque. Because of friction in the threads a significant proportion of the torque applied to the nut is transmitted into the bolt, causing a conventional distribution of torsional shear stresses in the bolt shank, i.e. uniform around the circumference and a linear function of radius. The resulting combination of stresses causes an earlier onset of yielding on the surface of the bolt and departure from linearity. If the torquing is continued to rupture, the same effect will lead to a reduction of the tensile capacity of the bolt compared to its strength under direct tension. It is worth noting that elongation to failure also diminishes: this is because limiting strain occurs at a lower elongation under combined displacements. However, if, after initial tightening, the bolt is then subject to direct tension it is found that the adverse interaction between torque and tensile strength is overcome and the bolt's tensile capacity is undiminished. (This result could be predicted from the normality principle of plastic deformations. Given the inevitable symmetry of the yield criteria for combined torque/tension about the tension axis, purely extensional displacement will inevitably cause a total relaxation of the shear stresses.) The deformation capacity also improves, although it is still less than that under pure tension, because the plastic shear strains do not relax and therefore contribute to the development of the limiting strain.

It is not necessary to take account of the torque/tension interaction in design because any external tensile loading will be in direct rather than torqued tension. Experience has demonstrated that it is generally possible to stress bolts to their proof load (related to their properties under direct tension) despite this adverse interaction *provided that the threads are undamaged*. If bolts with higher than average thread friction are used the ratio of maximum torsional shear stress to direct stress will increase, leading to a more severe interaction. In such circumstances it is clear that the bolt may not be stressed to its tensile proof load. Bolts with tight or damaged threads should clearly not be used for friction connections.

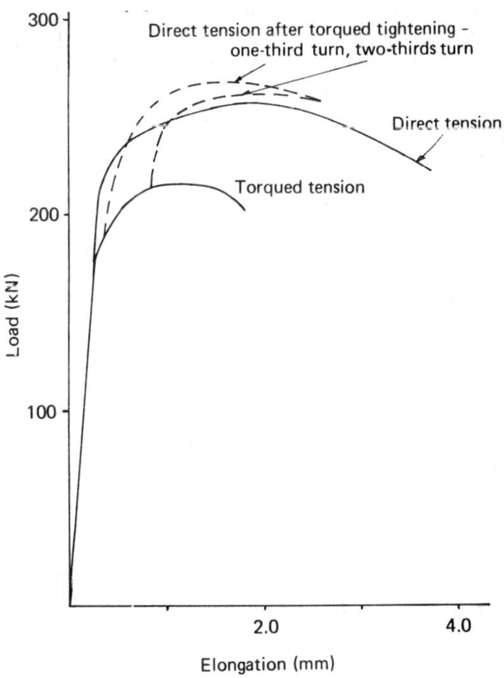

Figure 5.18 Torqued and direct tension behaviour of general grade HSFG bolt

5.5.4 Influence of bolt preload on post-slip shear capacity

Since designers are sometimes permitted to take advantage of the post-slip reserve of friction connections (and current design procedures might lead to slip at the serviceability limit state rather than the ultimate one) it is necessary to consider possible interaction between the tensile preloading of the bolts and their shear capacity. Figure 5.19 shows the results[9] of some tests on the shear capacity of high-strength bolts that have been prestrained in tension. Various degrees of prestrain have been applied, up to the point where the bolt has been taken past its maximum tensile capacity and is starting to neck prior to failure. In this case, where there is only a very short thread length within the stressed length, the critical tension region does not coincide with the shear planes and there is clearly no adverse interaction.

The situation is less clearcut where the shear plane cuts the threaded portion. Unfortunately there are no experimental results for this case. However, it seems likely that the adverse interaction will still be limited, at least until the bolt has started to neck prior to tensile failure. Once again, classical plasticity theory can be used as justification for this conclusion. If a bolt is subject to combined tension and shear to the point of plasticity and then to purely shear deformations the tensile preload will tend to relax (it is free to do so because it is not caused by any external continuing action). The bolt should then be able to sustain its full shear capacity, provided that the tensile prestrain has not reduced the bolt cross-section.

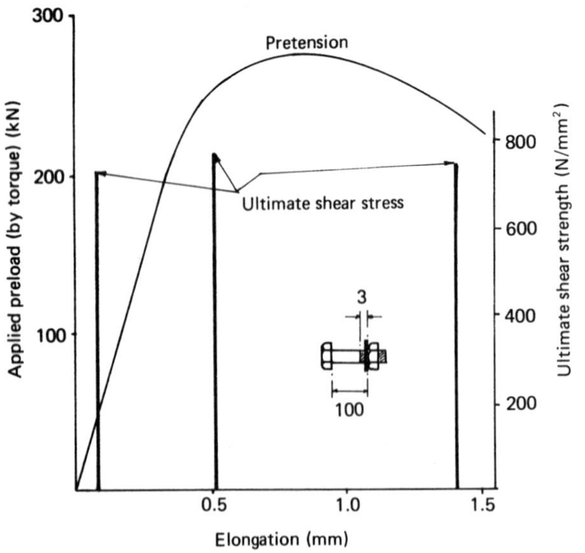

Figure 5.19 Effect of bolt prestrain on shear strength of higher grade HSFG bolts

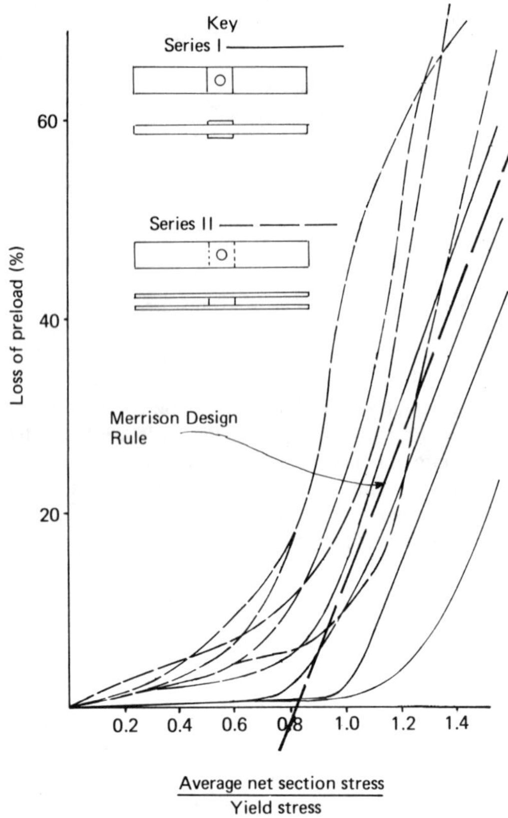

Figure 5.20 Relaxation of bolt preload due to connected plate stresses

5.5.5 Relaxation of bolt tensions

Any relaxation of bolt preload would clearly reduce slip capacity of the connection, and possible relaxation due to external tension is discussed in Section 5.6.2. However, Figure 5.20[20] shows another form of relaxation, that due to changes in thickness of the plies within the connection, arising from tensile stresses in the splice.

The most important parameter for any particular combination of plies appears to be the stress on the net section of the most heavily loaded ply. In the test programme summarized in Figure 5.20 two limiting conditions were studied. Series one represents the extreme end of a tension splice or the tensile side of a beam splice, where the main plate is fully stressed and the splice plates are unstressed. Series two represents the net sections nearest the centre of the splice where the main plate is now unstressed and the splice plates are fully stressed.

Two fairly distinct modes of behaviour can be identified. Initially, only modest relaxation of bolt tension occurred due to the elastic reduction in thickness of the stressed plies from Poisson's strain.

As the net section stress increased so the relaxation increased sharply as plastic thinning of the plies occurred. It is worth noting that the transition between the two modes took place at a lesser average net section stress than the uniaxial tensile yield one. This occurs because the stress concentrations from the bolt holes, together with the through-thickness compressions from the bolt tension, cause an early initiation of yield stress.

The effect on bolts at sections where the net section exceeds nominal yield can be quite significant, with losses of bolt tension of up to 40% at the maximum stresses permitted by some codes. However, this extreme reduction will only occur on the end row of bolts; those nearer the centre of the half-splice will suffer losses of less than 10%. It is worth noting that only one code – the Merrison Appraisal Rules – took account of the effect. It is clearly a secondary effect in most cases, but designers should be aware of the result for critical situations where slip could have particularly severe consequences.

5.5.6 Multibolt connections

In a similar way to bearing bolt connections, there are particular aspects of the behaviour of practical, multibolt HSFG connections under shear loading which need to be considered. There are many similarities between the two classes of bolted connection. Discussion in this section is therefore limited to a commentary on Section 5.2.6, where the principal aspects of multibolt connection behaviour were presented for bearing-bolt connections.

Design of *critical net sections* is similar for both classes of connection – reference should be made to Section 7.3.

Hole misalignment does not influence the distribution of shear between the bolts. Plastic methods of analysis for bolt groups under shear and torsion may therefore be used when considered to be appropriate in Chapter 8 and when permitted by the codes. However, bearing in mind the low factors against slip in practice, the use of elastic analysis can provide a useful additional reserve of capacity. Once

Table 5.4 Design data for general grade HSFG bolts to BS 5950: Part 1

Basic equations (using notation of BS 5950: Part 1)

Tension	P_t	$0.9P_0$
Slip	P_{SL}	$1.1K_s\mu P_0$
Combined shear and tension		$\frac{F_S}{P_{SL}} + 0.8\frac{F_t}{P_t} \leq 1.0$

Design capacities

Nominal dia. (mm)	16	20	22	24	27	30	36
Proof load (kN)	92.1	144	177	207	234	286	418
Tension capacity (kN)	82.9	129.6	159.3	186.3	210.6	257.4	376.2
Slip capacity (kN)	45.6	71.3	87.6	102.5	115.8	141.6	206.9

Slip capacity in single shear in presence of external tension (kN)

External tension (kN)							
25	34.6	60.3	76.6	91.5	104.8	130.6	195.9
50	23.6	49.3	65.6	80.5	93.8	119.6	184.9
75	12.6	38.3	54.6	69.5	82.8	108.6	173.9
82.9	9.1/0						
100		27.3	43.6	58.5	71.8	97.6	162.9
125		16.3	32.6	47.5	60.8	86.6	151.9
129.6		14.3/0					
150			21.6	36.5	49.8	75.6	140.9
159.3			17.5/0				
175				25.5	33.8	64.6	129.9
186.3				20.5/0			
200					27.8	53.6	118.9
210.6					23.2/0		
Change in interpolation interval							
250						31.6	96.9
257.4						28.3/0	
300							74.9
350							52.9
376.2							41.4/0

Assumes a coefficient of friction of 0.45.

slip has occurred, the same considerations of unequal load distribution apply with bearing-bolt connections and the greater conservatism of elastic methods of analysis would seem prudent.

Bearing in multibolt connections in the post-slip condition is subject to the same considerations as for bearing bolts.

Large connection effects are equally relevant for both HSFG bolted and bearing-bolted connections. The reduction factors should be applied to both.

5.5.7 Design

Tables 5.4 and 5.5 summarize the design strengths for the current codes of practice. Examination of the codes will show that there is a post-slip reserve for all but the thinnest plies. Most codes do not permit outer plies to be less than 10 mm in thickness in order to ensure satisfactory dispersion of the through-thickness stresses at the faying surface. This ensures a post-slip reserve for these plies for all but

Table 5.5 Design data for general grade HSFG bolts to BS 5400: Part 3
Basic equations (using notation of BS 5400: Part 3)

Tension $P_T = \dfrac{\sigma_t . A_{et}}{\gamma_m \gamma_{f3}} = \dfrac{0.7 \times 827}{1.2 \times 1.1} A_{et}$ $\ngtr$M24

$= \dfrac{0.7.725}{1.2 \times 1.1} A_{et}$ >M24

Slip $P_D = \dfrac{F_V . \mu}{\gamma_m \gamma_{f3}} = \dfrac{587.0.45}{1.3 \times 1.1} A_{et}$ $\ngtr$M24

$= \dfrac{512 \times 0.45}{1.3 \times 1.1} A_{et}$ >M24

Combined shear and tension $F_D = \dfrac{(F_0 - F_t)\mu}{\gamma_m \gamma_{f3}}$

Design capacities

	Nominal diameter (mm)	16	20	22	24	27	30	36
	Proof load (kN)	92.1	144	177	207	234	286	418
	Tension capacity (kN)	68.9	107.4	132.9	154.8	176.5	215.7	314.1
	Slip capacity (kN)	29.0	45.3	56.0	65.2	74.0	90.4	131.6
Slip capacity in presence of external tension (kn)	External tension (kn)							
	25	21.1	37.4	48.1	57.3	66.1	82.5	123.7
	50	13.3	29.6	40.2	49.5	58.2	74.6	115.9
	68.9	7.3/0						
	75		21.7	32.3	41.6	50.3	66.7	108.0
	100		13.8	24.4	33.7	42.4	58.8	100.1
	107.4		11.5/0					
	125			16.5	25.8	34.5	50.9	92.2
	132.9			14.0/0				
	150				18.0	26.6	43.0	84.4
	154.8				16.4/0			
	175					18.7	35.1	76.5
	176.5					18.2/0		
	200						27.2	68.6
		Change in interpolation interval					22.2/0	
	215.7							
	250							52.9
	300							37.2
	314							32.7/0

Assumes a coefficient of friction of 0.45.

Table 5.6 Reduction factors for slip resistance of HSFG bolted joints with non-standard clearance holes

	BS 5950: Part 1	*BS 5400: Part 3*
Oversize hole	0.85	0.85
Short-slotted hole	0.85	0.85
Long-slotted hole loaded perp. to slot	0.85	0.70
Long-slotted hole loaded parallel to slot	0.60	0.70

the largest bolts. It is usually not difficult to limit the ratio of bolt diameter to minimum ply thickness to ensure that all plies have a post-slip reserve.

Bearing strengths are generally higher than for bearing bolts. As discussed in Section 5.2.7, these latter are reduced to ensure that deformations at the serviceability limit state are acceptable. This is not a consideration for HSFG bolted connections which have not slipped at this stage in loading; the full safe bearing strength may therefore be mobilized for the collapse limit state.

Table 5.6 summarizes the reductions imposed by the various codes for oversize and slotted holes. These are partly to take account of possible reductions in slip capacity arising from the less optimum distribution of bearing stresses that is discussed in Section 5.5.2. They are also imposed in recognition of the probably more severe consequences of the greater slips that can occur with such connections.

5.6 HSFG bolts under external tension

5.6.1 Behaviour

The behaviour of an HSFG bolt under direct and torqued tension is discussed in Section 5.5.3. That of a pair of HSFG bolts and its associated Tee stub (the simplest form of connection which can be discussed in this way) will closely resemble that of preloaded bearing bolts under applied tension which is considered in Section 5.3.2. Indeed both illustrations in Figure 5.11 show bolt preloads which are close to their proof loads. The principal effect of the preload is to delay the onset of prying action. Any prying force near the edge of the stub implies a moment with associated curvatures in the portion of the Tee stub outside the bolt. These curvatures cannot develop to a significant extent while the Tee stub is clamped to the base at the bolt positions by the preload. In practice, each half of the flexible end plate of Figure 5.11(b) would still bend into double curvature between the bolt and web. The associated restraining moments at the bolts would initially be resisted by non-uniform bearing pressures under the bolt heads and between the Tee stub and the base. These would cause the bolts to flex, hastening separation and the onset of conventional prying action. Thus the 'prying factor' (Q/F) varies with applied load; the relevant factor is that close to ultimate load, when the effect of preload is virtually lost.

5.6.2 Relaxation of preload

By definition, the application of its proof load to a bolt takes it slightly into the elasto-plastic part of its response curve. Thus any external preload which causes separation, or significant flexing, as discussed above, will cause plasticity in the bolt. Although, as outlined in the previous section, the true situation is more complex, it appears that significant separation will not develop if the external tension, together with an adequate allowance for amplification due to prying, does not exceed the proof load. If separation is limited in this way little relaxation will occur. However, if this limit is not observed (for example, by not taking account of prying action) it is clear from an examination of Figure 5.11(b) that significant relaxation may take place.

A particular problem exists with bolts where load-indicating washers or bolts have been used to monitor bolt tightening. These devices are designed to deform plastically once a certain bolt load has been attained; a residual gap of up to 0.4 mm will remain after tightening. The *tangential* stiffness of such a system will be considerably less than that of a conventional HSFG bolt where plasticity is confined to the bolt. With a grip length of 100 mm, say, closure of a gap of 0.4 mm will lead to total relaxation of the bolt preload. If such devices are used for connections which are required to resist slip under shear loading but may also be subject to external tension loading it is clearly important to take a conservative view of the maximum load that may be applied to each bolt, with proper account of prying.

5.6.3 Design to codes of practice

Tables 5.4 and 5.5 summarize the design strengths of HSFG bolted connections subject to external tension for current codes of practice. It can be seen that these fasteners may generally be subjected to higher external tensile loading than the corresponding bearing bolt. (A general grade HSFG bolt is made of materials similar to those of an 8.8 bearing bolt.) This is probably appropriate when it is considered that:

1. Each bolt is proof tested during tightening;

2. Nuts are larger and of greater material strength, so that nut-related modes of failure are precluded;
3. Pretensioning will prevent any significant separation of plies under design tensile loading.

5.7 HSFG bolted connections under combined shear and tension

Any external tension will produce a corresponding reduction in clamping force between the plies. If any variation in coefficient of friction with bearing pressure is discounted (it is, in any case, a second-order effect) there will be a linear reduction in friction capacity of the connection. A linear interaction formula is therefore used for the design of bolts under combined shear and tension.

If the external tension arises because of an applied moment (say, on an end-plate beam/column connection) there will be no net change in clamping force between the end plate and the column. In such circumstances it is not theoretically necessary to make any reduction in slip capacity provided that the prepared faying surfaces extend over the whole region of the end plate. However, because of possible relaxation of preloads under the tensile component of the applied moment (for the reasons outlined in Section 5.6.2), it would be prudent to make some allowance for a reduction in slip capacity in critical situations.

References

1. Unpublished test data, Imperial College, University of London, 1981.
2. Owens, G. W., Dowling, P. J. and Algar, R. J., 'Bolted connections: bearing stresses in connections using Grade 8.8 bolts', *CESLIC Report 3+3A*, Civil Engineering Department, Imperial College, University of London, 1983.
3. Unpublished test data, Imperial College, University of London, 1983.
4. Back, J. de and Jong, A. de, 'Measurements on connections with high strength bolts particularly in view of the permissible arithmetical bearing stress', *Report 6-68-3*, Slevin Laboratory, Delft, the Netherlands, 1968.
5. Hirano, M., 'Bearing stresses in bolted joints', *Society of Steel Construction of Japan (Tokyo)*, **6**, No. 58 (1970).
6. Owens, G. W. and Hargreaves, A.C., 'Bolted connections: bearing stresses in connection with Grade 8.8 bolts. Stage 2', *CESLIC Report BC6*, Civil Engineering Department, Imperial College, University of London, 1980.
7. Unpublished test data, Imperial College, University of London, 1984.
8. Owens, G. W., 'Bolted connections: the influence of pitch on the bearing strength of connections using Grade 8.8 bolts', *CESLIC Report BC5*, Civil Engineering Department, Imperial College, University of London, 1980.
9. Kulak, G. L., Fisher, J. W. and Struik, J. H. A., *Guide to Design Criteria for Bolted and Rivetted Joints*, 2nd edn, Wiley, New York, 1987.
10. Rumpf, J. L. and Fisher, J. W., 'Calibration of A325 bolts', *Journal ASCE*, **89**, 5 and 6 (1963).
11. Unpublished test data, Imperial College, University of London, 1984.
12. Goodley, M. H. R., 'Comparative strengths of HSFG general grade and high tensile grade 8.8 structural bolts', *Report 101*, Department of Construction, Oxford Polytechnic, 1980.
13. Unpublished test data, Imperial College, University of London, 1978.
14. Harding, J. E. and Dowling, P. J., 'Report on the testing of beam–column connections on the Drax completion', *CESLIC Report DR1*, Civil Engineering Department, Imperial College, University of London, September 1979.
15. Douty, R. T. and McGuire, W., 'High strength bolted moment connections', *Journal ASCE*, **91**, No. ST2 (1965).
16. Surtees, J. O. and Mann, A. P., 'End plate connections in plastically designed structures', Conference on Joints in Structures, University of Sheffield, July 1970.
17. Chesson, E., Faustino, N. L. and Munse, W. H., 'High strength bolts subject to tension and shear', *Journal ASCE*, **91**, No. ST5 (1965).
18. Shakir-Khalil, H. and Ho, C. M., 'Black bolts under combined tension and shear', *The Structural Engineer*, **57B**, No. 4 (1979).
19. Harding, J. E., MSc thesis, September 1970.

6

Fatigue of connections

6.1 Scope

A full coverage of the topics covered in this chapter is inappropriate in a general book on connection design, and the presentation is therefore limited to those aspects that are of greatest importance to the designer. A bibliography is provided at the end of the chapter for the reader who wishes to pursue a topic in greater detail. A list of references is given for each of the principal headings in the text. The order within each list indicates the relative standard of the references; more general and elementary texts are listed first, followed by more advanced ones.

6.2 Introduction

6.2.1 Basic concepts

Fatigue is the mechanism whereby cracks in a structure grow when subjected to fluctuating stresses. The sum of the mean plus fluctuating nominal stresses that cause crack growth may be very much less than the stress to cause static failure.

The fatigue life, or endurance, of a specimen without defects comprises a period for crack initiation followed by one of crack propagation to failure. However, if a cracklike defect exists, even on a microscopic scale, this will provide a natural point for crack propagation and the contribution to fatigue life from the period of crack initiation is lost. Effectively, all welds contain minute metallurgical discontinuities from which cracks will grow. Thus the fatigue life of a connection containing welds is solely related to crack growth. Final failure usually occurs in a tension region when the reduced section becomes insufficient to carry the peak load without rupture. Fracture mechanics theory shows that the rate of crack growth is proportional to the square root of its length for a given stress range. Thus a fatigue crack is relatively small and difficult to detect for most of its life.

The primary factor influencing rate of crack growth is the stress range in the immediate vicinity of the crack tip. Thus in design it is very important both to seek to reduce stress concentration and to carry out a sufficiently detailed elastic analysis to make a realistic estimate of the true stress range. Note that this implies a different approach to analysis to the force path methods that are generally proposed within this book. Because of their importance, a detailed discussion of the different types of stress concentration is presented in the following section. Figure 6.1 illustrates the importance of minimizing stress concentrations in design by considering alternative truss connections. Figure 6.1(a) shows a typical detail from a major connection in industrial steelwork. Here design is primarily governed by static strength and the structure can safely accommodate the discontinuities and stress concentrations by local yielding. It is likely that local stresses are five to ten times those that would be calculated by the simple theory used for static design. Figure 6.1(b) shows a similar scale of connection for a fatigue-sensitive structure. Considerable care in design and expense in fabrication is now justified to ensure that the stress ranges are minimized. The design will be more detailed and should consider secondary bending stresses as well as primary axial ones. The connection layout is designed to minimize large-scale discontinuities; sound workmanship is necessary to ensure that weld fit-up is good and weld profiles are smooth to minimize local discontinuities. In extreme circumstances grinding of the weld surface or other

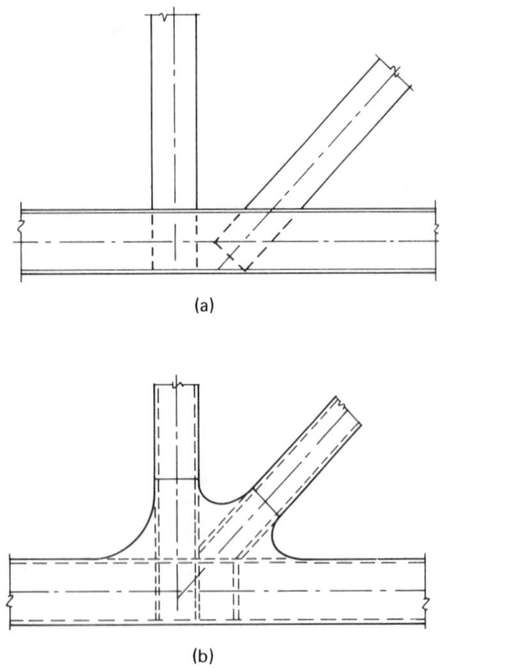

Figure 6.1 Truss details. (a) Node in industrial steelwork; (b) node in railway bridge

remedial action may be justified, as discussed in Section 6.6.

6.2.2 Stress concentrations

When designing against fatigue it is convenient to consider three levels of stress concentration.

Stress concentrations from structural action

Analysis for static design will typically only determine average stresses and, although all elastic analysis is based on compatibility concepts, detailed considerations of the relative deformation between neighbouring elements are frequently ignored. In reality, these local deformations cannot develop without additional strains and stresses. Frequently, load distribution between neighbouring elements is based on simple statics, discounting the effects of secondary elements such as bracing. Once again, the actual structural behaviour will differ from the static model. Differential displacements between neighbouring elements must imply strain and stress in the bracing elements and their supports.

These additional stress systems, which are effectively a form of stress concentration, are best illustrated by an example. In the truss whose connections are illustrated in Figure 6.1(a) the incompatibility that is discounted in simple static design is that of the restraint to end rotation of the individual elements that is provided by the connection, nominally pinned but in practice fully rigid. The ensuing bending stresses may well be of similar magnitude to the primary axial stresses in a truss of stocky proportions. When designing against fatigue these bending stresses must be analysed and accounted for.

Macroscopic stress concentrations

The second level of stress concentration with which the designer has to be concerned relates primarily to relatively large-scale geometric interruptions to stress flow. That is, geometric effects that are large in relation to the crack tip described in the introduction to this chapter. It is helpful to think in terms of stress trajectories and consider how these are modified by the presence of the changes in cross-section, notches, holes and other discontinuities. Figure 6.2 shows typical structural details containing such discontinuities. The stress concentration effect can clearly be seen from the closing down in the spacing of the stress trajectories.

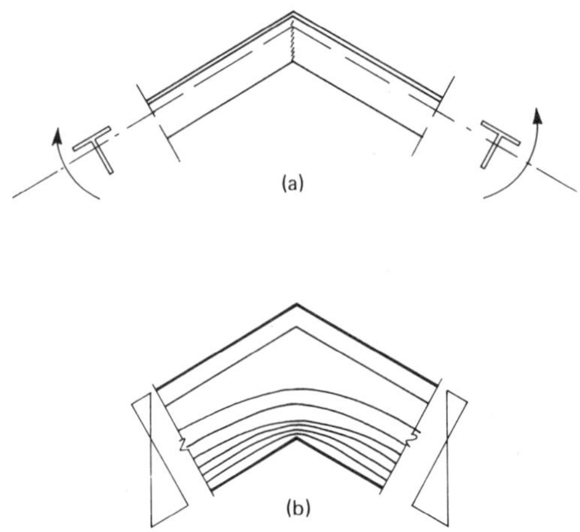

Figure 6.2 Bending stresses in a cranked beam. (a) General arrangement; (b) tensile stress trajectories at change in direction

Microscopic and local geometric stress concentrations

The final level of stress concentration relates primarily to the effects of the crack tip and any microscopic discontinuity or other defect. Such imperfections usually occur within the weld or heat-affected zone, as shown in Figure 6.3. Most of these defects will be very sharp and will therefore have a very considerable microscopic stress concentration effect.

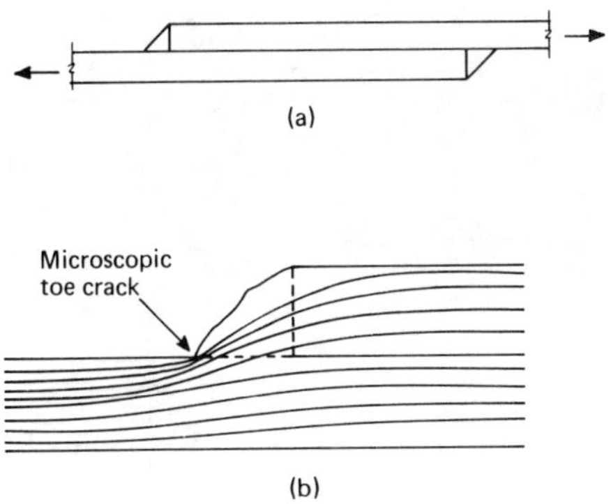

Figure 6.3 Stress concentration at the toe of a fillet weld. (a) General arrangement: (b) tensile stress trajectories

Stress trajectories through welds are, in any case, influenced by the structural action of the weld, by its geometry and surface roughness and by the relative position of the joined elements. These effects can also be seen, at least quantitatively, in Figure 6.3.

There is no clearcut dividing line between 'macroscopic' and 'microscopic' defined above. The overall discontinuity in Figure 6.3(b) is similar in both shape and scale to that of Figure 6.2. However, the distinction is still necessary to clarify the situation in design. In that context the essential difference relates to the basis of the fatigue design data. Thus the third category should include only those stress-concentrating effects that are modelled directly by the test specimens that were used to develop the fatigue data (see Section 6.4.1). It thus follows that the effects included in the second category are specifically those that are not covered by the basic design data – as such they may require additional actions by the designer.

6.3 Fatigue behaviour

6.3.1 Parent material

Most of the basic phenomena of behaviour for parent metal have been discussed in the introduction to this chapter and they are summarized as follows:

1. Fatigue life is the sum of the number of cycles for crack initiation plus the number of cycles for crack growth to the point where static failure occurs.
2. Fatigue life is a function of stress range plus mean stress, local to the point from which the crack grows. The crack itself, once formed, acts as a microscopic stress raiser.
3. For the common structural steels, fatigue life does not vary significantly with material composition and mechanical strength.
4. Surface defects, particularly at cut edges, act as stress raisers. Some cutting processes, notably shearing, cropping and punching (for holes) produce rough, work-hardened edges that are very prone to cracking. These should be avoided in fatigue situations.

6.3.2 Welds and welded structures

The following summary covers the most significant aspects of the fatigue behaviour of welds and welded structures:

1. Welds always contain small cracklike defects so that the crack-initiation portion of the fatigue life is lost. Fatigue life is simply the number of cycles for the crack to grow to the point where static failure occurs.
2. Regions in the vicinity of welds, and the weld metal itself, are likely to be subject to residual stresses of yield in tension. Stress cycling is from yield stress downwards (at least for constant amplitude of stress range), irrespective of the externally applied mean stress. Fatigue life is therefore only a function of stress range.
3. Fatigue life varies with type of welded detail. Different details have varying probabilities of cracklike defects; local stress flow past these defects varies with joint configuration and structural function.

6.3.3 Bolts and bolted joints in tension

A bolt is clearly a poor element to resist fatigue. The change in section at the junction of the head and shank will cause a concentration of the axial tensile stress. This will be exacerbated by radial bending of the head due to the eccentricity between the bearing surface under the head and the bolt shank. At the other end of the bolt the threaded portion will inevitably extend into the stressed length. The irregular profile of the threads will induce stress concentrations, and this is worsened by the local bending of the threads as they transfer load from the bolt into the nut. For both regions of stress concentration any eccentricity of seating under the nut or head is going to induce bending action, which will further increase stress range.

The only way to improve the fatigue resistance of the bolt itself is to seek to ensure that the profiles in the immediate vicinity of the stress concentrations are as smooth as possible. Thus a small radius is specified where the shank joins the head. The root profile of the threads is likely to be smoother if they are rolled rather than machined; probably of greater significance, the rolling process is likely to induce compressive residual stresses at the thread root.

However, in the context of the fatigue resistance of the bolted joint much the most important consideration is to recognize that the bolts are poor fatigue details and must therefore be protected from fluctuations in stress. This is simply achieved by ensuring that the bolts are reliably tightened to a load that is greater than the maximum external load to which they will be subjected in service. Examination of Figure 5.11(a) shows that in this state, provided that a robust end plate is used, fluctuations in load are primarily resisted by variations in the interface compression. For example, an increase in joint load per bolt from 50 to 130 kN only increases bolt tension by 7 kN. Figure 5.11(b) demonstrates the importance of taking proper account of prying action when ensuring that the bolt preload is not exceeded in service. Thus with a relatively flexible end plate, i.e. in this instance with a thickness of only 85% of the bolt diameter, an increase in joint load from 50 to 130 kN induces an increase in bolt load of 45 kN. In addition, the flexing of the end plate would be very likely to induce bending stresses in the bolt, further increasing the true stress range at the critical section. Premature failure of this detail would be very likely to occur in a fatigue environment. Thus, when carrying out a fatigue-sensitive design one of the methods given in Section 5.4.3 should be used to determine the magnitude of the prying action, the objective of the design being to ensure that the sum of the maximum external plus prying forces in the bolts is suitably less than their preload.

In practice, a very reliable way of ensuring that the bolts are suitably tightened is simply to specify HSFG bolts with the torque control method of tightening. Load-indicating devices should be used with caution, because any overstress can lead to a significant loss of preload. Turn-of-nut methods will strain the bolt into the plastic region and are also to be avoided.

Bearing in mind that the reduced stress range in the bolts is achieved by prestressing the connection, care must be taken during fabrication and erection to ensure that the reduced stress range in the bolts is not increased by any lack of fit between the plates.

6.3.4 Bolted joints in shear

For obvious reasons, ordinary bolts in shear connections will not behave satisfactorily under load reversal, or indeed during significant repeated fluctuations in load. Discussion in this section is therefore limited to joints with HSFG bolts.

Such joints have a particular type of stress concentration that has some significance on their fatigue performance. As shown in Figure 6.4, the concentration occurs at the outer limit of the region of high bearing contact around each bolt. Within each region, significant shear transfer between plies occurs but friction ensures that there is no relative movement. Outside and between these regions there must be relative movement under fluctuating load because of variations in stress and strain in neighbouring plies (the implication of this variation on the static strength of long joints in shear is discussed in Section 5.2.6). At the boundary between these two regions the change from relative to zero movement causes a stress concentration. Immediately outside this boundary there will still be significant bearing pressure and this, in the presence of movement, gives rise to fretting corrosion. In time, pitting of the surfaces occurs, and this provides a further stress raiser, as well as being the likely region for crack initiation. Thus, when such a joint fails in fatigue it is likely to crack through the gross section with cracks running through the edges of the regions of high bearing pressure.

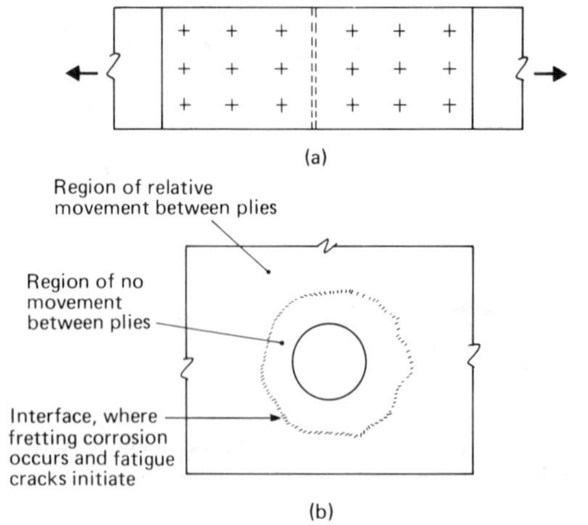

Figure 6.4 Stress concentrations and fatigue behaviour of HSFG splices under fluctuating load. (a) General arrangement; (b) details of stress concentration

Fortunately, the implications of this rather unusual mode of fatigue failure on practical design prove not to be particularly severe, and they are discussed in the following section.

In considering the performance of HSFG bolted joints in shear it should be recognized that conventional design procedures (for example, BS 5950: Part 1, Clause 6.4.2.1) leave little margin against slip at the serviceability limit state. Where the load cycles are close to this load it would seem prudent to introduce some additional conservatism into the design. Without this, any difficulties of fit-up or lack of control in tightening could easily lead to slip.

6.4 Design data

6.4.1 BS 5400: Part 10

For conventional structural steelwork, by far the most comprehensive summary of fatigue data is presented in BS 5400, Steel, Concrete and Composite Bridges: Part 10, Code of Practice for Fatigue.

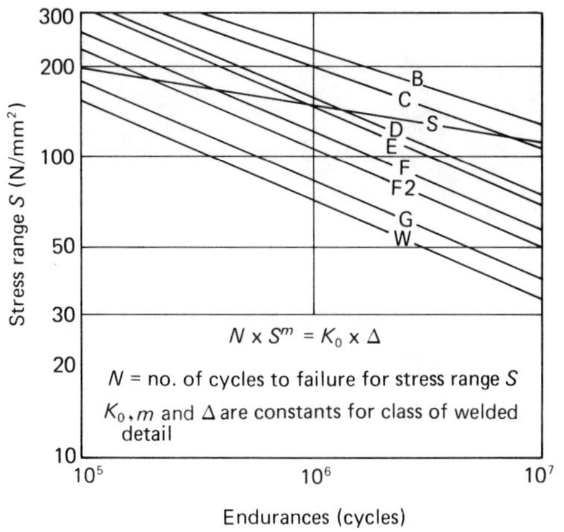

Figure 6.5 Mean *S/N* curves from BS 5400, where *S* is stress range and *N* is number of cycles to failure. (*Note*: not for use in design)

S–N curves relating stress range (*S*) to fatigue life (*N*) are presented for a wide range of commonly occurring details, and these are reproduced in Figure 6.5. (Note that this figure presents mean curves; design curves are normally based on mean minus two standard deviations. The stress ranges are unfactored.) The details are classified in BS 5400, Table 17, Parts (a)–(c). These tables are reproduced and discussed within this section as Figures 6.6, 6.7 and 6.8, to highlight the behavioural aspects that underlie the classifications. Further discussion of the background and guidance on usage are presented in Appendix H to the Code. Types 1.1–1.5 cover parent metal away from connections. The concern over discontinuities can be seen in the move from Class A to B, even when well-radiused corners are introduced in 1.3, together with the requirement to use stress concentration factors (s.c.f.) to determine stress range. The influence of edge preparation shows in the reduction from B to C between types 1.3 and 1.4 when the hardness and roughness of the flame cut edges are not machined away. Type 1.6 specifically relates to the fretting corrosion failure on the gross section that was discussed in Section 6.3.4.

Types 1.8–1.11 record the deteriorating fatigue resistance of the net section of bolted splices. Note that punched holes are not permitted. HSFG bolts and rivets provide a clamping action which helps to reduce the stress concentration at the net section. Precision bolts are likely to be carefully fitted and so share the load equally without causing additional stress concentration on any net section. Ordinary or black bolts in clearance holes are clearly undesirable – they are given a G classification, with caution! Type 1.12, the bolt in tension, is given a high fatigue classification, but this is only if the bolt is properly tightened. The reasons for this are discussed in Section 6.3.3.

Section (b) of Table 17 classifies the welded details on the surface of the member. The twin concerns of the proximity of likely defects to the stress concentration and the severity of the discontinuity can clearly be seen. Thus the Class C of the longitudinal fillet weld of 2.2 reduces to E in Class 2.4 because the discontinuity of the weld concentrates the stress at the end toe, where defects are likely to occur in any case. Stress flow past the defect site is particularly severe at the end of the cover plate in Class 2.7 leading to a G classification. Transverse attachment welds (for example, type 2.9) are Class F where the attachment is not transferring load into the plate. This is because the mere presence of the attachment draws stress to it, with the concentration occurring at the toe of the weld where defects are most likely to occur. The reason for the G classification of Class 2.11 is bound up with the stress concentration that occurs at that point in a lap splice, as discussed in Section 4.5.2.

Section (c) of Table 17 classifies the welded details at the end connections of a member. The concern to limit the severity of the discontinuity is clearly seen in the geometric limitations of types 3.2–3.5. The butt welded joint in the rolled section of type 3.6 is given a lower rating (F_2) than the basic plate splice of 3.1 (c) because of the near-impossibility of avoiding defects at the web/flange junction. The basic geometric problems are compounded by the increased proportions of impurities and larger grain size that are likely to exist there in the parent sections. Where there is lack of continuity on the primary tension path, as in type 3.10, a very low fatigue resistance results, i.e. Class F_2 and separate consideration of s.c.f.s is required. This lack of continuity can either arise from the use of partial penetration welds (not recommended for connections subjected to significant tension) or the lack of any stiffener behind the attached flange. As can be seen from type 3.7, removal of this discontinuity raises the classification to F, without any s.c.f. if a stiffener is provided. Finally, note that types 3.12–3.15 relate to shear connectors and other elements buried in concrete that are beyond the scope of this text.

A detailed examination of these tables will show that it is difficult to avoid at least some Class F details in a welded, stiffened structure. Designers would be prudent to recognize this fact at the initial design stage.

6.4.2 Standard stress concentration factors

Because of the importance of fatigue in the manufacturing and aerospace industries there is a substantial body of data on stress concentration factors for commonly occurring geometries.

Table 17. Classification of details

BS 5400 : Part 10 : 1980

(a) Non-welded details

Product form	Rolled steel structural plates and sections											Threaded fasteners
Location of potential crack initiation	Away from all welding											
	Away from all structural connections					At a lapped or spliced connection fastened with:						In a butt joint, fastener axis parallel to σ_r
	On a member of constant or smoothly varying cross section		At any external or internal edge		At a small hole (may contain bolt for minor fixtures)	high strength friction grip bolts			rivets	precision bolts	black bolts	
						Away from hole	At a hole					In thread root
Dimensional requirements	No holes		Any aperture or re-entrant corner radius > t		Hole diameter $\leqslant 3t$	Double covered symmetrical joints only				Close tolerance hole		
	No re-entrant corners											
Manufacturing requirements (see also Part 6)	All surfaces fully machined and polished	Edges as rolled or machined smooth	Any flame cut edges subsequently machined or ground smooth		Hole drilled or reamered							Bolts to BS 3692 or BS 4395 Screw threads to BS 3643 : Part 2
						Tighten to BS 4604 : Parts 1, 2 and 3				Torque to full capacity or use lock nuts		
	No flame cutting			Any cutting of edges by planing or machine flame cutting with controlled procedure								
Special inspection requirements												
Design stress area			Net cross section			Gross	Net cross section					Core area (minor diameter)
Special design stress parameter			Use stress concentration factor for apertures or re-entrant corners			Designed for no slip at ultimate load (see Part 3)						See 7.5
Type number	1.1	1.2	1.3*	1.4*	1.5*	1.6*	1.7	1.8	1.9	1.10	1.11	1.12*
Class	A‡	B‡	B‡	C‡	D†	C‡	C†	D†	D†	E†	G‡	B†

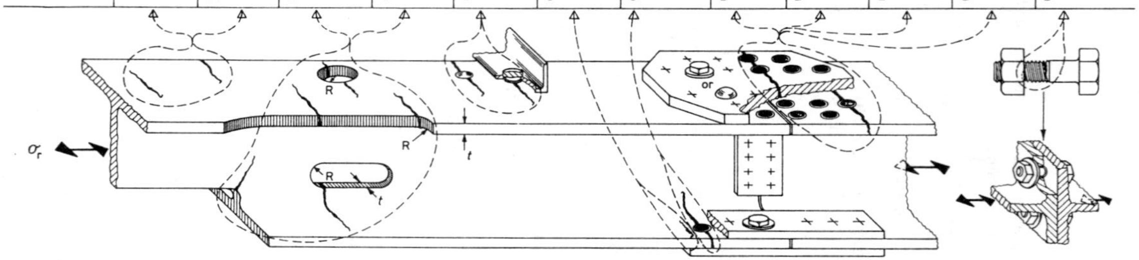

Key. Typical crack location — Cut edge — Fastener — Direction of stress fluctuation

*See H.2.3.
†Important features that change significantly from one type to another.
‡Classifications that should be used with caution.

Figure 6.6 Classification of details (a) non-welded details (Table 17 of BS 5400: Part 10: 1980)

BS 5400 : Part 10 : 1980

(b) Welded details on surface of member

Product form	Rolled steel structural plates, sections and built-up members											Reinforcing steel in concrete
Location of potential crack initiation	At a long welded attachment (in direction of σ_r)								At a short welded attachment		At any attachment	At welded intersections in fabric or between hot rolled bars
	Away from weld end			At an intermediate gap in a longitudinal weld	At a cope hole	At a weld end						
					Narrow attachment		Wide attachment				Close to edge of member	
							On one side only	On both sides symmetrically				
Dimensional requirements	Butt weld full penetration			Fillet weld		Weld toe not less than 10 mm from member edge					Weld toe within 10 mm of member edge	
				Intermittent $\frac{m}{h} \leq 2.5$		Weld length (parallel to σ_r) $l > 150$ mm			$l < 150$ mm			
					Attachment width $w \leq 50$ mm		$w > 50$ mm					
Manufacturing requirements (see also Part 6)	Grind smooth any undercut on member edges										Grind any undercut	Resistance or manual plus grind smooth undercut
	Dress flush reinforcement	Automatic no stop-starts									Avoid weld returns round laps (see z)	
Special inspection requirements	Proved free of all significant defects											
Design stress area	Minimum transverse cross section of member at location of potential crack initiation											
Special design stress parameter									$\tau \leq 0.5\sigma$	$\tau > 0.5\sigma$		
Type number	2.1*	2.2*	2.3	2.4*	2.5	2.6	2.7*	2.8*	2.9	2.10*	2.11*	2.12
Class	B‡	C‡	D†	E†	F†	F2†	G†	F2†	F†	E‡	G†	D†

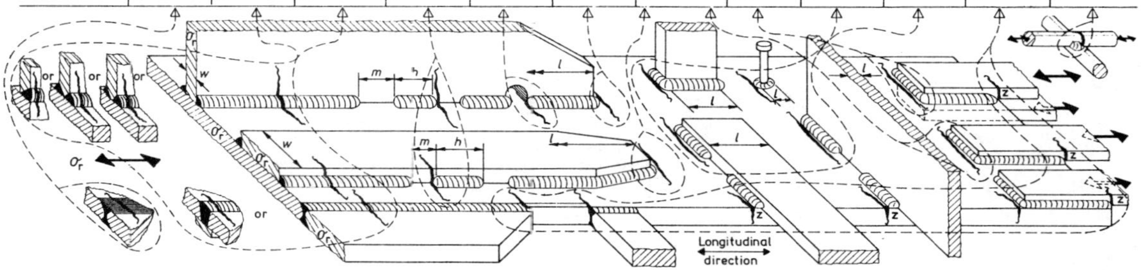

Key. Typical crack location — Surface grinding — Direction of stress fluctuation ↔

NOTE. Weld throat cracks are type 3 (see table 17(c)).

*See H.3.3.

†Important features that change significantly from one type to another.

‡Classifications that should be used with caution.

Figure 6.7 Classification of details (b) welded details on surface of member (Table 17 of BS 5400: Part 10: 1980)

(c) Welded details at end connections of member

Product form	Rolled steel plates only					Rolled steel sections of built-up members						Shear connectors in concrete	Rolled steel reinforcing bars in concrete		
							(including plates)								
Location of potential crack initiation	At transverse weld joining:										In weld throat		At transverse weld		
	two single plates end to end					two members end to end			end of one member to side of another		In any incompletely fused joint	Between encased connector and any member	Between ends of bars	Between end of bar and surface of plate	
							with third member transverse through joint								
Dimensional requirements	Longitudinal axes in line								Full penetration butt weld	Partial penetration butt or fillet weld			Axes in line	Axes perpendicular	
	Full penetration butt weld							Partial penetration butt or fillet weld					Full penetration butt weld		Fillet weld leg size $\geqslant 0.4\,D$
	Equal width	Any width change $\leqslant$ 1 in 4 slope			Abrupt width change	Similar profile							Slope at any diameter change $\leqslant$ 1 in 4	Plus reinforcing fillets with leg size $\geqslant 0.25\,D$	
	Equal thickness	Any thickness change $\leqslant$ 1 in 4 slope													
			[also includes plug welds (see footnote)]				[also applies if third member is narrow and slots through first (see appendix H)]								
Manufacturing requirements (see also Part 6)	Welded from both sides			On permanent backing strip	Build-up corners to radius $\geqslant 1.25t$	Misalignment slope $\leqslant$ 1 in 4	If width permits make weld continuous round joint otherwise grind weld ends flush (see X)						Flash butt or manual welding from both sides		
	Misalignment slope $\leqslant$ 1 in 4	Downhand shop welds, not submerged arc		No permanent tack welds within 10 mm of edge			Maximum misalignment of main members $\leqslant 0.5t_3$								
	Dress flush reinforcement				Grind corners within $2t$	Dress flush reinforcement									
	Temporary run-on and run-off plates used, weld and ground smooth														
	Grind smooth any undercut particularly on external corners												Grind smooth any undercut		
Special inspection requirements	Proved free of all significant defects						All regions stressed in through-thickness direction to be free from lamellar defects and tears							Check plate for lamellar defects and tears	
Design stress area	Minimum transverse cross section of member at location of potential crack initiation										'Effective' weld throat area		Minimum area of bar		
Special design stress parameter							Use the stress concentration factor unless third member is plate or has continuity plating		Stress concentration factor shall be used		See 6.3	See 6.4			
Type number	3.1*	3.2*	3.3*	3.4*	3.5*	3.6*	3.7*	3.8*	3.9*	3.10*	3.11*	3.12*	3.13*	3.14	3.15
Class	C‡	D†	E†	F†	F2†	F2†	F†	F2†	F†	F2†	W†	S†	D†	F†	F2†

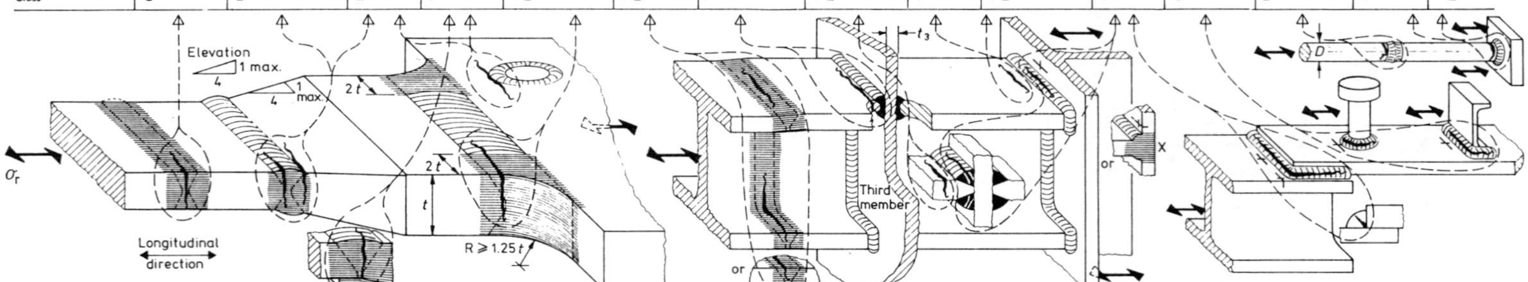

Key. Typical crack location — Surface grinding — Direction of stress fluctuation — Permitted variation to joint dimensional requirements [ ]

NOTE. Potential crack locations marked ⤫ are type 2 (see table 17(b)).

*See H.4.3.

†Important features that change significantly from one type to another.

‡A classification that should be used with caution.

Figure 6.8 Classification of details (c) welded details at end connections of member (Table 17 of BS 5400: Part 10: 1980)

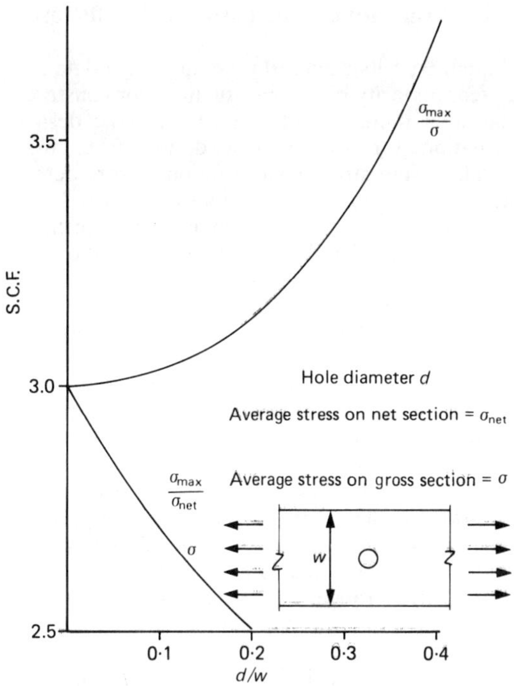

Figure 6.9 Stress concentration factors for a plate with a single, central hole

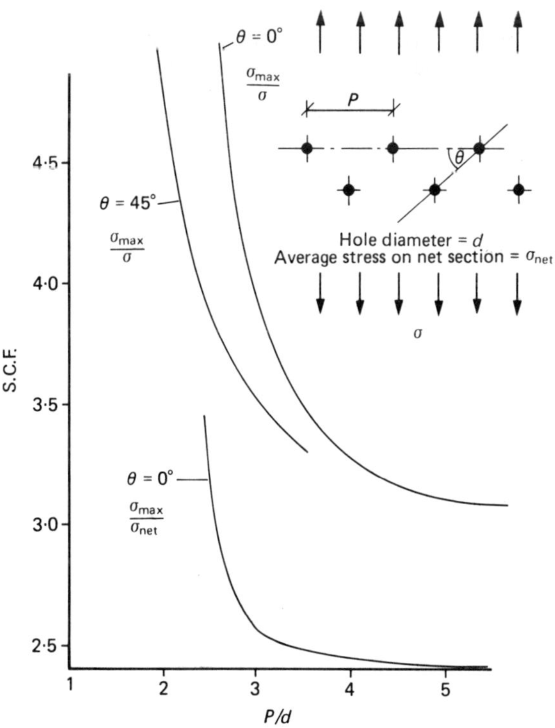

Figure 6.10 Stress concentration factors for a wide plate with a series of holes

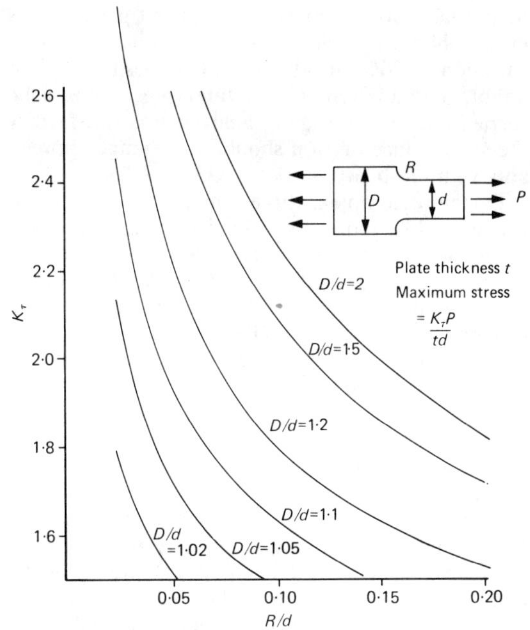

Figure 6.11 Stress concentration factors for a flat bar with a change in width and re-entrant corners with radius R

Although not often required for the design of constructional steelwork, the designer should be aware of the most useful summaries of these data that are available. They are listed in the Bibliography and Figures 6.9–6.11 show examples of their output. Incidentally, these examples have been chosen to illustrate the interaction between severity of discontinuity and s.c.f. The value of introducing even small radii to smooth out a discontinuity can clearly be seen.

6.4.3 Determination of non-standard stress concentration factors

Where stress concentration factors are required for important but non-standard situations it may be necessary to determine them directly by numerical analysis or experimental model. Numerical analysis is certainly feasible for two-dimensional studies. The most important point to appreciate is that, almost by definition, stress concentrations imply rapidly varying stress and strain fields. For accuracy, it is essential that the mesh size local to the point of greatest interest is small in relation to the local geometric changes. Figure 6.12 shows the mesh of higher-order elements (i.e. those with at least quadratic displacement formulations and mid-side models) that would typically be necessary to study a simple welded connection by this technique. As a retrospective guide on the adequacy of the mesh, the critical stress should certainly not vary by more than a factor of two between adjacent gauss points in neighbouring elements. In critical cases a variation of 30% or 40% would be desirable. If in doubt, a re-analysis with a finer mesh should be carried out; if the original analysis was satisfactory the stress concentration should not change appreciably with mesh size.

For more complex, three-dimensional, situations it may well be more cost-effective to use a physical model to determine s.c.f. This has certainly proved to be the case in the offshore sector. An appropriate plastic scale model is manufactured and strain gauged. Acrylic material (Perspex) is readily available and with careful choice of scale factor may frequently be used in standard component form (for example, tubes and sheets): it can readily be glued. However, its major disadvantage is that it creeps significantly, even at low stresses. Considerable care is necessary in testing to ensure accuracy of results. Araldite is a much more satisfactory material for testing, being effectively creep-free up to 500 μs. However, no standard components are available and considerable effort must be expended in order to cast components of the model with adequate control on element thickness. Being a thermoplastic, it may readily be formed to complex shapes from sheet form.

Strain gauging has to be planned carefully if adequate accuracy is to be achieved. Since the point of greatest interest is likely to occur at a change of direction of the surface, it will not usually be possible to gauge precisely at that point. Instead, a strip of at least three (and preferably more) gauges is laid down at uniform spacing in a line away from the discontinuity. It is then possible to extrapolate into the discontinuity with reasonable accuracy.

6.4.4 Basic fatigue data onshore and offshore

Although BS 5400: Part 10 presents design data for the great majority of design situations for constructional steel there is still a need for basic design information for use with a design from first principles using stress concentration factors determined in accordance with Sections 6.4.2 or 6.4.3.

The most convenient presentation for onshore design is that given in BS 5500 (the specification for unfixed fusion welded pressure vessels), reproduced in Figure 6.13. This design curve is based on strain-controlled fatigue tests of specimens containing a ground flush butt weld. This is to recognize that virtually any location in a pressure vessel (and in practice many other fabrications) may contain a ground flush repair weld. The design curve is based on mean strength minus four standard deviations. The code gives further guidance on the practical application of this approach. One very important point to note when using this approach is that the basic data are presented as a stress amplitude, i.e. half the stress range. (In Figure 6.10 the more conventional presentation of total stress range is used.)

For offshore design the more convenient presentation is given in the Department of Energy Guidance Notes for Offshore Structures. The basic design curve, the T curve, is also presented in Figure 6.13. This is primarily for use on the design of tubular joints where the stress concentration factor is determined by one of the methods described in Sections 6.4.2 and 6.4.3. It is the same as that document's design curve for a Class D detail.

These Guidance Notes also contain valuable information on two aspects of fatigue design. The

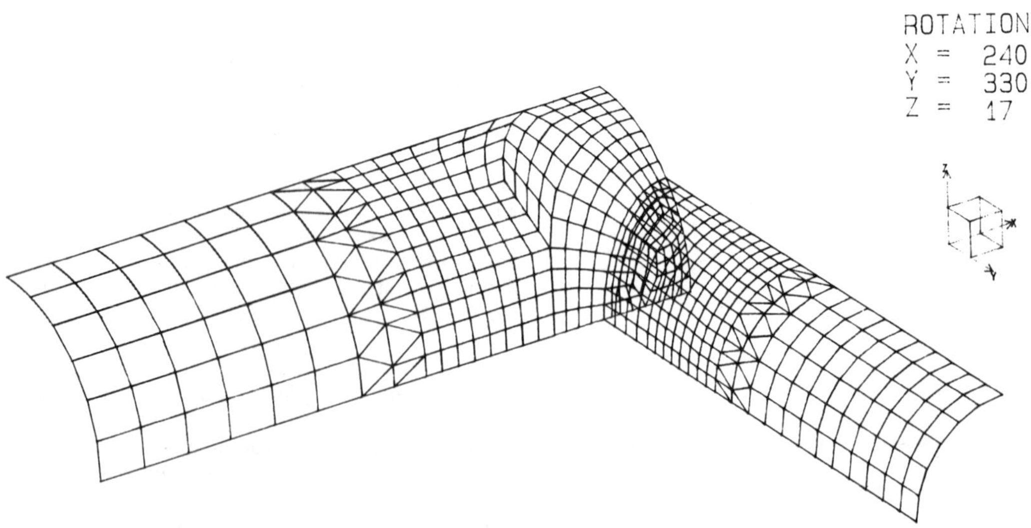

Figure 6.12 Finite element mesh to estimate stress concentration factors in a tubular X joint

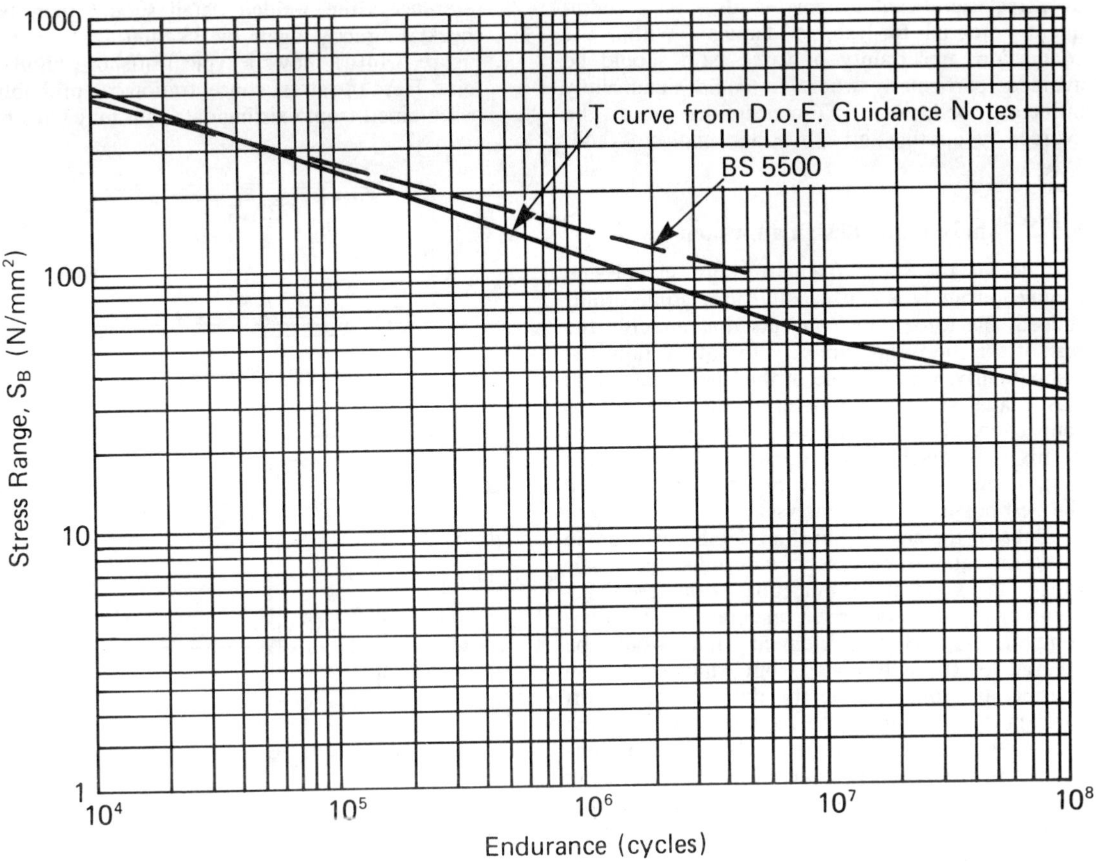

Figure 6.13 Basic design S-N curve for Nodal joints

first (only applicable to offshore structures) is that for unprotected joints exposed to seawater the basic *S–N* curves should be reduced by a factor of 2 on life. The second concerns the effect of plate thickness. This effect arises because the thicker the plate, the more the crack tip is subject to plain strain conditions, with consequent increase in true stress range. The datum thickness for the T curve is 32 mm; for other curves in that document it is 22 mm. For joints of greater thickness a correction factor on stress range of $(t_B/t)^{1/4}$ should be applied, where t is the actual thickness of the member under consideration and t_B the thickness relevant to the basic *S–N* curve.

6.5 Design

6.5.1 General

The earlier sections of this chapter have described the main aspects of fatigue behaviour and the data that are available to assist in design against fatigue. In doing so, they may have given the impression that design against fatigue is an exact applied science. The reality is very different. In the laboratory fatigue results are notorious for their scatter. In the design office analysis can rarely be carried out to the level required to predict stress ranges accurately; there is likely to be great uncertainty about the frequency of occurrence of loading; the detail under consideration is very likely not to fit neatly into one of the classes. On site, the actual stress range for a particular loading occurrence is likely to be influenced strongly by detailed fit of the joint and overall fit of the structure.

Faced with this practical uncertainty, the most important attribute that the designer can develop is an instinct for good, fatigue-resistant design. The overall form should be such that load paths are as smooth as possible. Additional unintended load paths should be avoided; for example, some bracing systems in multigirder bridges are inevitably going to attract significant primary loads – with consequent potential for fatigue damage. Within the connection, uncertainty of force path should be avoided, particularly where fit could significantly influence behaviour. Discontinuities must be avoided by tapering and appropriate choice of radii.

6.5.2 Analysis and design approach

For design for static strength, both global and connection analysis can usually be a crude affair. Indeed, the force path concepts that underlie the rest of this book are unlikely to satisfy fully the compatibility, which is one of the ingredients of an elastic analysis of an indeterminate structure or connection. However, fatigue life for most welded details is inversely proportional to (elastic stress range)^3 and it is clearly very important to carry out the best possible elastic analysis.

Globally, the analysis must model the structure in sufficient detail to be able to determine the elastic force ranges on the connection with acceptable accuracy. The model must include all structural components, including elements that would be ignored for strength analysis but which may have considerable influence on the detailed load paths through the structure. If this is not carried out the analysis will not reproduce the stress concentrations from structural action that were considered in Section 6.2.2. An example of this unintentional structural action is given by some forms of stability bracing in multigirder bridges, as referred to in Section 6.5.1.

It is difficult to give specific guidance on connection analysis because the most effective approach will depend, on the one hand, on the connection and its structural environment and, on the other, on the format of design data that is being used. Examples are given below:

1. Figure 6.14(a) shows a simple bolted splice. Here the distribution of tension between the flange and web would simply be proportioned by their respective contributions to the overall moment of inertia and BS 5400 data, types 1.6 or 1.7, applied directly to the resultant flange stress.
2. In Figure 6.14(b) account needs to be taken of the discontinuity of tension load path brought about by the lack of stiffening to the column web. The flange tension, obtained as in 1 above, is assumed to be resisted only by the effective weld area that is given by the dispersion rules of Section 7.4.2. The resultant stress range in the fillet weld is checked by BS 5400 data, type 3.11.
3. Figure 6.14(c) shows a situation where some determination of the macroscopic stress concentration is necessary. This could be obtained by careful interpretation of one of the standard references; the welded detail should then be checked directly, again by BS 5400.
4. Figure 6.14(d) shows a typical offshore tubular joint. Here the stress concentration can probably be obtained from existing data, but may have to

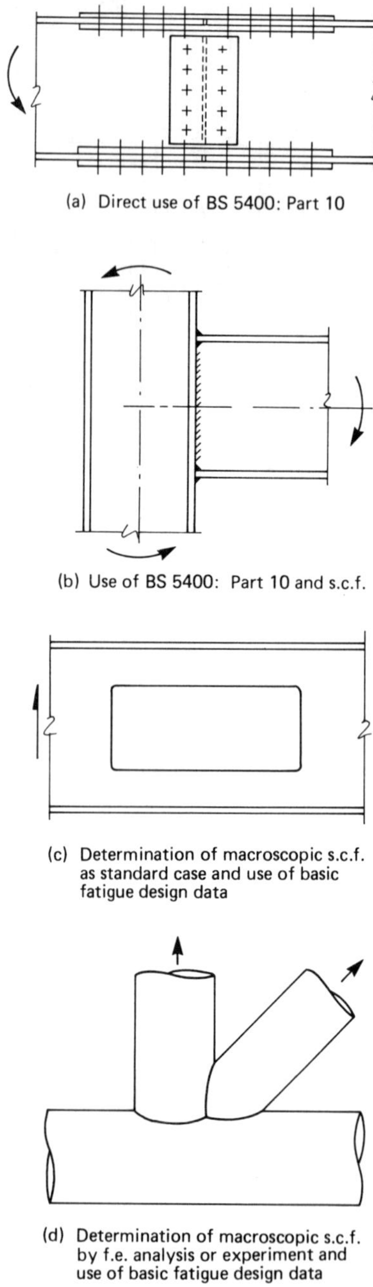

(a) Direct use of BS 5400: Part 10

(b) Use of BS 5400: Part 10 and s.c.f.

(c) Determination of macroscopic s.c.f. as standard case and use of basic fatigue design data

(d) Determination of macroscopic s.c.f. by f.e. analysis or experiment and use of basic fatigue design data

Figure 6.14 Different methods of design for (a) HSFG bolted splice, (b) unstiffened moment connection, (c) web opening in beam and (d) tubular joint

be determined experimentally if the joint proportions are not adequately represented in that data. The basic fatigue strength data will be applied as in Section 6.4.4, using the T curve of the Guidance Notes for Offshore Structures.

6.5.3 Variable amplitude loading, cumulative damage and Miner's Rule

All the previous discussion in this chapter has been in the context of a single fluctuating load of constant amplitude producing a constant stress range. In practice, any particular detail may well be stressed by more than one type of loading and each type of loading may well vary in intensity.

When faced with such a variable stress history the first step is to break the sequence into a stress spectrum as shown in Figure 6.15, using a cycle counting method. The two most popular methods are the Reservoir Method, intended for hand calculation of short stress histories and described in BS 5400: Part 10, and the Rainflow Method, intended for analysing long stress histories by computer. Alternatively, a load spectrum may be determined initially, as is the case for bridge design, from which the stress spectrum may be determined directly.

Under variable amplitude of stresses the life has to be estimated by calculation of the total damage arising from each band of the stress spectrum. The damage done by each band is defined as n/N, where n is the number of cycles in the band during the design life and N is the endurance under that stress range. If failure is to be prevented during the design life, Miner's Rule must be satisfied. This states that the damage done by all the bands must not sum to more than unity, i.e.

$$\frac{n_1}{N_1} + \frac{n_2}{N_2} + \frac{n_3}{N_3} + \ldots + \frac{n_n}{N_n} \leqslant 1.0$$

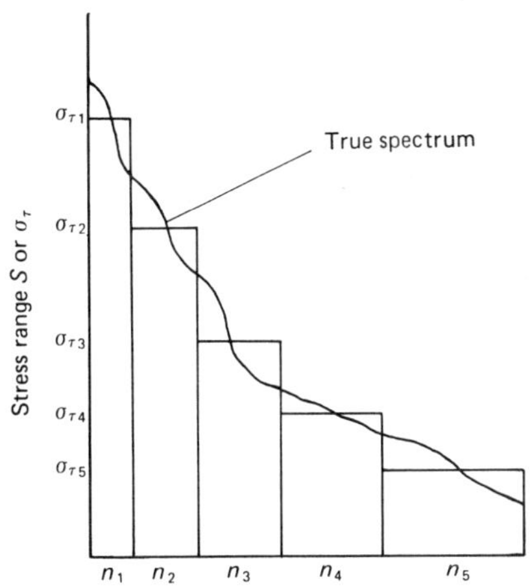

Figure 6.15 Simplification of stress spectrum for design

Note that under variable amplitude, loading stress ranges less than the non-propagating stress range (σ_0) will still cause damage. This is because the larger amplitude cycles will start to propagate the crack and, once it is formed, the microscopic stress range at its tip may still be sufficient to continue the propagation even for very low nominal stress ranges. In such circumstances the horizontal cutoff at σ_0 is replaced by a sloping line with a log gradient of $1/(m + 2)$.

6.6 Improvement and remedial techniques

There are several recognized techniques for improving fatigue performance, but it is not generally recommended that advantage should be taken of these at the initial design stage. However, the designer should be aware of them in case either the severity of the design fatigue condition worsens as the design develops or the in-service behaviour is less satisfactory than anticipated and remedial action is required.

6.6.1 Weld toe grinding

Where fatigue cracks are going to grow from the toes of welds, or the junction between beads in multipass welds, as is usually the case, an improvement in strength of at least 30% can be achieved by controlled local machining or grinding of the toe for single-pass welds or the entire surface for multipass ones. The purpose of this is to remove the very small cracks that will almost inevitably exist in this region and act as initiators for fatigue crack growth. Although these cracks are sharp, they are usually not deeper than 0.5 mm. The treatment should therefore seek to produce a smooth concave profile to the weld toe with a depth of at least 0.5 mm below the bottom of any visible undercut. Where toe grinding is used to improve the fatigue life of fillet-welded connections care should be taken to ensure that the required throat area is maintained. Of course, where the fatigue crack may grow from the weld root (as, for example, with a partial-penetration weld) no improvement in fatigue life can be anticipated from this technique.

6.6.2 Overall weld machining and profiling

The benefits of dressing off the reinforcement to butt welds is already recognized in BS 5400, where such details are given a higher fatigue classification than undressed welds. A similar improvement is to be expected where the profile of a fillet weld is smoothed by dressing. Unfortunately, there are still insufficient data to quantify the benefits of this technique, but its use is to be encouraged where the design is very fatigue-sensitive and the costs of repair of fatigue damage would be sufficiently high to justify the additional initial cost. The technique has found some practical use in the offshore sector.

6.6.3 Peening

Peening is the application of repeated hammering, usually with a round-headed punch or hammer, to cause local yielding of the material. It is applied to the weld toe or other region where a fatigue crack may initiate and has the effect of reducing locally the tensile residual stresses. This decreases the mean stress to which the critical region is subjected and therefore improves fatigue life. Once again, there are insufficient data on which to offer general advice for design. An examination of the traditional fatigue data in BS 153 suggests that a reduction in mean stress by a factor of two (which ought to be attainable on the surface if the peening is carefully carried out) may improve stress range for a particular design life by up to a factor of two. This seems most encouraging; however, the real difficulty is to ensure that the peening procedures can reliably achieve this sort of reduction in mean stress.

Where this is proposed as part of some remedial action to improve the design life of a weldment it would be appropriate to carry out a simple test series to check the efficacy of a proposed detailed peening procedure for the particular joint in question.

6.6.4 Repairs to cracked welds

Detailed repair requirements for welds that have cracked from fatigue will depend very much on circumstances, and it is not possible to make specific recommendations. However, the following points should be borne in mind when determining detailed procedures:

1. If any portion of the crack remains it will continue to act as a stress raiser.
2. The repair weld is likely to be carried out in more difficult circumstances than the original one, particularly if the latter was laid down in the shop, probably in the downhand position. It is therefore more likely to contain defects than the original weld and to have a lower fatigue life.
3. It follows that, if possible, the repair weld should be to a revised detail having a better fatigue classification than the original one.
4. If at all possible, some additional stiffening/strengthening should be added to reduce the stress range on the detail where the crack has occurred.

6.6.5 Repairs to cracks remote from welds

Where the crack has propagated into, or originated from, some region that is unwelded it is again possible to offer some general considerations for designers when they specify remedial action:

1. If any portion of the crack remains it will continue to act as a stress raiser. Metallurgical damage will have occurred some distance in front of the visible crack tip and should also be removed, preferably by drilling a hole with a radius of at least the plate thickness, just touching the visible crack tip and lying in the direction of propagation.
2. Any repair to the crack back from this hole should preferably be a full-penetration weld; this will involve grinding a weld preparation onto the two faces of the crack.
3. If possible, the repair design should incorporate some strengthening of the structure, or other modification, to reduce the stress range. Where the fatigue damage arises from some unconsidered structural action it may well be more economic and more effective to modify the structure to remove the unconsidered structural action, provided, of course, that static strength is not diminished to an unacceptable extent.

Bibliography and References

General and welded structures

1. Richards, K. G., *Fatigue Strength of Welded Structures*, The Welding Institute, Abington, Cambridge, 1969.
2. Gurney, T. R., *Fatigue of Welded Structures*, 2nd edn, Cambridge University Press, Cambridge, 1979.

Bolts

See reference 9 of Chapter 5 for a general discussion.

3. Spiers, R. and Cullimore, M. S. G., 'Fretting fatigue failure in friction grip bolted joints', *Journal Mech. Eng. Science*, **10**, No. 5 (1968).
4. 'Fatigue test on high strength friction grip bolts and Coronet Load Indicators', *Technical Report 26*, Cooper and Turner, Sheffield.

Stress concentration factors

5. Peterson, R. E., *Stress Concentration Design Factors*, Wiley, New York, 1953.
6. Roark, R. J., *Formulas for Stress and Strain*, 5th edn, McGraw-Hill, New York, 1975.

Design codes

7. BS 5400, Steel, Concrete and Composite Bridges. Part 10: 1980, Code of Practice for Fatigue, British Standards Insitution.
8. BS 5500: 1985, Specification for Unfired Fusion Welded Pressure Vessels, British Standards Institution.
9. Department of Energy, *Offshore Installations: Guidance on Design and Construction*, 3rd edn, HMSO, London, 1984.

7

Other components in the connection

7.1 Introduction

In Chapters 4–6 the behaviour and design of bolts and welds were summarized. However, the complete connection consists of much more than these. The members being connected are usually subjected to some local forces within the connection that require special consideration. It is frequently necessary to add some additional components, such as stiffeners or gusset plates to complete the connection. This chapter is devoted to the behaviour and design of these other components. Such topics are generally given much less attention than the bolts and welds themselves, both in codes of practice and in design textbooks. However, a reference back to Section 1.4 on the necessity for complete connection design should emphasize the importance of the material presented in this chapter.

The comments in Section 1.2 concerning the uncertainty of connection behaviour and the influence that this should have on sound connection design have particular relevance in the design of these elements. For example, quite arbitrary partitions of load are sometimes made between parallel elements on a particular load path. In such circumstances it is essential that the individual elements behave in a ductile way in order to permit any redistribution that may be necessary.

7.2 Slenderness limitations

7.2.1 General

The local geometric limitations on outstands and plate panels that should be adhered to in connections can usually be related to the criteria that are used in general element design. These criteria are in turn based on different performance requirements for different types of stress resultant or loading. Thus outstands in tension are limited by requirements of minimum robustness against accidental damage and the need to ensure that shear lag effects are sufficiently subdued that they may be ignored in design.

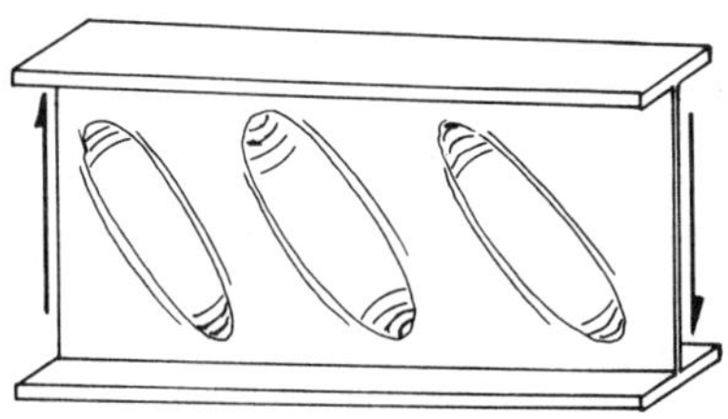

Figure 7.1 Local buckling in steel elements

Outstands in compression and plate panels in shear or compression are governed by the need to prevent, or at least limit, the local buckling indicated in Figure 7.1. Limiting geometric criteria will be a function of the requirements for strain capacity. For example, an outstand that is just able to remain stable in the presence of extreme fibre yielding would not be suitable for use in plastic design, with its requirements for stability in the presence of strains that are many times yield strain.

7.2.2 Classification of local geometric criteria for compressive or shear stresses

Figure 7.2 summarizes the generally accepted classification of geometric criteria for elements subject to instability and Figure 7.3 illustrates the influence of these criteria on behaviour by a comparison of moment/rotation curves for a restrained I-beam. Definitions and applications to element design are listed below.

Slender elements are those whose local proportions are such that they will buckle locally before attaining nominal yield stress. These slendernesses rarely occur in rolled section structures but plated structures may operate in this range (for example, compressive plating with widely spaced stiffeners where a reduced effective breadth is used).

Semi-compact elements are those which can attain nominal compressive yield strain on their most unstable fibres but will buckle shortly afterwards. An example would be a plate girder designed elastically for flexure.

Compact sections are those where a limited redistribution can be permitted within the cross-section; the extreme fibres will remain stable under strains of two or three times yield strain. This is sufficient, for example, to mobilize the effective reserve of a cross-section in flexure beyond first yield of extreme fibres; many modern codes permit the design bending strength of such sections to be based on the plastic rather than the elastic modulus.

Plastic sections are those which are of such robust proportions that they will not be prone to local buckling even when subject to strains that are many times yield strain. They may therefore be safely used in structures that are designed plastically, i.e. where redistribution of moments is assumed to occur due to the formation of plastic hinges.

7.2.3 Application to connection design

Similar concepts are relevant in connection design although their application differs in detail. The primary concern is with the degree of uncertainty of any compressive load paths in the connection. The extreme robustness of plastic geometric criteria is not generally required. However, if there is any uncertainty about the compressive load which any element is going to be subject to, either of magnitude or distribution over the cross-section, then compact criteria should be used. It is only if both the magnitude and the distribution of any compressive load are known that semi-compact criteria can be used with confidence in conjunction with truly elastic design. Slenderness criteria are rarely used in connections, and should only be applied if the stresses are very low and if proper account is taken of destabilizing actions. An example is illustrated in Figure 7.5 and is discussed later in this section.

Table 7.1 summarizes these geometric criteria in

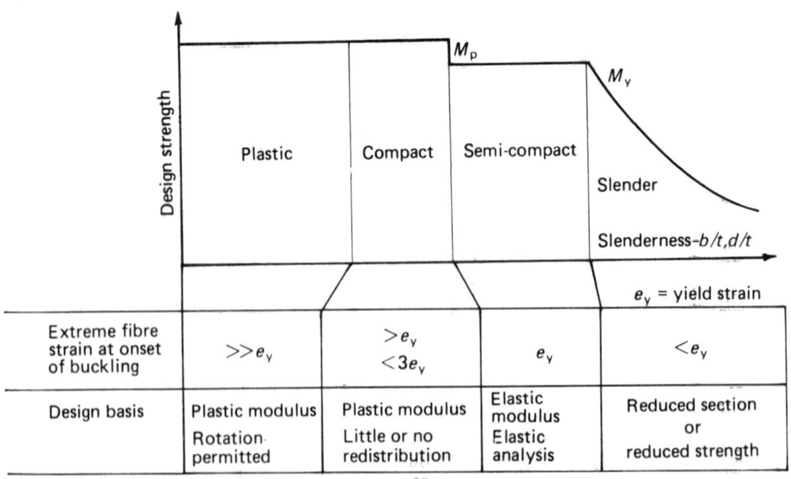

Figure 7.2 Classification of slenderness limitations for local buckling and their implications on design strength in flexure

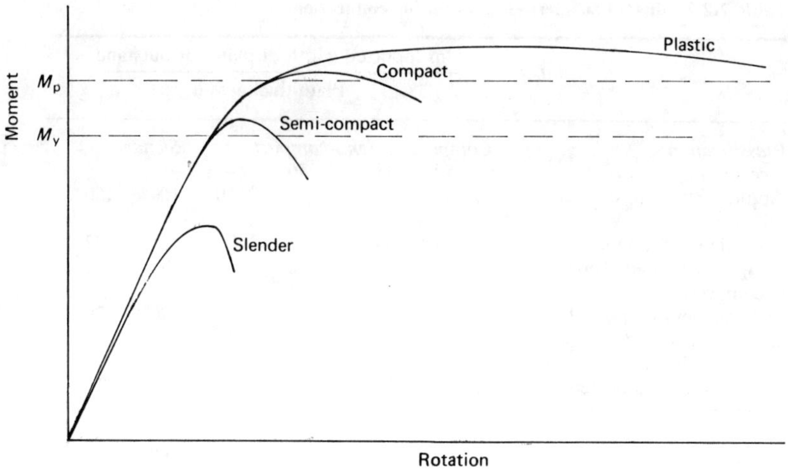

Figure 7.3 Moment/rotation curves for restrained I-beams of different cross-section proportions

general terms[1] and Table 7.2 demonstrates their application to the common structural steels. Alternatively, general local slenderness criteria from the relevant code may be adopted. These tables also present the criteria for gusset plates, which only occur in connections and therefore cannot be related to the general treatment for elements. The figures presented are derived from traditional detailing rules. These elements are permitted to be considerably more slender than others: traditional approaches are only suitable if the connection geometry is such that unsupported edges of the gusset plate are not subject to significant compression. If they are, they should comply with the general criteria for compressive elements.

The application of these criteria is best illustrated by examples. Figure 7.4 shows a fully welded interior beam-to-column connection where the column web requires both compression and tension stiffening. Suppose the unstiffened web has a capacity of $0.4P$ in compression. Designers have two choices. They may design the stiffener (A) for $0.6P$, thus making an arbitrary partition of load P; in this instance they should use compact criteria for proportioning the stiffener, since the partition in practice will tend to put a greater load than they have assumed into the stiffener and it is required to deform in a stable manner as load is redistributed into the web. Alternatively, they may decide to design the stiffener to carry the whole of P, in which case they can be certain that the stiffener design load cannot be exceeded by adverse distribution within

Table 7.1 Slenderness criteria for connections

	Compact	*Semi-compact*	*Slender*
Plate element supported along both unloaded edges, in compression	$b/t \leqslant 1.00\sqrt{\frac{E}{p_y}}$	$b/t \leqslant 1.38\sqrt{\frac{E}{p_y}}$	$b/t \leqslant 1.38\sqrt{\frac{E}{f_a}}$
Plate element supported along both unloaded edges in shear or in-plane bending	$b/t \leqslant 1.67\sqrt{\frac{E}{p_y}}$	$b/t \leqslant 2.30\sqrt{\frac{E}{p_y}}$	$b/t \leqslant 2.30\sqrt{\frac{E}{f_a}}$
Plate element supported along one unloaded edge only	$b/t \leqslant 0.35\sqrt{\frac{E}{p_y}}$	$b/t \leqslant 0.5\sqrt{\frac{E}{p_y}}$	$b/t \leqslant 0.5\sqrt{\frac{E}{f_a}}$
Gusset plates	Not applicable	$b/t \leqslant 2.07\sqrt{\frac{E}{p_y}}$	As semi-compact

b: Unsupported width of plate or outstand, measured transverse to load.
t: Plate thickness.
E: Young's modulus.
p_y: Yield stress.
f_a: 1.5 × extreme fibre stress at collapse (i.e. 2.25 × extreme fibre stress at working load).

Table 7.2 Evaluated slenderness criteria for connections

$$b/t = \frac{\text{Unsupported width of plate, or outstand}}{\text{Plate thickness}}$$

Classification	*Compact*		*Semi-compact*		*Slender*		
Steel grade	*43*	*50*	*43*	*50*	–	–	–
Applied ext. fibre stress at u.l.s.	–	–	–	–	50	100	150
Plate element supported along both unloaded edges in compression	27	24	38	33	70	51	42
Plate element supported along both unloaded edges in shear or in plane bending	45	40	63	55	70[a]	70[a]	70
Plate element supported along one unloaded edge only	9.5	8.5	13.5	12	26	18.5	15
Gusset plates			57	50	60[a]	60[a]	60[a]

[a]These slendernesses are restricted, on practical grounds.

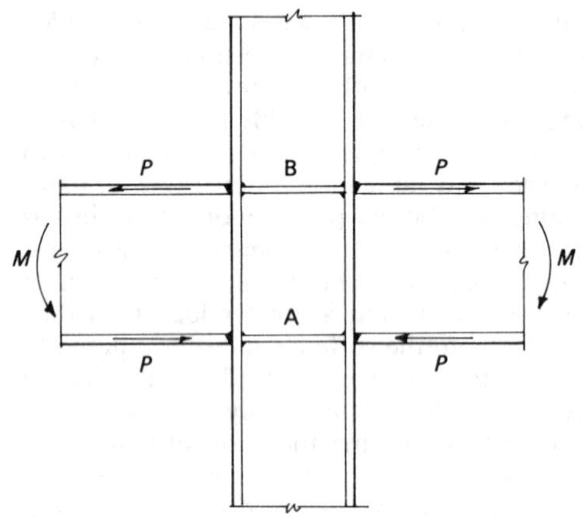

Figure 7.4 Stiffening for an interior beam-to-column connection

the connection. In this instance they may use semi-compact criteria.

The analogous consideration with the tension load is that of weld size. The tension stiffener (B) is not subject to any particular slenderness limitations other than those conventionally used. However, suppose the web tensile capacity is 0.5*P* and the tension stiffener is designed to resist the remaining 0.5*P*. This is an arbitrary partition of load and ductility must therefore be achieved. This necessitates providing welds for (B) that are at least as large as those required for ductility (see page 49).

Figure 7.5 shows a welded column bracket where designers are faced with the problem of proportioning the bracket stiffener. The load is low but has a considerable eccentricity. Because of the compressive stress on the extreme fibre it would not be appropriate to treat this element as a gusset plate. As the load paths are well defined it is not necessary to use compact criteria, but even semi-compact criteria can lead to an uneconomic stiffener operating at low stress. In this instance it would be appropriate to use slenderness criteria, ensuring that the ratio b/t does not exceed

$$0.5\sqrt{\frac{E}{f_a}}$$

in accordance with Table 7.1, where f_a is 1.5 × the extreme fibre compressive stress at collapse.

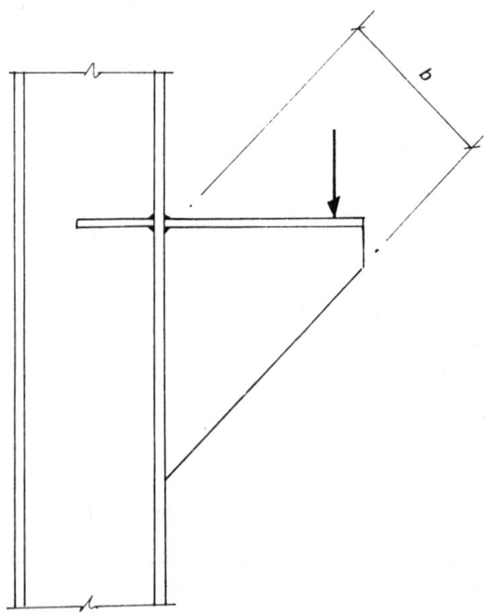

Figure 7.5 Welded column bracket

This point can be further illustrated by a numerical example. Suppose that b = 375 mm and the first trial design is based on semi-compact criteria. From Table 7.2 the maximum b/t is 14.5 for Grade 43 plate and the minimum stiffener thickness is 30 mm. When analysed, it is found that the factored extreme fibre stress is only 25 N/mm^2 – clearly an uneconomic design. As a second trial design a stiffener thickness of 15 mm is considered; this gives a factored extreme fibre stress of 50 N/mm^2. Table 7.2 indicates that this can be sustained with a b/t of up to 26. The actual b/t is 375 ÷ 15 = 25, a more stable value, and clearly the design is satisfactory.

7.3 Local in-plane loading: effective and critical sections

7.3.1 Presence of holes

Design was traditionally based on gross sections under compressive loading and net sections under tensile loading. The former was justified on the grounds that the rivets would fill the holes and therefore be able to transfer compressive loads. The latter restriction was imposed because it was not considered appropriate to permit any overstress on the net sections in tension. The argument about rivets is no longer valid for bolts in clearance holes, although in many situations the bolts in the first row in a compressive connection will remove sufficient load from the connected member for the critical section to be the neighbouring gross section.

It is now considered that a degree of overstress may be permitted on net sections in many circumstances without impairing structural safety. It is argued that a local exceedance of yield over a small proportion of the member length at the collapse limit state is unlikely to lead either to unserviceability of the structure at the appropriate lower load factors or to cause any major redistribution of forces at the collapse limit state. It is, of course, necessary to ensure that there is a suitable margin against rupture of the net section. Different codes apply this relaxation in different ways. It is most normally applied to beams. There is less readiness to do so to tension elements because so many of these are only partially connected. The derivation of the empirical

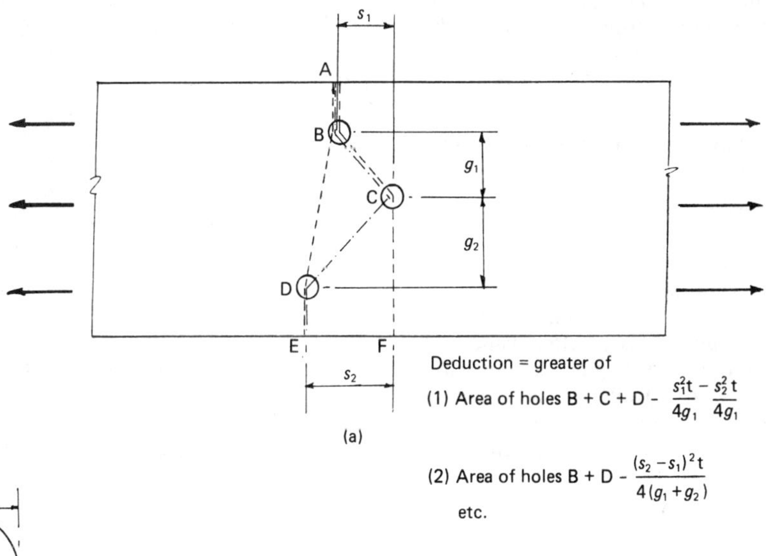

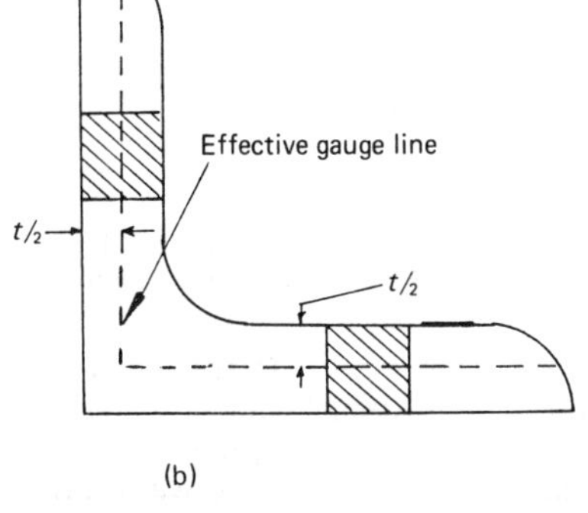

Figure 7.6 (a) Critical net sections: (b) effective gauge lines for elements under in-plane loading

rules for such elements, presented in Section 7.3.2, is obscure. Because of this lack of knowledge it is considered prudent not to permit any overstress.

In most codes the overstress is permitted by modifying the definition of net section. For example, the effective net section may be defined as k times the true net section but not greater than the gross section. The value of k might be 1.2 for Grade 43 steel and 1.1 for Grade 50. This variation in k with material grade relates to the changing ratio of yield stress to u.t.s. and ensures that there is a minimum factor against rupture.

A particular problem arises if bolt holes are staggered. As shown in Figure 7.6, there will generally be more than one potential critical section. The traditional, empirical, rule that is used in such circumstances is that the effective net section along any potential failure surface is the gross section minus the sum of the bolt hole cross-sections plus $s^2t/4g$ for each staggered line between bolt holes, where:

s is the bolt spacing in the loaded direction
g is the bolt gauge transverse to the loaded direction

The same figure illustrates the application of this empirical design rule. For non-planar sections, the gauge g should be the distance along the centre of thickness of the section between hole centres.

7.3.2 Connection of tension members through part of cross-section

It is common practice to connect angle, channel and Tee sections through part of the cross-section only. It is sometimes convenient to use partial connection for other elements. There are traditional, empirical rules for the more common situations that take account both of the partial loss of effectiveness of the cross-section and of the ensuing end eccentricity, and these are summarized in Table 7.3. Where there is no eccentricity (for example in an I-section connected by the flanges only, or where bending is prevented by other means) it is appropriate to use the dispersion rules of Section 7.4.

7.3.3 Critical sections

Apart from the particular cases described above, connection element and gusset plate design should be based on checks of critical sections, making full deductions for any holes and any reasonable assumption for stress distribution. The dispersion rules of Section 7.4 may often be used as the basis of this distribution. Designers should be on their guard for any unusual situations that may arise and lead to particular difficulty.

An example that has been the subject of considerable research is shown in Figure 7.7, the 'block shear problem'.[2] After a review of this research and other design rules it is found that the most appropriate design approach is to base the shear capacity of AB on its effective section $A_{e(AB)}$,

Table 7.3 Effective areas for partially connected members

	Effective area
Single angle connected through one leg	$\dfrac{3A_1^2 + 4A_1A_2}{3A_1 + A_2}$
Pair of angles connected together and attached to same side of gusset or equivalent by one leg of each angle	$\dfrac{5A_1^2 + 6A_1A_2}{5A_1 + A_2}$
Single channel connected through the web	$\dfrac{5A_1^2 + 6A_1A_2}{5A_1 + A_2}$
Single tee	$\dfrac{5A_1^2 + 6A_1A_2}{5A_1 + A_2}$
Double angles or tees, back to back + connected to each side of gusset or equivalent, and connected together along length	Full sectional area

A_1: Area of connected part.
A_2: Area of unconnected part(s).

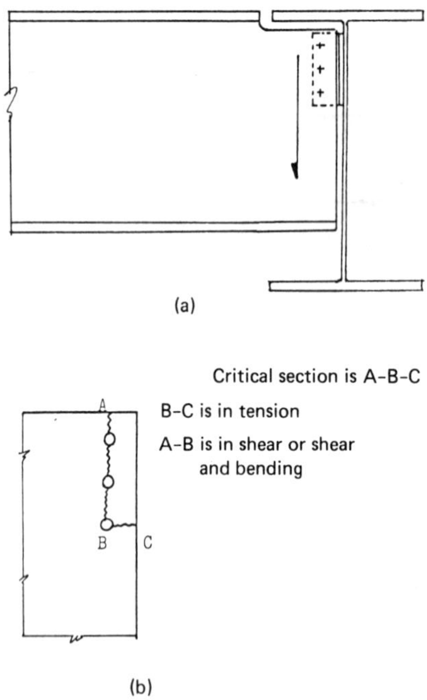

Figure 7.7 Design of shear block in notched beam. (a) General arrangement; (b) detail of web of minor beam

as defined in BS 5950: Part 1, Clause 3.3.3, and to assume that BC is carrying half the tensile capacity of its effective section $A_{e(BC)}$. The factor of half is necessary to provide some reserve to resist local eccentricity. These rules are satisfactory for one or more columns of bolts. Thus capacity is:

$$0.6p_y\, A_{e(AB)} + 0.5p_y\, A_{e(BC)}$$

7.4 Local in-plane loading: dispersion rules

7.4.1 Behaviour

This is an aspect of connection behaviour that has received relatively little attention in research. However, if the information that is available is combined with traditional design practice it is possible to put forward a systematic approach to this aspect of design.

One of the smallest dispersion angles put forward was that due to Whitmore (reference 3 and Chapter 5, reference 9), when he suggested that the maximum direct stress in a gusset plate from an individual member could be estimated adequately by ensuring that the member force was distributed uniformly over an effective area given by a 30° dispersion from the outer fasteners, as shown in Figure 7.8(a). In a different context, studies of short, side-fillet, welded lap joints suggested that

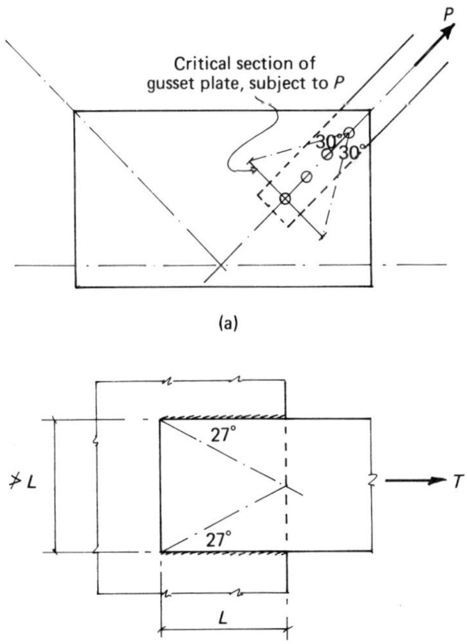

Figure 7.8 In-plane dispersion near a free boundary. (a) Gusset plate; (b) side-weld tension connection

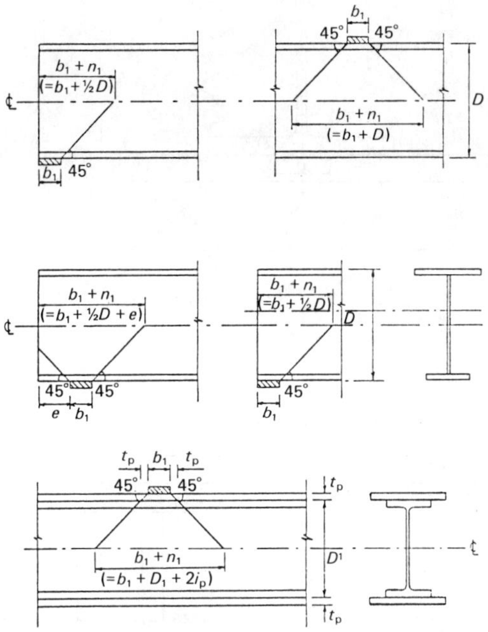

Figure 7.9 Effective width of equivalent column for buckling under local compressive load

adequate joint efficiency could only be achieved if the weld spacing was not greater than weld length.[4] As shown in Figure 7.8(b), this implies a dispersion of 27°. The common characteristics in these two examples are the presence of a free, unstiffened, boundary normal to the applied load, together with a discontinuity of cross-section in the load path. These cause local in-plane deformation, which limits the spread of the load.

Wider dispersion can be permitted, even for compressive loads which might lead to local instability, if the element concerned has a stiffened boundary along the loaded edge. Figure 7.9 shows the traditional approach used for checking unstiffened beam webs under local loads or support reactions. Where such elements are slender they are likely to reach their maximum load-carrying capacity before plasticity has spread to the mid-depth. The satisfactory use of a dispersion angle of 45° in this context suggests that such an angle can be achieved in the presence of only very limited plasticity. This is supported by the approaches for determining stiff bearing lengths that are shown in Figure 7.10.

Greater dispersion still may be permitted in situations where considerable plasticity can be tolerated and there is local symmetry, as shown in Figures 7.11(a) and (c). In tension this is the case where full-strength welding has been used that is capable of ductile deformation. In compression similar criteria may be used when checking local

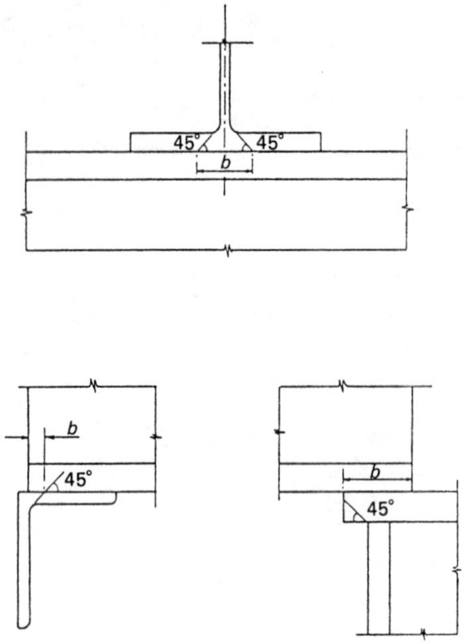

Figure 7.10 Stiff bearing length determined by dispersion through a boundary element. Note that no dispersion may be permitted through packs unless they are symmetric about the point of application of load. Dispersion depends on a plastic hinge forming in the flange, and that requires a balancing restraint

crushing. Both these figures show the established dispersion angle of 68°, which corresponds to a 1:2.5 spread. This has been widely used in other countries for many years, notably Europe and the USA. The traditional value in the UK used to be 60° but this is now regarded as unnecessarily conservative. BS 449 partly compensated for the smaller spread by using a higher stress. It is worth noting that still broader angles are permitted for some steels and particular situations in draft Eurocode 3, based on recent Dutch research.

However, where there is no local symmetry, as in Figure 7.11(b), this dispersion gives a significant overestimate of strength. In such circumstances a 45° spread should be adopted, as shown.

7.4.2 Design

Based on the above discussion, it is thus possible to make general recommendations for dispersion rules, and these are summarized in Table 7.4. The only intentional omission from this table is guidance on the dispersion of bolt tensions through end plates and flanges to their supporting members. Here, behaviour is governed by flexure of the end plates and flanges, and this aspect of dispersion is discussed in Section 7.7.

Section 7.7.6 offers additional guidance where the dispersion is taking place in the presence of high stresses in the plane of the dispersing flange in Table 7.4(c).

7.5 Local in-plane loading: strength assessment

The preceding two sections give guidance on the effective sections that should be used for strength assessment, and guidance is also needed on design stresses before the strength assessment can be completed. In tension the design strength is usually based on the yield stress, although there is an effective exceedence of this value because of modifications to the critical net section, in certain circumstances, as discussed in Section 7.3.1. In shear the design strength is usually based on shear yield. However, if the panel concerned has slender proportions (see Table 7.2) the design strength should be reduced in accordance with the general requirements for such elements. In compression the design strength may be a function of slenderness. Where local buckling near the centre of the element is being considered the radius of gyration is determined from the effective section and the effective length takes account of the boundary conditions, as shown in Figure 7.12. For local crushing, at the element boundary, the design strength is taken as the yield stress.

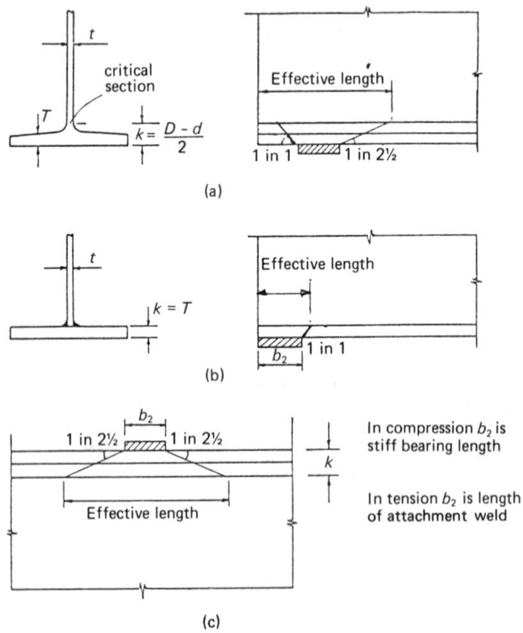

Figure 7.11 Effective length of web under local tensile or compressive loads

Table 7.4 Recommended dispersions of in-plane forces

	DISP'N ANGLE	CONDITIONS	EXAMPLES
A	30°	Near a free boundary where in-plane distortions will modify stress distribution	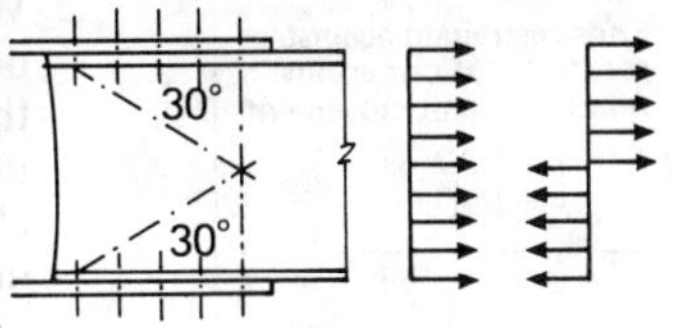 (i) Symmetric partial connection (as illustrated above) (ii) Gusset plates - Fig 7.8(a) (iii) Side welded tension connections - Fig 7.8(b)
B	45°	Near boundaries where boundary members inhibit in-plane distortion, but where strains have to be limited. I.e. In tension if partial strength welds are used; in compression if buckling can occur, or if deformations have to be limited, or if there is a lack of local symmetry	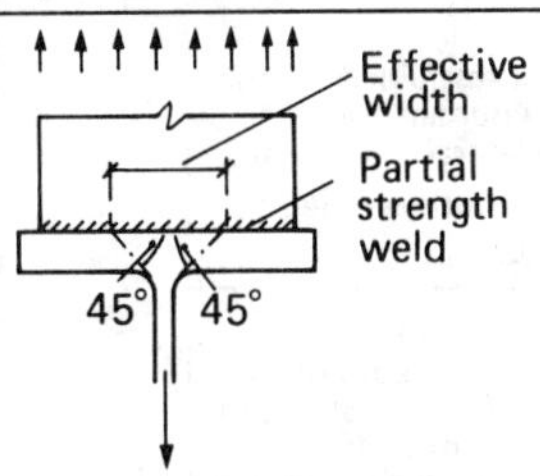 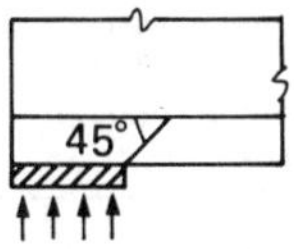 (i) Cruciform connection (as illustrated above) (ii) Local crushing without symmetry (as illustrated above) (iii) Equivalent column – Fig 7.9 (iii) Stiff bearing – Fig 7.10
C	68° $\tan^{-1}2.5$	Where ductile behaviour is guaranteed and where there is no restriction on local strains or deformations. I.e. In tension where full strength welds are used; in compression where instability is not a consideration	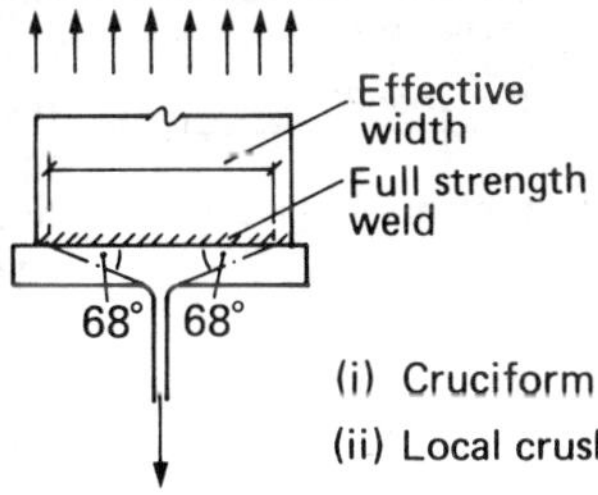 (i) Cruciform connection (ii) Local crushing - Fig 7.11(c)

7.6 Local in-plane loading: stiffener design

7.6.1 Stiffener choice

There is some confusion about the appropriate choice of stiffener. This at least partly stems from some rather ambiguous wording in BS 449, which required that:

> The *ends* of load bearing stiffeners shall be fitted to provide a tight and uniform bearing upon the loaded flange unless welds designed to transmit the full reaction or load are provided between the flange and stiffener.

In practice the designer should have considerable freedom in his choice of stiffening. Where the stiffener is only required to overcome a local overstress close to the point of application of the load, stub stiffeners, as shown in Figure 7.13(a), may be used. Their length should be based on two considerations. First, they should be attached to the web by a sufficient length of weld for their load to be transferred to the web in shear. Second, they should have sufficient depth to be able to resist in-plane bending as the distributed load is transmitted to the web. (This latter consideration is only significant for short, wide stiffeners.)

Long, singly fitted, stiffeners, as shown in Figure 7.13(b), should be used to prevent buckling of the web. These should be fitted to the flange through which the load is transmitted but need not necessarily be fitted to the other flange. Subject to the comments below, it is generally satisfactory and considerably cheaper to curtail these stiffeners short of the unloaded flange, using the detailing rules for intermediate stiffeners.

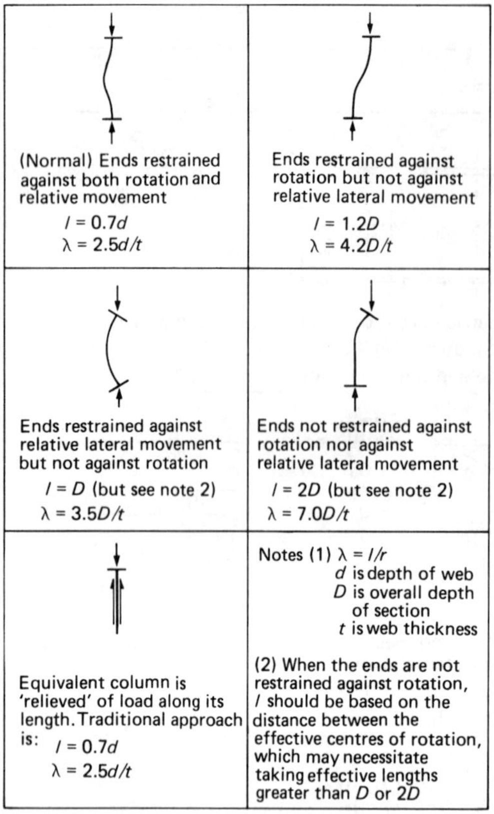

Figure 7.12 Slenderness of equivalent column for buckling under local compressive load

Fully fitted stiffeners, as shown in Figure 7.13(c), need only be used in the following circumstances:

1. Where significant load is transmitted through the member (as illustrated);
2. Where the stiffener is over a support and the beam is subject to lateral–torsional buckling. In these circumstances the stiffener has an important role in controlling overall instability. In the absence of detailed guidance about how curtailing stiffeners could modify behaviour, curtailment should be avoided;
3. Where the stiffener is required to resist any distortion of the unloaded flange;
4. Where the stiffener is required to transmit any torsional moments, including any restraint action, into the beam. This is particularly important where fatigue is a consideration; any gap between the end of the stiffener and the flange could lead to unacceptable local bending stresses in the web.

7.6.2 Effective section and design stress

Unless covered separately under some aspect of dispersion, it is customary and reasonable to assume that part of the stiffened element is acting in conjunction with the stiffener when determining resistance to buckling. Figure 7.14 shows typical proportions. Design stresses are based on considerations similar to those for the unstiffened plate, once again taking account of support conditions in order to determine the effective length of the equivalent strut for compressive loading, in accordance with Figure 7.12.

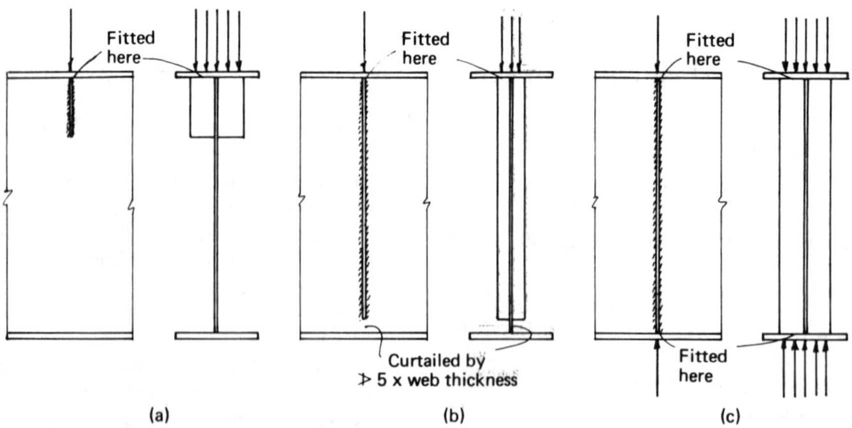

Figure 7.13 Load-bearing stiffeners. (a) For local crippling in compression; (b) for web buckling where load is dispersed into web; (c) for web buckling where load is transferred through element

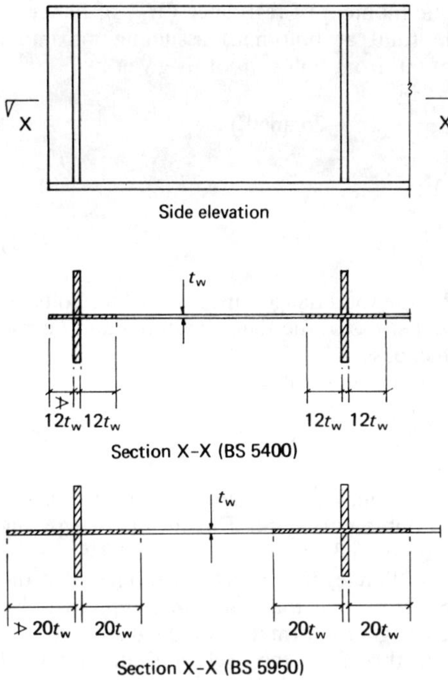

Figure 7.14 Effective cross-sections for stiffeners

7.7 Local out-of-plane loading

7.7.1 General

In marked contrast to most other topics discussed in this chapter, plates and other elements that are locally loaded in shear and flexure by bolts in tension have been the subject of substantial research activity in recent years. Most of this work has been in the context of moment-resisting beam-to-column connections, and it is well summarized in reference 5 and Chapter 5, reference 9. The method of theoretical study most widely used has been that of yield line analysis, and the yield line pattern has frequently been determined empirically. Because of variations in proportions, different experiments have resulted in different yield line patterns, which in turn have led to different formulae for 'design'.

Complete design of this portion of the connection has, in any case, to consider much more than just a simplified distribution of local hinges with severe deformations at the collapse limit state. As shown in Figure 7.15, plasticity and its associated non-linearity of behaviour can frequently commence at a low proportion of the load which produces the final yield line pattern. It is therefore important to consider, at least implicitly, the serviceability condition. It is probably unrealistic and indeed unreasonable to insist that a separate calculation has to be carried out, which demonstrates that there is no local surface yielding at a load factor of 1.0. However, some means of limiting plasticity and non-linearity should be built into the design. In addition, the designer needs information on the distribution of shear among the elements supporting the plate in flexure if the attachment welds and the elements themselves are to be designed properly.

7.7.2 Long Tee stubs

The most straightforward situation occurs where there are two rows of bolts arranged symmetrically about the stalk of a long Tee stub. The traditional

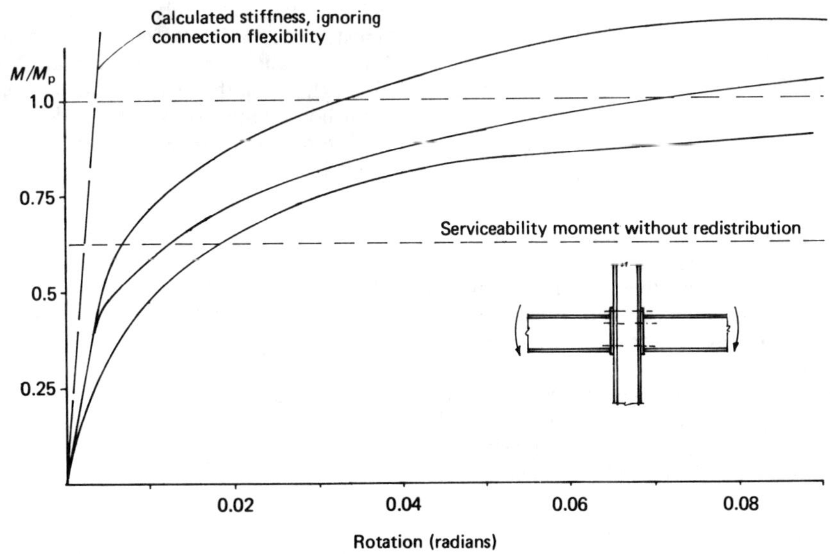

Figure 7.15 Moment/rotation curves for interior beam-to-column connections

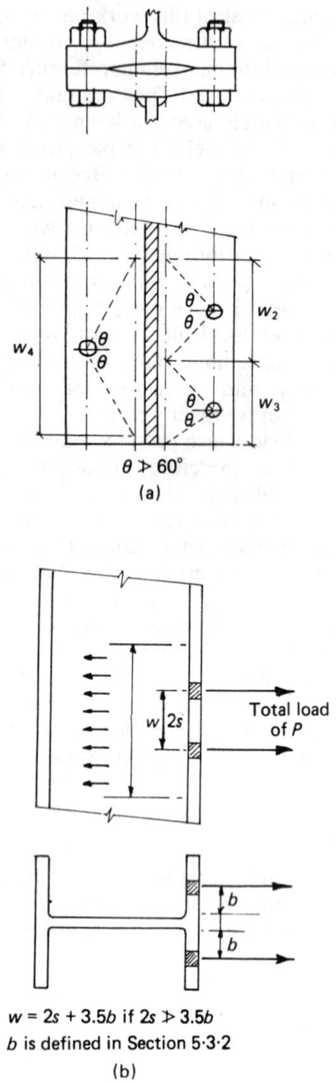

Figure 7.16 Traditional approach to effective breadth of flanges under flexure from local bolt loads

approach to design was to use a simple dispersion angle from the bolt centreline, as shown in Figure 7.16. This defined the effective width of flange in flexure, which would be considered to act in single or double curvature as appropriate, as $2s + 3.5b$ (provided that $2s \not> 3.5b$).

If, in the four-bolt example shown in Figure 7.16(b), the flange is assumed to be in double curvature and the element is designed using the elastic modulus the capacity would be calculated as follows:

$$\text{Moment at root line (and at bolt line)} = M = \frac{P}{2} \times \frac{b}{2}$$

The elastic moment of resistance (M_E) of flange at root line (and at bolt line) assuming maximum spread of 60° from bolt to root, is given by:

$$M_E = \frac{\sigma_y t^2}{6}(2s + 2b\tan 60°)$$

If $M = M_E$,

$$P = \frac{2\sigma_y t^2}{3b}(2s + 3.5b) \qquad (7.1)$$

where P is the total flange capacity for four bolts, σ_y is the yield stress of the flange material and t is the flange thickness.

If the plastic moment of resistance is used:

$$P = \frac{\sigma_y t^2}{b}(2s + 3.5b) \qquad (7.2)$$

In the above analysis, the flange has been assumed to be in double curvature. This means that prying forces (Section 5.3.2) must occur at or near to the edge of the flanges to balance the moments at the bolt lines. Note that this traditional approach also gives the designer information on the distribution of tension in the Tee stub web and in any weld attaching the web to the flange.

Much research activity has been devoted to investigating the yield line patterns that occur in the flange in such circumstances, and a comparative study of this research has recently been carried out.[5] This is summarized in Figure 7.17; Figure 7.17(a) shows a yield line pattern for a doubly symmetric four-bolt flange connection. The remainder of the figure shows different yield line patterns with the effective lengths for the same four bolts as that shown in (a), although advantage has been taken of symmetry to show only one bolt in the sketches. The most conservative pattern, and therefore the one most suitable for design, is that shown in Figure 7.17(g). Using this pattern, the tension capacity of a four-bolt connection would be derived as follows.

The moment (M) at the root line (and at the bolt line) is given by:

$$M = \frac{P}{2} \times \frac{b}{2}$$

The plastic moment of resistance of the flange (M_p), for an effective length $L_y = 2s + 2a + 2.8b$, is given by:

$$M_p = \frac{\sigma_y t^2}{4}(2s + 2a + 2.8b)$$

If $M = M_p$,

$$P = \frac{\sigma_y t^2}{b}(2s + 2a + 2.8b) \qquad (7.3)$$

Once again, this analysis assumes double curvature and therefore requires prying forces for

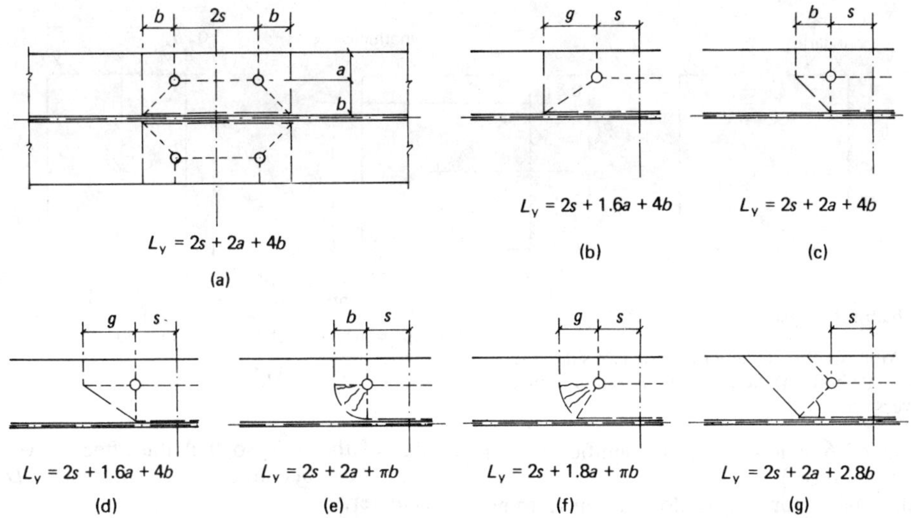

Figure 7.17 Yield line patterns for long-flange Tee connections with yield lines between bolt holes. L_y is effective length of equivalent cantilever (i.e. w in Figure 7.16(b))

equilibrium. If the distance from the bolt line to the edge of the plate is greater than $1.7b$ then use $1.7b$ instead of a in the formula. This is to cover the possibility of a local yield line forming around the bolt. (See also Section 7.7.6, when there are significant axial stresses in the flange.)

For a typical arrangement where, say, $s = b$ and $a = b$, the above formula gives an increase in design capacity of 84% compared to the traditional approach using elastic design or 24% if plastic bending capacity were used.

With the pattern shown, the boundary shear per unit length is $\sigma_y t^2/2b$ between the bolts. Thus the load per unit length on the web is $\sigma_y t^2/b$. (This shear is determined from the moment gradient in the plate between the two parallel yield lines. Elsewhere, where yield lines of opposite sign converge to a point, infinite shear is implied, which shows the limitations of these simple plastic analyses.) In neither method of design is it usually necessary to take direct account of the presence of bolt holes, for the reasons given in Section 7.7.4.

If the flange is stronger than necessary, due to σ_y or t being greater than the minimum required, the complete hinge system will not form and there may be a higher concentration of reaction adjacent to the bolt. If the flange is connected with fillet welds, care must be taken to ensure that they are not critical, bearing in mind their limited ductility.

It is more difficult to give precise guidance on means of ensuring that early yielding of such elements does not lead to unserviceability of the structure. Figure 7.15 showed the overall response of a connection containing such flexural elements, and the early departure from linearity can clearly be seen. The appropriate action in design should depend on circumstances. Thus in a plastically designed portal frame, where deflection does not govern the particular design and cladding is tolerant of limited frame movement, it seems unlikely that the degree of non-linearity shown would cause any difficulty in practice. No additional design checks would seem appropriate, whichever method of design for strength is used. However, in any frame where either deflections govern design or the finishes are intolerant of frame movement it would seem important to ensure that significant plasticity did not occur at working load levels.

The yield line pattern that ultimately leads to collapse of the flange or end plate forms progressively as the applied load is increased. It would normally start to form at the root opposite the bolt, and surface yielding is likely to occur before working load is reached. This is not normally considered to be detrimental. (The possibility of yielding of the cleats in simply supported end connections at working load has long been accepted.) For connections of normal proportions (i.e. where the bolts are reasonably close to the web) major movement will only occur as the development of the full yield line pattern is approached. Up to this stage the movements are small compared to the 'lack of fit' due to fabrication and erection tolerances. A satisfactory design for most multistorey frames should be obtained if it is based on a yield collapse of the plate with an additional γ factor of 1.15. It is only if very stiff connection behaviour is required that design need be based on the more conservative, elastic approach. Thus equation (7.3) is modified to:

$$P = \frac{\sigma_y t^2}{1.15b}(2s + 2a + 2.8b) \tag{7.4}$$

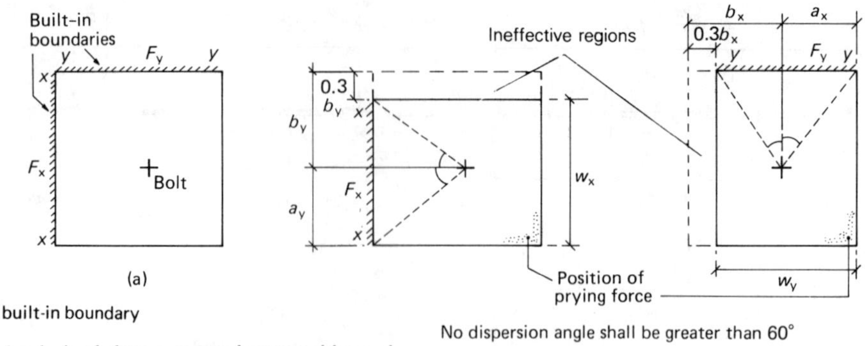

Figure 7.18 Analysis of plate supported on two sides and loaded by a single bolt. (a) Actual plate; (b) simulation as two cantilevers

(See Section 7.7.6, where there are significant axial stresses in the flange.)

The additional factor of 1.15 does not need to be applied for the design of the bolts if the tension capacity is taken as:

$$\frac{Y_f A_t}{\gamma_m} \text{ but} \ngtr \frac{0.7U_f}{\gamma_m} A_t$$

and prying forces are included.

It is, of course, most important that the design of any plate in flexure is carried out in a manner consistent with the determination of the prying action of the plate on the bolt. The above discussion relates to the limiting case where the plate is in symmetric double curvature, giving an upper bound on the prying force. In such circumstances the prying force may be calculated in accordance with Section 5.3.2. This value should be compared with that implied by the moment in the plate at the bolts, assuming that the prying force is acting at distance n from the bolts. If there is any significant disagreement the more conservative (i.e. greater) prying force should be used when determining total bolt forces. Where, in order to limit the total bolt force, lesser prying forces are required these should be calculated in accordance with the same section. These should then be used in conjunction with the traditional effective width approach outlined above to determine the compatible distribution of moments in the flange or end plate. It is *not* correct to use these lesser prying forces with the above yield line approaches because they can only develop with full prying action.

Finally, a note of caution should be sounded about the traditional 60° spread from the bolt centreline for the determination of the effective breadth of flange in flexure. It becomes unsafe where the bolts are not reasonably close to the web, i.e. if $b > 3a$. (Above this limit it gives a higher value than the $2a + 2.8b$ from yield line analysis for a single bolt.) For such situations it is better to adopt US practice, i.e. to assume a spread of 45° from the edge of the hole so that the effective width for a single bolt becomes $D + 2b$, where D is hole diameter.

7.7.3 Short Tee stubs

Where the Tee stub is too short for the full yield line to develop the element will simply behave as a beam. It may therefore be treated by the traditional method, with the effective breadth replaced by the true one.

7.7.4 Plates supported on more than one side

It is possible to idealize a plate supported on two adjacent sides, as shown in Figure 7.18(a), as two overlaid Tee stub flanges. Insufficient research has been carried out to justify any yield line analysis for general use (although results are available for particular cases), and it is therefore appropriate to use an extension of the traditional, effective-width approach in general design. The two effective cantilevers are shown in Figure 7.18(b), including various restrictions on geometry. Note that in the corners the effective breadths of the cantilevers are only taken as 70% of the distance from the bolt line to the adjacent support. This allows for the formation of fan type yield lines in the corners, i.e. it recognizes that not all the plate can deform in flexure. Partition of the load between the beams can usually be based on their relative plastic capacities. Thus:

$$F_x = \frac{w_x/b_x}{(w_x/b_x) + (w_y/b_y)} \times F;$$
$$F_y = \frac{w_y/b_y}{(w_x/b_x) + (w_y/b_y)} \times F \qquad (7.5)$$

where F_x and F_y are the loads transmitted to the xx and yy boundaries and F is the total external load on the bolt, i.e. without any increase for prying action. It is then possible to proceed as before, in accordance with the traditional method of Section 7.7.2.

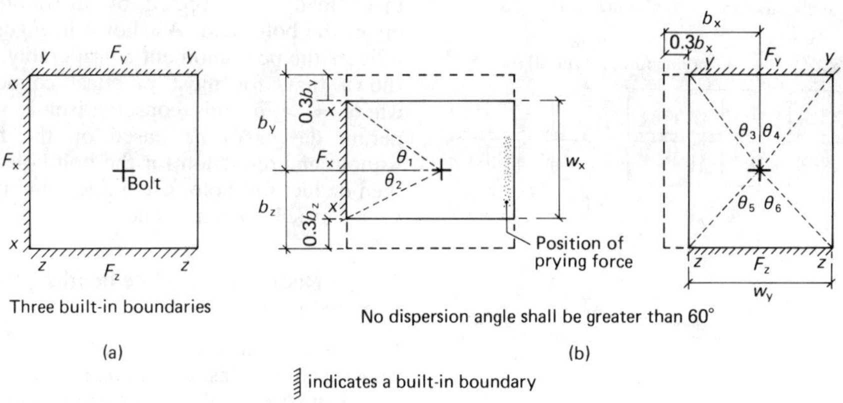

Figure 7.19 Analysis of a plate supported on three sides and loaded by a single bolt. (a) Actual plate; (b) simulation as a cantilever and beam

If a plate is supported on three sides, as shown in Figure 7.19(a), the same broad principle of superposition may be applied. The plate should now be considered as a Tee stub flange and a built-in beam, as shown in Figure 7.19(b). Once again, a conservative general bound on effective widths may be used. Load partition is now given by:

$$F_x = \frac{w_x/b_x}{(w_x/b_x) + (w_y/b_y) + (w_y/b_z)} \times F;$$

$$F_y = \frac{w_y/b_y}{(w_x/b_x) + (w_y/b_y) + (w_y/b_z)} \times F; \qquad (7.6)$$

$$F_z = \frac{w_z/b_z}{(w_x/b_x) + (w_y/b_y) + (w_y/b_z)} \times F$$

In both the above cases it is necessary to ensure that the built-in boundary conditions shown in Figures 7.16 and 7.17 are justified by the actual support conditions. This criterion can usually be satisfied by inspection, that is, by ensuring that the plastic moment capacity of the plate can be developed by the supporting element and any extension of the plate. Where this condition is not satisfied the capacity of the boundary hinge must be based on the flexural capacity of the supporting element. In all cases the reduction in plate bending capacity described in Section 7.7.6 should be allowed for in the presence of significant axial stresses in the plate. It is frequently convenient to apply the reduction coefficient simply to the width of plate that is transverse to the direction of the axial stress.

The formulae given above for the partition of F to the supporting elements are based on plastic flexural action. This is reasonable for such a ductile element which could accommodate any variation from this distribution. However, the supporting elements may be less tolerant. Elastic distributions of load to the supported elements are given for the two-sided case by:

$$F_x' = \frac{w_x/b_x^3}{(w_x/b_x^3) + (w_y/b_y^3)} \times F$$

$$\text{and} \quad F_y' = \frac{w_y/b_y^3}{(w_x/b_x^3) + (w_y/b_y^3)} \times F \qquad (7.7)$$

For the three-sided case the corresponding distributions are:

$$F_x' = \frac{w_x/b_x^3}{(w_x/b_x^3) + (w_y/b_y^3) + (w_y/b_z^3)} \times F;$$

$$F_y' = \frac{w_y/b_y^3}{(w_x/b_x^3) + (w_y/b_y^3) + (w_y/b_z^3)} \times F; \qquad (7.8)$$

$$F_z' = \frac{w_z/b_z^3}{(w_x/b_x^3) + (w_y/b_y^3) + (w_y/b_z^3)} \times F$$

In designing the attachment welds to the loaded plate it would be prudent to take the more severe value (of F_x and F_x', etc.) of the elastic and plastic distributions. Generally, the reactions should be assumed to apply on the line of the bolts.

Prying forces may be evaluated separately for each of the Tee stub flanges. (The built-in beam, i.e. the three-sided case, is unlikely to produce significant prying unless b_x and b_z differ considerably – a case that is discussed below.) In the two-sided case, where the two Tee stub flanges are overlaid, the greater prying force should govern bolt design. Once again, any design based on less than maximum prying action and other than full double-curvature bending should be consistent between bending action in the plate and assumptions for bolt tensions. The calculated prying force should be applied at the positions shown in Figures 7.18 or 7.19 and the calculation be carried out for each of the overlaid elements.

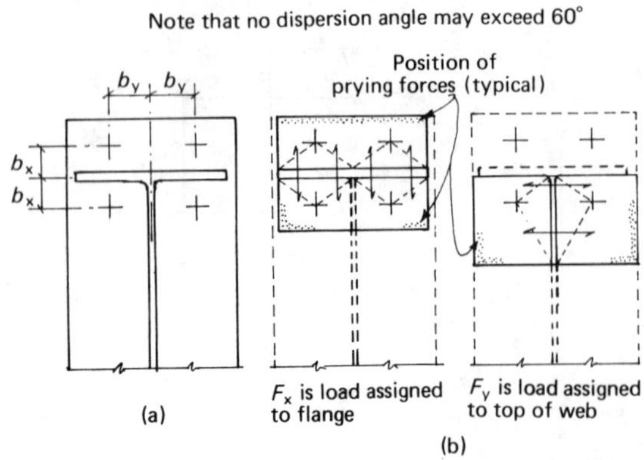

Figure 7.20 Design approach for a plate supported on two sides

Where the proportions (and therefore the contributions to strength from the alternative load paths) differ considerably, little is to be gained from considering the combined action. It is suggested that, for the two-sided case, if $b_y > a_{min} + 1.4b_x$ assume that all the load is supported on edge XX and vice versa, where a_{min} is the lesser of a_x and a_y. For the three-sided case, if $b_X > 1.4\sqrt{b_y b_z}$ assume that $F_X = 0$ and design the plate as supported on two opposite sides; if $b_z > 2b_y$ ignore the support on edge ZZ and if $b_y > 2b_z$ ignore the support on edge YY.

Figure 7.20 shows an example of the application of the above principles to a practical, end-plate connection. The two superimposed beams in Figure 7.20(b) are considered separately in accordance with the above recommendations.

7.7.5 Reductions in effective section due to holes

Calculations of moments on critical sections are based on the assumption that bolt loads are applied as point loads on the bolt centreline. In practice, these loads are applied by distributed pressure under the bolt head. As shown in Figure 7.21, this reduces the peak moment considerably. It has been shown that, for most practical connections, i.e. where $w > b$, this conservatism is sufficient to permit design to be based on the full section, without any reduction for the bolt holes. If $w < \frac{1}{2}b$ then deduct full hole; use a linear interpolation for values of w between $\frac{1}{2}b$ and b.

7.7.6 Reduction in plate bending strength from axial stresses

By simple application of plastic analysis it can be shown that the plastic bending moment capacity per unit width (m_p') of a flat plate in the presence of axial stresses is given by:

$$m_p' = \left[1 - \frac{f_a^2}{p_y^2}\right]\frac{p_y t^2}{4} = \mu\, m_p \qquad (7.9)$$

where f_a is the average stress in the plate due to axial load,

$$\mu = 1 - \frac{f_a^2}{p_y^2}$$

This adverse interaction only occurs where the stresses are in the same direction (for example, in a transverse hinge in a column flange). It is usually convenient to apply the reduction factor to the width of plate in bending. Thus a and b should be replaced by:

$$a\left[1 - \frac{f_a^2}{p_y^2}\right] \quad \text{and} \quad b\left[1 - \frac{f_a^2}{2p_y^2}\right]$$

in the expression for L_s in equations (7.3) and (7.4) in the presence of significant axial stresses. (The reason for the coefficient 2 in the denominator of the reduction factor for b is that only half of the yield lines associated with the 2.86 length are transverse to the flange.) A similar reduction in

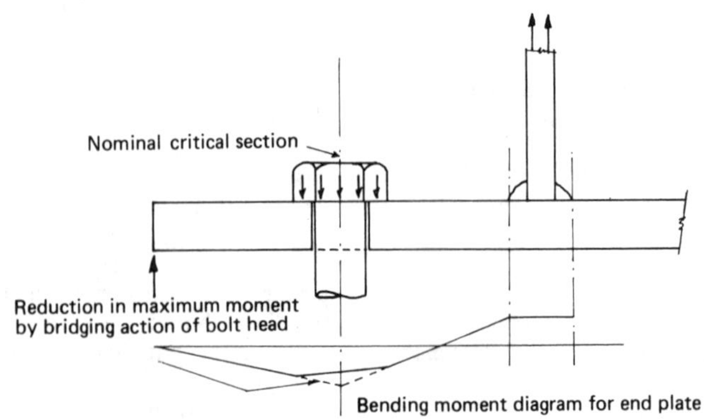

Figure 7.21 Bridging action of bolt head on plate flexure around bolt hole

Table 7.5 Flowcharts for selection of methods of analysis for end plate beam/column connections in buildings

Overall analysis		*Use Section 8.3*
Analysis of end plate/ column flange		
Multistorey sway frame with low critical load ratio where high connection stiffness under factored load is necessary for stability	*Y* →	Use elastic analysis and traditional dispersion (e.g. equation (7.1) for 4 tension bolts)
↓ *N*		
Multistorey sway frame where high connection stiffness is only necessary to ensure serviceability (most practical frames)	*Y* →	Use plastic analysis with some conservatism (e.g. equation (7.2) or equation (7.4))
↓ *N*		
Single storey or low-rise multistorey building designed plastically, high connection stiffness not crucial for serviceability or stability	*Y* →	Use plastic analysis and yield lines (equation 7.3)
Consideration of prying forces		
Is connection of conventional proportions, i.e. (1) End plate and flange thickness > bolt diameter (Grade 8.8) (2) a and b sensible minimum values, $\ngtr 2d$ (3) $a \nless b$?	*Y* →	Carry out initial design without direct consideration of prying. Subsequently, check that Q is not significantly greater than notional 20% allowance in BS 5950 (see Section 5.3.4)
↓ *N*		
Design from first principles to equations (5.12), (5.13) and (5.21)		

effectiveness can also occur when a concentrated load is dispersing through a flange. Here the most convenient approach in design is to replace the conventional 1:2.5 spread by $1{:}2.5\sqrt{\mu}$.

7.7.7 Design

For bridges, Clause 14.3.6 of Part 3 of BS 5400 gives quite specific guidance on methods of design for local out-of-plane loading from bolts in tension. The principal design freedom concerns the selection of prying force. A low force will cause the plate to be in the largely single-curvature bending, increasing the governing moment and hence plate thickness. Such a design would be appropriate where the designer wished to minimize the design bolt forces and was prepared to compensate by increasing end-plate and column flange thickness. A higher value of prying force would clearly increase the bolt design forces, but there could be compensating reduction in end-plate and column flange thickness.

For buildings, there is much less detailed advice, and therefore much greater freedom of choice for the designer. This choice can be bewildering, as the designer seeks to match connection stiffness requirements with selection of method of analysis and to reconcile structural economy with a minimizing of design effort. The flowchart in Table 7.5 is an attempt to rationalize the designer's principal choices for moment-resisting beam-to-column connections, the most common situation where elements are subjected to local out-of-plane loading.

References

1. Owens, G. W., Private communication, based on comparative code study in preparation of Brazilian bridge code, 1982.
2. Yura, J. A., Birkemoe, P. C. and Rieles, J. M., 'Web shear connections: an experimental study', *Journal ASCE*, **108**, No. ST2 (1982).
3. Whitmore, R, E., *Experimental Investigations of Stresses in Gusset Plates*, University of Tennessee Engineering Experiment Station Bulletin 16, May 1952.
4. Freeman, F. R., 'The strength of arc-welded joints', *Proc. ICE*, **23** (1931). Discussion by Dr C. H. Jobban.
5. Cheal, B. D., *Interim Report on Project 306*, CIRIA, London, 1983.

8

Analysis

8.1 Introduction

It is a recurrent theme of this book that, for many connections, an approximate method of analysis is at least as appropriate as traditional, 'rigorous' analysis based on questionable assumptions. However, in many situations these traditional analytical approaches are as convenient as any other to use and probably no less correct. Their usage is likely to continue in appropriate circumstances for the foreseeable future.

In this chapter their derivation from first principles is first summarized. Guidance is then given on their application, including any special considerations that may be necessary if computer-based solutions are utilized. Finally, because all these methods of analysis are more or less tedious to use in practice, a series of interaction diagrams is presented. It is hoped that these will be a useful complement to existing, tabular, design aids.

8.2 Bolt groups subject to shear and moment in their shear plane

8.2.1 General

Figure 8.1 shows the most general case of an arbitrarily distributed bolt group subject to shears P_x and P_y acting with eccentricities e_x and e_y, respectively, about the group centroid. (The eccentricity e_y is defined negative as shown in order to ensure that clockwise moments are positive.) All methods of analysis assume that:

1. Deformation of the connected parts may be ignored;
2. The relative movement of the connected parts may thus be considered as the relative rigid body rotation of the two parts about some centre of rotation;
3. This movement defines the deformations of the individual bolts which are tangential to this centre of rotation;
4. This deformation induces reactive bolt forces which are also tangential to the centre of rotation.

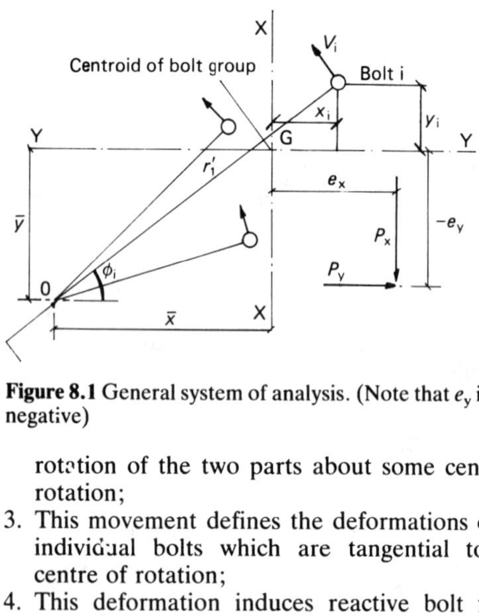

Figure 8.1 General system of analysis. (Note that e_y is negative)

Note that the magnitude of the bolt forces is a function of the assumptions made about bolt behaviour.

These assumptions lead to three basic equations of equilibrium, where the sums of the appropriate components of the reactive bolt forces are equated with the applied loading:

$$\sum_n V_i \cos \varphi_i = P_x \tag{8.1}$$

$$\sum_n V_i \sin \varphi_i = P_y \tag{8.2}$$

$$\sum_n V_i r_i' = P_x(\bar{x} + e_x) + P_y(\bar{y} + e_y) \tag{8.3}$$

where n is the number of bolts in the group. No solution can be achieved without some assumption about the relative magnitude of V_i, that is, the load/deformation relationship of the bolts.

8.2.2 Elastic analysis using superposition

The most common assumption is that of elastic behaviour, where bolt forces are assumed to be linearly proportional to bolt deformations. Because of the preceding assumptions, they are also proportional to distance from the centre of rotation, r_i'. It is possible to use this assumption to solve equations (8.1)–(8.3) directly, and this is demonstrated in the following section. However, it is more convenient to use the principle of superposition, which is applicable to any elastic analysis.

This principle (or, more precisely, its converse, that of decomposition), may be used on the kinematics of a bolt group, as shown in Figure 8.2. The rigid body rotation ψ about the centre of rotation is replaced by the displacement components of the bolt group centroid. The two translations are given by:

$$\delta_x = \psi \times \bar{x} \tag{8.4}$$

$$\delta_y = \psi \times \bar{y} \tag{8.5}$$

The rotation remains constant as ψ.

It is now possible to consider the various actions at the group centroid separately. Thus shear P_x is associated with displacement δ_x, P_y with δ_y and the moment at the centroid $(P_x e_x + P_y e_y)$ with ψ. The constant displacement δ_x imposed on all bolts implies that the vertical shear per bolt V_{xi} is uniform for all bolts and is given by:

$$V_{xi} = \frac{P_x}{n} \tag{8.6}$$

Similarly, the horizontal shear per bolt V_{yi} is given by:

$$V_{yi} = \frac{P_y}{n} \tag{8.7}$$

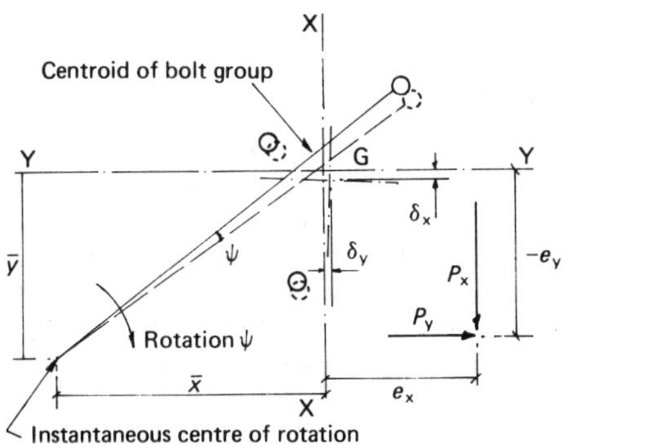

Figure 8.2 Kinematic superposition for elastic analysis

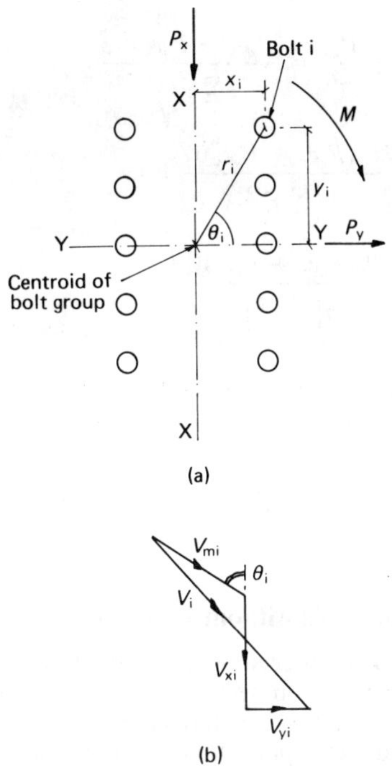

Figure 8.3 Elastic analysis of bolt group subject to torsion and shear. (a) General notation; (b) vector force diagram for bolt i

The moment is resisted by bolt shears V_{mi} acting tangentially to radii r_i, as defined in Figure 8.3. From the preceding assumptions, V_{mi} is proportional to r_i or:

$$V_{mi} = k\, r_i \tag{8.8}$$

Hence

$$M = P_x e_x + P_y e_y = \sum_n V_{mi} r_i = \sum_n k\, r_i^2$$

From which

$$k = \frac{P_x e_x + P_y e_y}{\sum_n r_i^2} \tag{8.9}$$

which may be substituted into equation (8.8) to give:

$$V_{mi} = \frac{(P_x e_x + P_y e_y).r_i}{\sum_n r_i^2}$$

The total bolt force V_i is the vector sum of V_{xi}, V_{yi} and V_{mi}:

$$V_i = \sqrt{[(V_{xi} + V_{mi}\cos\theta_i)^2 + (V_{yi} + V_{mi}\sin\theta_i)^2]}$$

$$= \sqrt{(V_{xi}^2 + V_{yi}^2 + V_{mi}^2 + 2V_{mi}V_{xi}\cos\theta_i + 2V_{mi}V_{yi}\sin\theta_i)} \tag{8.10}$$

$$\text{or } V_i = \sqrt{\left[\frac{P_x^2}{n^2} + \frac{P_y^2}{n^2} + \frac{(P_x e_x + P_y e_y)^2 r_i^2}{\left(\sum_n r_i^2\right)^2} + \frac{2P_x(P_x e_x + P_y e_y)x_i}{n\left(\sum_n r_i^2\right)} + \frac{2P_y(P_x e_x + P_y e_y)y_i}{n\left(\sum_n r_i^2\right)}\right]} \quad (8.11)$$

(noting that $r_i \cos\theta_i = x_i$ and $r_i \sin\theta_i = y_i$).

An alternative format which is easier to remember is:

$$V_i = \sqrt{\left\{\left[\frac{P_x}{n} + \frac{Mx_i}{\Sigma(x_i^2 + y_i^2)}\right]^2 + \left[\frac{P_y}{n} + \frac{My_i}{\Sigma(x_i^2 + y_i^2)}\right]^2\right\}}$$

8.2.3 Elastic analysis without superposition

Readers who find the preceding application of the principle of superposition too tenuous for their tastes may use the following derivation, which is based on the direct application of the elastic assumption of bolt behaviour to the basic equations (8.1)–(8.3).

The reactive bolt shear V_i is proportional to the displacement of bolt i arising from the rigid body rotation ψ. Thus, with the notation of Figures 8.1 and 8.2:

$$V_i = Kr_i'\psi \quad \text{or} \quad K = \frac{V_i}{r_i'\psi} \quad (8.12)$$

where K is the elastic bolt stiffness. From vertical equilibrium:

$$P_x = \sum_n V_i \cos\varphi_i \quad (8.13)$$

but

$$\cos\varphi_i = \frac{\bar{x} + x_i}{r_i'}$$

Substituting into equation (8.13) gives:

$$P_x = \sum_n K(\bar{x} + x_i)\psi = K\psi n\bar{x} + K\psi\sum_n x_i$$

However, $\Sigma_n x_i = 0$, by definition. Thus:

$$P_x = K\psi n\bar{x} \quad (8.14)$$

Similarly:

$$P_y = K\psi n\bar{y} \quad (8.15)$$

Taking moments about the centre of rotation:

$$P_x(\bar{x} + e_x) + P_y(\bar{y} + e_y) = \sum_n V_i r_i' = K\psi\sum_n r_i'^2$$

$$= K\psi\sum_n[(\bar{x} + x_i)^2 + (\bar{y} + y_i)^2]$$

$$= K\psi\sum_n[\bar{x}^2 + 2\bar{x}x_i + x_i^2 + \bar{y}^2 + 2\bar{y}y_i + y_i^2]$$

Thus, noting that $\sum_n x_i = \sum_n y_i = 0$

$$P_x(\bar{x} + e_x) + P_y(\bar{y} + e_y) = K\psi n(\bar{x}^2 + \bar{y}^2) + K\psi\sum_n r_i^2$$

Substituting from equations (8.14) and (8.15):

$$P_x(\bar{x} + e_x) + P_y(\bar{y} + e_y) = P_x\bar{x} + P_y\bar{y} + K\psi\sum_n r_i^2$$

Hence:

$$P_x e_x + P_y e_y = K\psi\sum_n r_i^2 \quad (8.16)$$

Eliminating $K\psi$ from equations (8.12) and (8.16):

$$V_i = \frac{(P_x e_x + P_y e_y)\, r_i'}{\sum_n r_i^2}$$

$$= \frac{(P_x e_x + P_y e_y)}{\sum_n r_i^2} \sqrt{[(\bar{x} + x_i)^2 + (\bar{y} + y_i)^2]}$$

However, from equations (8.14) and (8.16):

$$\bar{x} = \frac{nP_x \sum_n r_i^2}{(P_x e_x + P_y e_y)}$$

and from equations (8.15) and (8.16):

$$\bar{y} = \frac{nP_y \sum_n r_i^2}{(P_x e_x + P_y e_y)}$$

$$V_i = \sqrt{\left[\frac{P_x^2}{n^2} + \frac{P_y^2}{n^2} + \frac{(P_x e_x + P_y e_y)^2 r_i^2}{\left(\sum_n r_i^2\right)^2} + \frac{2P_x(P_x e_x + P_y e_y)x_i}{n\left(\sum_n r_i^2\right)} + \frac{2P_y(P_x e_x + P_y e_y)y_i}{n\left(\sum_n r_i^2\right)}\right]}$$

This is the result obtained in the previous section.

8.2.4 Plastic analysis

The bolts may be considered to act as rigid/plastic connectors. That is, up to a certain load (for example, the slip load for HSFG bolts) they do not deform significantly; at this they deform while sustaining that load. In such circumstances a plastic analysis may be used to determine the capacity of the bolt group, setting V_i equal to a constant value V.

Equations (8.1)–(8.3) can only be solved iteratively if V_i is made constant because of the form of the trigonometric relationships, and this is far too tedious for design office use. Superposition cannot be used for plastic analysis. A possible iterative procedure, suitable for use on a microcomputer, is outlined below.[1]

With initial values of V_i, $\bar{x}$ and $\bar{y}$, equations (8.1)–(8.3) may be rewritten as shown below. Sensible starting values might be those derived from an elastic analysis.

$$\Sigma V \cos \varphi_i - P_x = A \quad (8.18)$$

$$\Sigma V \sin \varphi_i - P_y = B \quad (8.19)$$

$$\Sigma V r_i - P_x(\bar{x} + e_x) - P_y(\bar{y} + e_y) = C \quad (8.20)$$

Variations in A, B and C are given by:

$$\delta A = \frac{\partial A}{\partial V}.\delta V + \frac{\partial A}{\partial x}.\delta\bar{x} + \frac{\partial A}{\partial y}.\delta\bar{y} \text{ etc.}$$

This may be expressed in matrix form as:

$$\begin{bmatrix} \delta A \\ \delta B \\ \delta C \end{bmatrix} = \begin{bmatrix} \frac{\partial A}{\partial V} & \frac{\partial A}{\partial \bar{x}} & \frac{\partial A}{\partial \bar{y}} \\ \frac{\partial B}{\partial V} & \frac{\partial B}{\partial \bar{x}} & \frac{\partial B}{\partial \bar{y}} \\ \frac{\partial C}{\partial V} & \frac{\partial C}{\partial \bar{x}} & \frac{\partial C}{\partial \bar{y}} \end{bmatrix} \begin{bmatrix} \delta V \\ \delta\bar{x} \\ \delta\bar{y} \end{bmatrix} \quad (8.21)$$

Note that all the partial derivatives $\partial A/\partial V$ etc. can be derived from the geometry of the bolt group and equations (8.18)–(8.20). If equations (8.18)–8.20) were linear, setting $\delta A = -A$, $\delta B = -B$ and $\delta C = -C$ would give, from equation (8.21), values of δV, δx and δy that lead to exact solutions to the original equations. With non-linear equations the same approach can be used as the basis of a rapidly convergent iteration scheme. It is essentially a three-dimensional Newton–Raphson approach.

Another difficulty with plastic methods of analysis is that, in their simplest form, equations (8.18)–(8.20) cannot yield solutions for many combinations of P_x, P_y, e_x and e_y. This can be demonstrated by considering the reactive forces arising from different positions of the centre of rotation in the simple, two-bolt, connection shown in Figure 8.4(a). If the centre of rotation is on the horizontal axis and outside both bolts, either to the right or left as in Figure 8.4(b), then both bolt forces are acting in the same direction and the connection is resisting pure shear. If the centre of rotation is between the bolts, as in Figure 8.4(c), then the bolts are acting in opposite directions and the connection is resisting pure moment. Thus no solution is possible for any vertical, eccentric load which subjects the connection to a combination of shear and moment.

The problem arises because of the unrealistic way in which the reactive force in any one bolt maintains its full value but changes direction as the centre of rotation passes through the bolt, centroid. One solution to the problem[2] is to change to a modified plastic analysis assuming that the bolt only attains its full reactive shear if it is more than a certain distance (R_0) (say, three times the bolt diameter) from the centre of rotation. Bolts at a closer distance (R) to the centre of rotation may be assigned an 'elastic' value. Hence $V_i = R_i V/R_0$ but not greater than V.

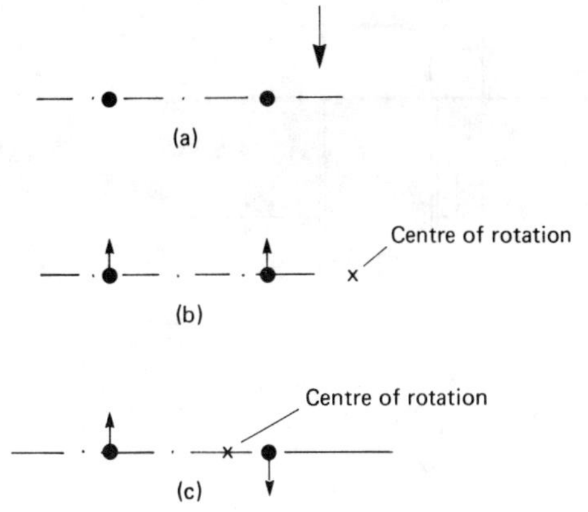

Figure 8.4 Simple plastic analysis of a two-bolt group

Another solution, only applicable to friction connections, is to recognize that the frictional resistance of any one bolt does not all develop at the bolt centroid but is distributed throughout the annular ring of bearing contact surrounding the bolt. Thus, referring to Figure 8.5:

$$\delta V_i = F_s r \mathrm{d}\theta \mathrm{d}r \quad (8.22)$$

where F_s is some uniform frictional 'shear' stress and equals:

$$\frac{V}{\pi(r_2^2 - r_1^2)}$$

Integrating over the annular ring of contact and taking account of symmetry yields expressions for the shear (V_{si}) and contribution to moment resistance (V_{mi}) for bolt i:

$$V_{si} = 2F_s \int_{r_1}^{r_2} \int_{-\pi/2}^{\pi/2} \frac{r(r_i + r \sin\theta)\mathrm{d}\theta.\mathrm{d}r}{\sqrt{(r_i^2 + 2r_i r \sin\theta + r^2)}} \quad (8.23)$$

$$V_{mi} = 2F_s \int_{r_1}^{r_2} \int_{-\pi/2}^{\pi/2} \times r\sqrt{(r_i^2 + 2r_i r \sin\theta + r^2)}.\mathrm{d}\theta.\mathrm{d}r \quad (8.24)$$

Evaluation of these integrals may appear tedious, but this method has been used successfully on a microcomputer.[1]

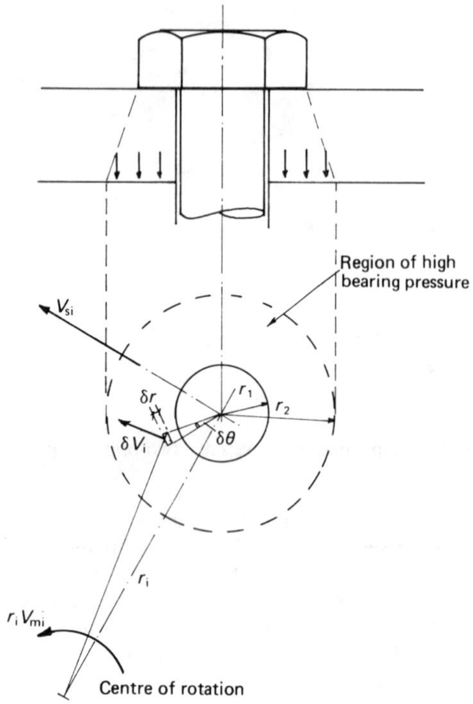

Figure 8.5 Slip resistance of HSFG bolt close to centre of rotation

By spreading the frictional resistance of a bolt over this annular ring in this way the centre of rotation may approach a bolt without introducing any discontinuities into the solution.

8.2.5 Applicability of analytical methods

With groups of bearing bolts the actual (as opposed to theoretical) distribution of shear between the bolts is largely influenced by the fit of the connection. Unless the components were clamped together prior to drilling there will be significant mismatch between the holes. Bolts will not start to carry any load until they are bearing against both sets of plies, irrespective of their position in relation to the theoretical centre of rotation. The effect of this on carrying capacity is discussed in Section 1.2. Because of the adverse effect of lack of fit on strength it would seem prudent to use a conservative method of analysis. Thus the elastic method is to be preferred for such connections. It is reassuring to note that, even in the presence of the uncertainties outlined above, it has led to satisfactory connections when used with conventional values for shear strength.

Friction connections are not subject to any such uncertainty. The behaviour of a single-bolt connection is more accurately represented by a rigid-plastic than by an elastic model. In most circumstances slip can be considered as an unserviceability rather than an ultimate condition because the connection will have a considerable post-slip reserve. Because of these factors it would seem reasonable to use plastic methods of analysis for such bolt groups. Note, however, that this is not permitted in some codes, notably BS 5400: Part III; this is presumably because the code drafters wished to retain the greater conservatism of elastic analysis.

8.2.6 Interaction diagrams

It is not feasible to present interaction curves for all four variables P_x, P_y, e_x and e_y. Instead, the simpler (and more widely occurring) case is considered where the bolt group is subject to a shear which is parallel to one of its principal axes. Two forms of interaction curves are presented.

Figure 8.6 shows a non-dimensional plot of shear against moment.[3] Before such a graph can be used directly the position of the loading vector on the interaction diagram must be determined from the following empirical relationships. Using the notation of Figure 8.6, it has been shown that, for a wide variety of bolt groups, if P_0 is capacity in pure shear, the moment capacity (M_0) is generally equal to, or greater than $(P_0 \times d)/3$ for plastic analysis or $(P_0 \times d)/4$ for elastic analysis. (The denominator will tend to values of 4 and 6, respectively, for very long, narrow bolt groups.) Thus the loading vector direction in the interaction diagram:

$$\tan^{-1}\left(\frac{P}{P_0}\cdot\frac{M_0}{M}\right)$$

is given by:

$$\tan^{-1}\left(d/3\,\frac{P}{M}\right)$$

for plastic analysis and

$$\tan^{-1}\left(d/4\,\frac{P}{M}\right)$$

for elastic analysis, where P and M are any pair of maximum coincident values of shear and moment, respectively.

Since M/P is the eccentricity of loading, it is possible to evaluate the relevant vector simply by selecting a connection size. Thus a convenient procedure for preliminary design is:

1. Estimate connection size and hence determine the diagonal d;
2. Calculate the load vector direction;
3. Read off from the appropriate curve in Figure 8.6 the value of P/P_0;

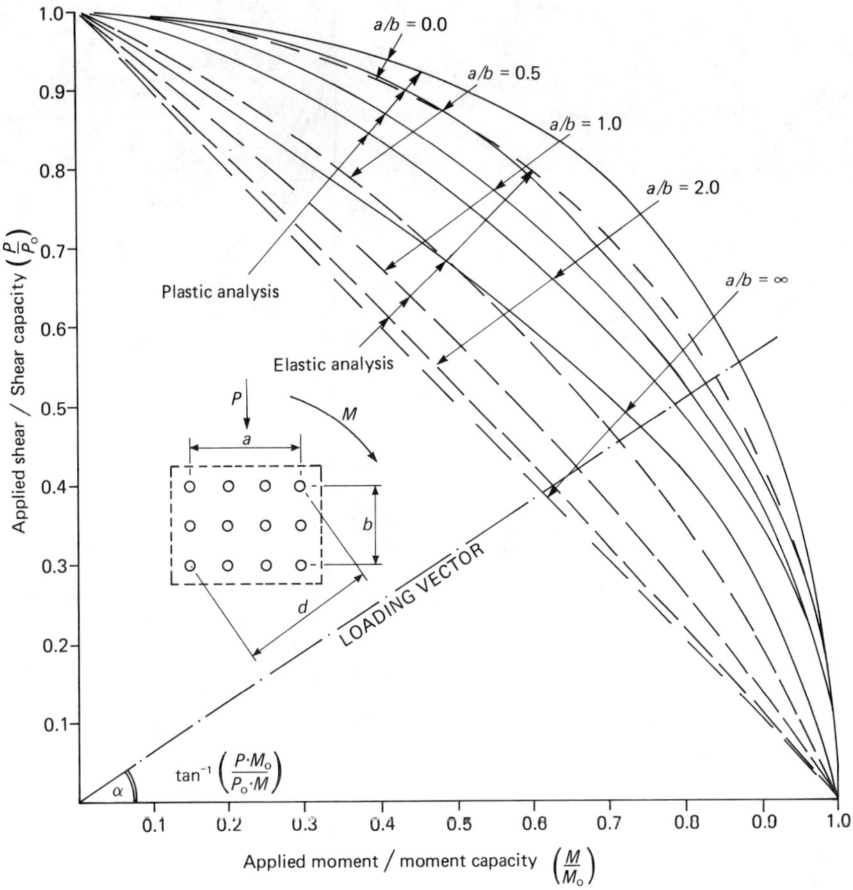

Figure 8.6 Interaction curves for bolt groups subject to torsion and shear

4. Design the connection for the determined value of P_0, checking that this is compatible with the original assumption for the connection size.

This preliminary design may then be checked in the conventional way, with every confidence that it will prove to be satisfactory.

Figure 8.7 presents alternative forms of interaction diagrams where non-dimensionalized shear is plotted against eccentricity. The two extremes are given, i.e. where a single line of bolts is alternatively parallel or normal to the line of action of the applied shear. Intermediate bolt groups may be designed initially using linear interaction between these extremes, with acceptable accuracy. For example, the nine-bolt group shown in Figure 8.8 has a shear capacity (P_0) of 9 V and an elastic moment capacity (M_0) of 3 $V.D$, where V is the shear capacity of a single bolt and D is the bolt group diagonal.

Consider the situation where we are required to find the maximum eccentricity at which a load of 50% of the concentric load can be sustained. Thus $P/P_0 = 0.5$ and the limiting elastic values of e/D for a vertical and horizontal three-bolt line are 0.57 and 0.33, from Figures 8.7(i) and (ii), respectively. The mean value (appropriate for a square bolt group) is 0.45. Rigorous elastic analysis of the bolt group gives an eccentricity of 0.39D when $P = 0.5P_0$.

Alternatively, consider the same bolt group for the particular case where it is subject to a P/P_0 of 0.9. (This implies a plastic centre of rotation 2p from its centroid, a value which may be determined iteratively.) Figures 8.7(iii) and (iv) suggest values of e/D of 0.19 and 0.06, respectively, with an average of 0.125. Exact plastic analysis gives a value of $M/M_0 = 0.35$, which may be confirmed from Figure 8.6. This corresponds to an e/D of 0.15.

Thus in both cases, appropriate interpolation gives reasonable values for initial design. Alternatively, it would be easy to construct appropriate diagrams corresponding to Figure 8.7 for different bolt groups from Figure 8.6.

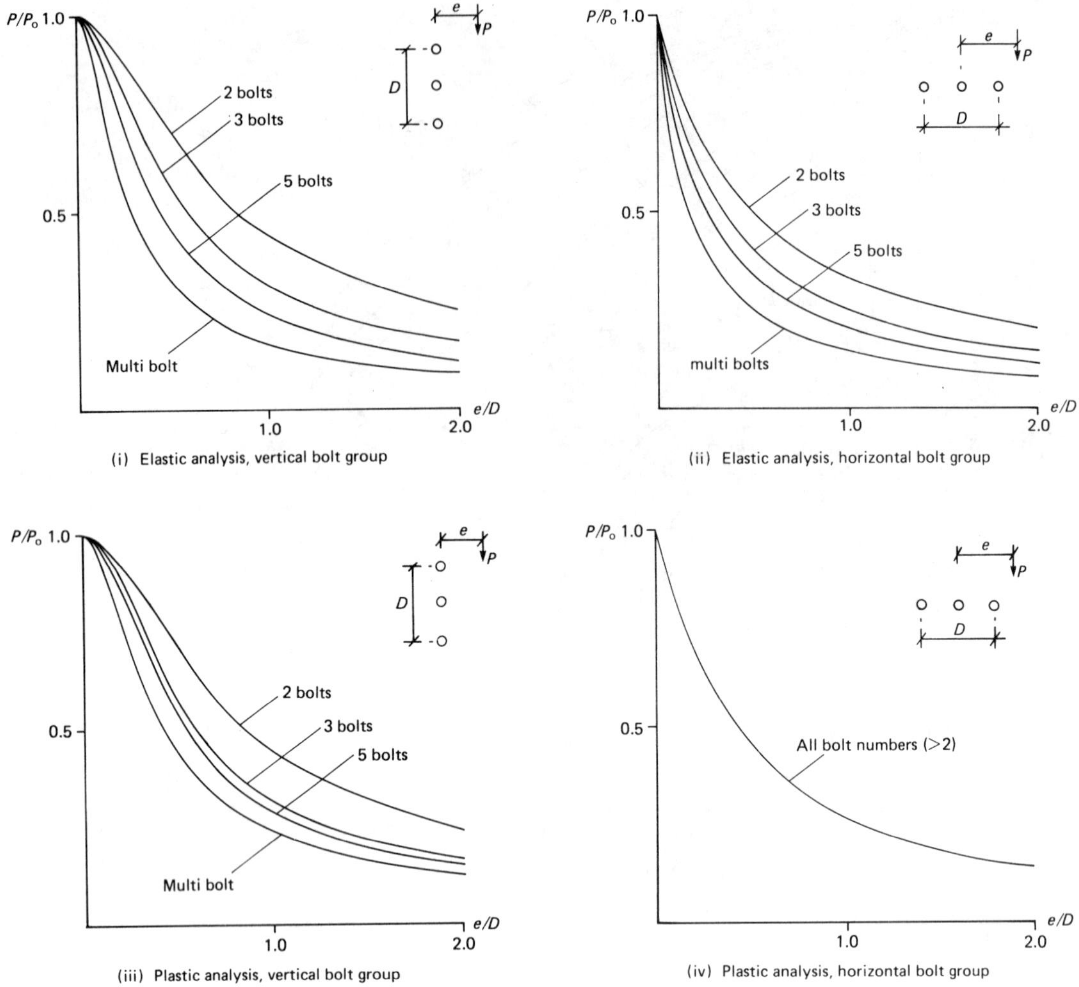

Figure 8.7 Shear:eccentricity diagrams for vertical and horizontal bolt groups

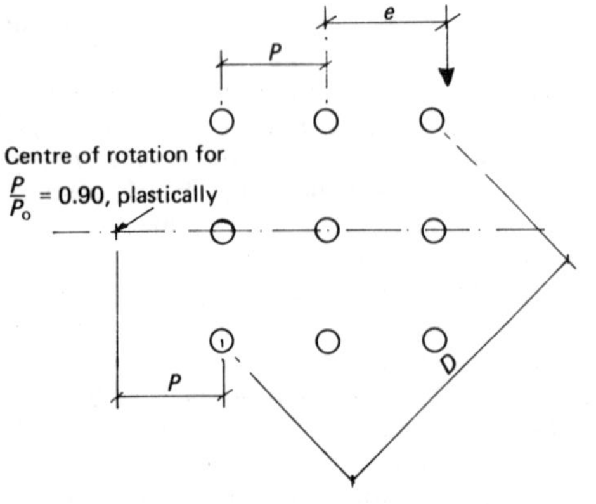

Figure 8.8 Bolt group for interaction diagram example

8.3 Bolt groups subject to loading eccentric to the shear plane

As shown in Figure 8.9, the eccentricity of loading induces a couple in the connection. In addition, the connection has to resist the shear force. Connection response to this form of load is very much a function of detailed design, and it is no longer possible to present definite methods of analysis.

Consideration of the compressive component of the couple is most straightforward if there is a well-defined 'hard spot' on the load path. This would be the case in the examples shown if stiffeners were provided to the column webs opposite the lower bracket flanges. In such cases it seems logical to assume that the compression acts at the mid-depth of the hard spot.

Without a well-defined hard spot the problem is less easy to resolve. In practice, the position of the

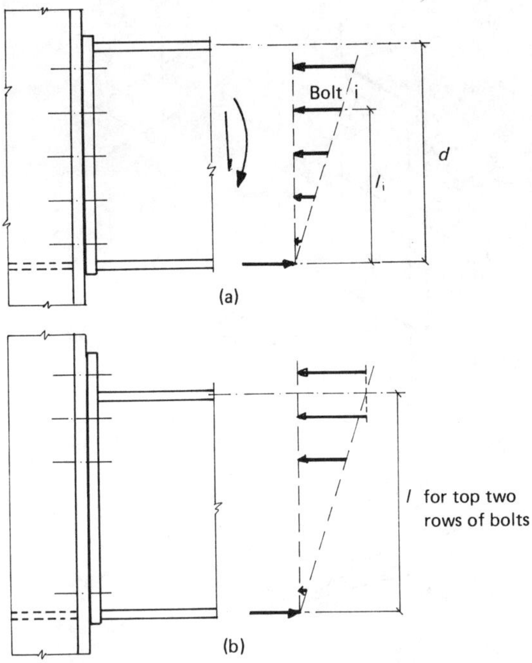

Figure 8.9 Bolt and compressive force distributions for end-plate connections. (a) Short end plate; (b) extended end plate

neutral axis for the connection would be a function of the stiffness of both tensile and compressive regions. Deformation of the column flange and end-plate flexure would have a significant influence on behaviour, and it is unrealistic to attempt to define these accurately. In addition, stress flow in elements coming into the connection will also influence force distributions within the connection. Fortunately, the magnitude of the tensile forces in the critical bolt is relatively insensitive to the distribution of bearing stresses. Unless consideration of the elements coming into the connection suggests otherwise, it is generally acceptable to assume a neutral axis position that is one-sixth of the depth of the connection up from the bottom. This is the value that has traditionally been used; its deviation from one-half (which would be appropriate for a symmetric bending action) is in recognition of the greater stiffness of the compression portions of the connection. A triangular distribution of bearing stress was assumed, leading to a position of the line of action of the compressive force that is $d/18$ up from the bottom. For connections of conventional proportions this will give a line of action that is close to the lower flange centroid that was proposed for analysis in the presence of stiffeners.

Where such connections are required to develop moments greater than approximately 80% of the yield moment of the beam it will be found that any of the procedures outlined above will lead to flange forces that are greater than their design capacity. In such circumstances, the 'load path' methods, discussed in Section 12.4.1, should be adopted, mobilizing the bending capacity of the web directly.

The distribution of nominal bolt tensile forces is generally assumed to be a linear function of distance from the neutral axis. Where there is a defined hard compressive point this distance can be replaced by that from the line of action of the compressive force with negligible loss of accuracy. Where extended end plates are used research has indicated that the top portion of the plate behaves as a Tee stub that is symmetric about the tension flange. In such circumstances the forces in the top two rows of bolts may be assumed to be constant and to be based on the distance from the top flange centroid to the neutral axis or line of action of compressive force, as appropriate.

Based on the assumptions outlined above, the nominal tension in the ith bolt (T_i) is given by:

$$T_i = kl_i \tag{8.25}$$

where k is some elastic constant,
l_i is distance from centre of rotation.

For moment equilibrium:

$$M = \sum T_i L_i = k \sum_n l_i L_i$$

where L_i is the lever arm of the ith bolt. (Where the compressive force is distributed over a depth $d/6$, $L_i = l_i + d/g$; where the centre of rotation coincides with the line of action of the compressive force, $L_i = l_i$.)

Substituting for k in equation (8.25) gives:

$$T_i = \frac{Ml_i}{\sum_n l_i L_i} \tag{8.26}$$

Note that this analysis only determines nominal bolt tensions; appropriate allowance should subsequently be made for prying action.

There are different approaches to the distribution of shear between the bolts. The more conservative one is to assume that all bolts carry equal shear. On this basis, the critical bolts are always those at the top of the connection, and these must be checked under combined shear and tension. The more optimistic approach is to assume that all bolts can be working at their design capacity under varying ratios of tension and shear. On this basis the residual shear capacities of all the bolts are determined, taking account of coincident bolt tension. Adequacy in shear is then checked by ensuring that the sum of these residual shear capacities is greater than the applied shear. As in the previous section, it seems logical to consider the practicalities of fit before selecting design procedure. Unless matched drilling can be ensured it seems unduly optimistic to assume that the redistributions implied in the latter method can safely take place in connections with bearing bolts. Thus without matched drilling, the usual situation for general fabrication, the more conservative assumption of an equal distribution of bolt shears should be made. Only if matched drilling is ensured, or HSFG bolts are used, should the latter method of assessment of shear distribution be adopted.

Where a bolt group is subject to a load with eccentricities in and out of the shear plane the same general methods of analysis apply. The bolt tension forces are determined in accordance with the initial part of this section. The bolt shears are determined in accordance with Section 8.2; individual bolts are checked under their combined loading.

8.4 Weld groups subject to shear, moment and torsion

8.4.1 Traditional analysis

Figure 8.10 shows a weld group under a combination of vertical shear, moment about a horizontal axis in the plane of the weld group and torsion about a horizontal axis normal to the weld group. Elastic analysis is generally used for weld groups, generally ignoring all deformations other than those of the welds themselves. It is assumed that weld response is linear and invariant with direction of the load vector. (If deformation of the connected components will significantly influence the distribution of forces on the weld group then either some other method of analysis must be used or welds loaded by flexible elements must be regarded as ineffective.)

These assumptions lead to the following derivations for the three components of total loading vector on any portion of the weld. The shear load per unit length (F_s) is constant throughout the weld and is given by:

$$F_s = \frac{P}{\int_{\text{weld}} \text{d}s} \tag{8.27}$$

where ds is an element of the weld.

Taking moments about XX gives:

$$Pe_m = \int_{\text{weld}} x\, F_m\, \text{d}s$$

where F_m is the force vector per unit length resisting the moment,
x is the distance from the XX axis.

However, $F_m = k.x$ from elasticity.

Substituting in equation (8.28):

$$Pe_m = k \int x^2\, \text{d}s$$

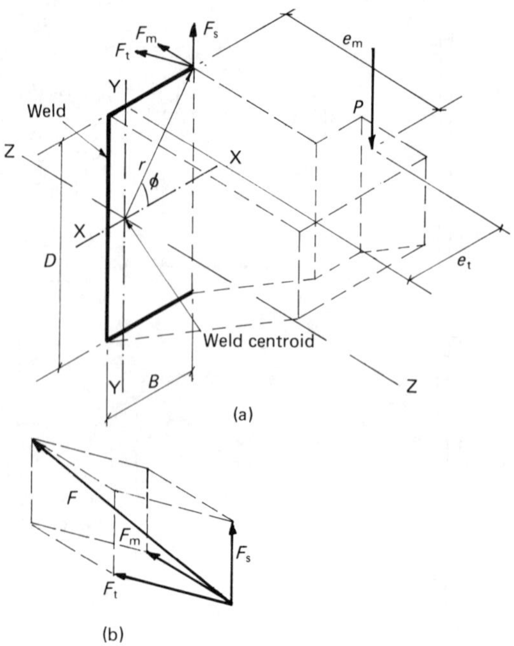

Figure 8.10 Weld group under combined shear torsion and moment. (a) Overall analysis; (b) vector summation

Eliminating k gives:

$$F_m = \frac{Pe_m\, x}{\int_{\text{weld}} x^2\, \text{d}s} \tag{8.29}$$

Taking moments about ZZ:

$$Pe_t = \int_{\text{weld}} F_t\, r\, \text{d}s \tag{8.30}$$

where r is distance from weld centroid. However, $F_t = k^*.r$ from elasticity, where k^* is another elastic constant.

Substituting in equation (8.30):

$$Pe_t = k^* \int r^2\, \text{d}s$$

Eliminating k^* gives:

$$F_t = \frac{Pe_t.r}{\int_{\text{weld}} r^2 \text{d}s}$$

(For convenience, note that $\int r^2 \text{d}s = \int x^2 \text{d}s + \int y^2 \text{d}s$. The latter is usually easier to evaluate.) The total vector F at any point in the weld group is the vector sum of F_s, F_m and F_t.

8.4.2 Approximate analysis

Elastic analysis is based on the assumption, *inter alia*, that weld response does not vary with

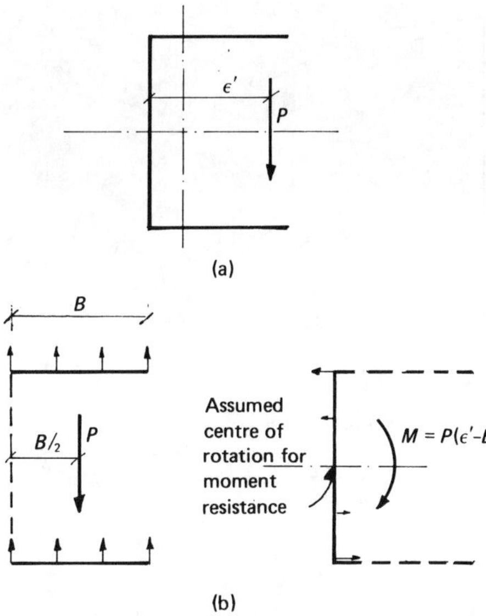

Figure 8.11 Approximate analysis of a weld group. (a) Original system; (b) equivalent system

orientation of the load vector with respect to weld axis. In reality, as discussed in Section 4.2 and illustrated in Figure 4.5, weld response varies considerably with load direction. For example, welds are both stiffer and stronger in the end-fillet condition than in the side-fillet one. For this reason, a weld group will respond in a way that will maximize end-fillet loading where the load vector is normal to the longitudinal weld axis and will minimize side-fillet loading. This can sometimes be exploited to carry out an approximate analysis which is less tedious than a rigorous one and probably more accurate. An example is shown in Figure 8.11.

Before using such an approach a cautionary note should be sounded. Such an analysis should lead to a distribution of stresses very different to that which is normally associated with a particular shape, where subconsciously an engineer thinks in terms of the distribution of stresses that are associated with flexure. For example, in the three-sided weld that looks like a channel in Figure 8.11, the 'web' does not carry any 'shear'. The horizontal portions of weld carry all the vertical load because of their greater stiffness.

8.4.3 Applicability of analytical methods

It should be clear from the above discussion of both the conventional and approximate methods of elastic analysis that they are only appropriate in situations where deformation of the weld group is the dominant flexibility in that part of the structure.

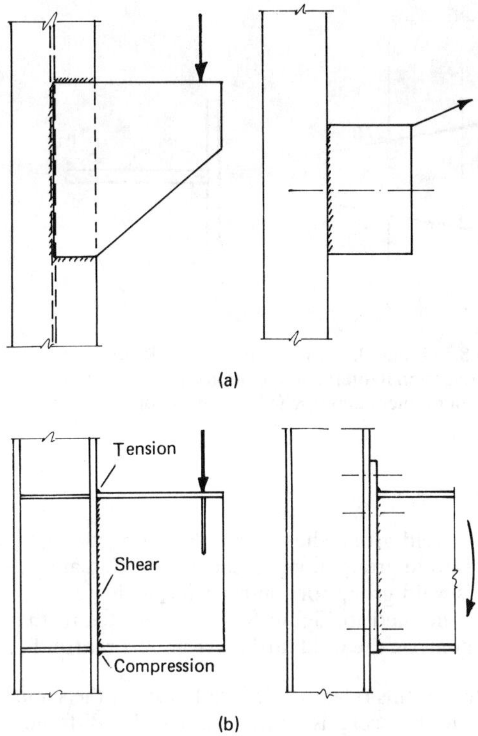

Figure 8.12 Classification of analytical methods for welded connections. (a) Weld deformation and conventional weld group analysis; (b) force paths through the connection

In most situations where welds are used it is more sensible to distribute the load to the welds in accordance with the stress distributions in the connected parts. Figure 8.12 shows examples where weld group analysis is appropriate and contrasts it with situations where weld forces are best determined from considerations of overall behaviour.

In any situations where real uncertainty about analysis exists then minimum sized welds, in accordance with those described on page 49, should be used in order to ensure that the connection is capable of ductile redistribution.

8.4.4 Interaction diagrams

It can readily be shown that the interaction relationship for any weld group subject to shear P, moment M and torsion T is generally given by:

$$\left(\frac{P}{P_0}\right)^2 + \left(\frac{M}{M_0}\right)^2 + \left(\frac{T}{T_0}\right)^2 + 2\left(\frac{T}{T_0}\right)\left(\frac{P}{P_0}\right) \cos \varphi = 1.0$$

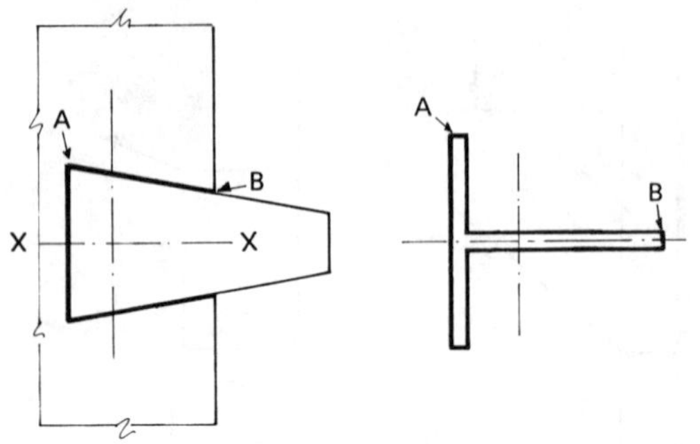

Figure 8.13 Examples of weld groups that do not conform to a conventional interaction relationship. A – critical points for moment about XX; B – critical points for torsion

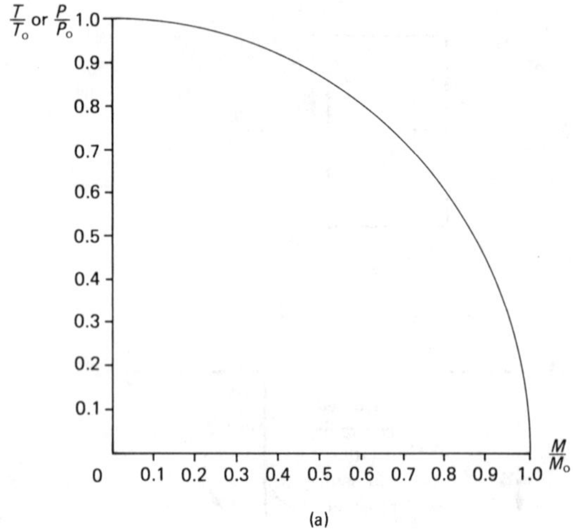

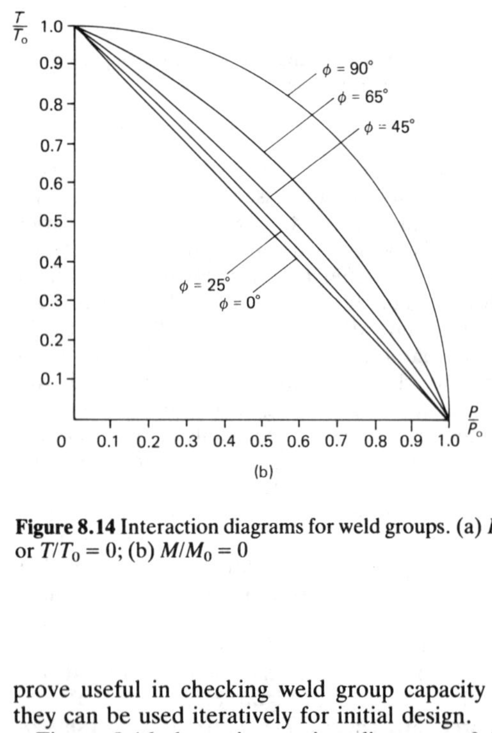

Figure 8.14 Interaction diagrams for weld groups. (a) P/P_0 or $T/T_0 = 0$; (b) $M/M_0 = 0$

where

P_0 is weld group shear capacity, acting alone,
M_0 is weld group moment capacity, acting alone,
T_0 is weld group torsion capacity, acting alone,
φ is defined in Figure 8.10, and relates to the point on the weld furthest from 'the centroid'.

Note that this relationship is only valid if the same point on the weld is critical under the different modes of loading. This is generally the case, but Figure 8.13 shows counter-examples where different points are critical under moment and torsional loading. In such circumstances the interaction relationship given above will underestimate design capacity under combined loading.

Unfortunately, this relationship is not very suitable for graphical presentation in a form that is suitable for direct design. It is always difficult to present design charts for more than two variables and the term in φ creates an added complication. However, it is still worth presenting some general diagrams in order to have a better understanding of interactive design capacity. In addition, for direct use in design it is feasible to present a few diagrams of particular cases which commonly occur in practice.

Figure 8.14 shows interaction diagrams for the situations where one of the three force components is absent. The influence of variations of φ on strength can clearly be seen. Figure 8.15 presents four sets of interaction diagrams for all three components of loading and for varying values of φ. Examination of various weld groups shows that it is not possible to derive sensible bounding relationships between P_0, M_0 and T_0, which would enable loading vectors to be determined for these diagrams in a similar way to bolted connections. Without these, it is not possible to use these diagrams for direct initial design. However, they will prove useful in checking weld group capacity and they can be used iteratively for initial design.

Figure 8.16 shows interaction diagrams of P/P_0 against non-dimensionalized eccentricity for the limiting cases of no torsion and no moment. For the reasons outlined above, it is not possible to plot general design charts. In lieu, the response of a range of common weld groups is presented. These may be used directly for these particular geometries. For other cases the charts may be used to give realistic estimates of P/P_0 by interpolation, for use in initial design.

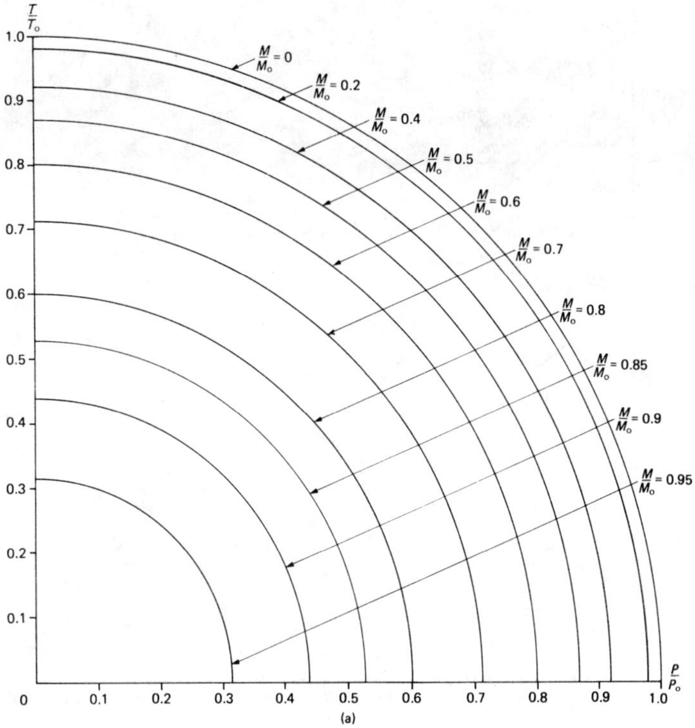

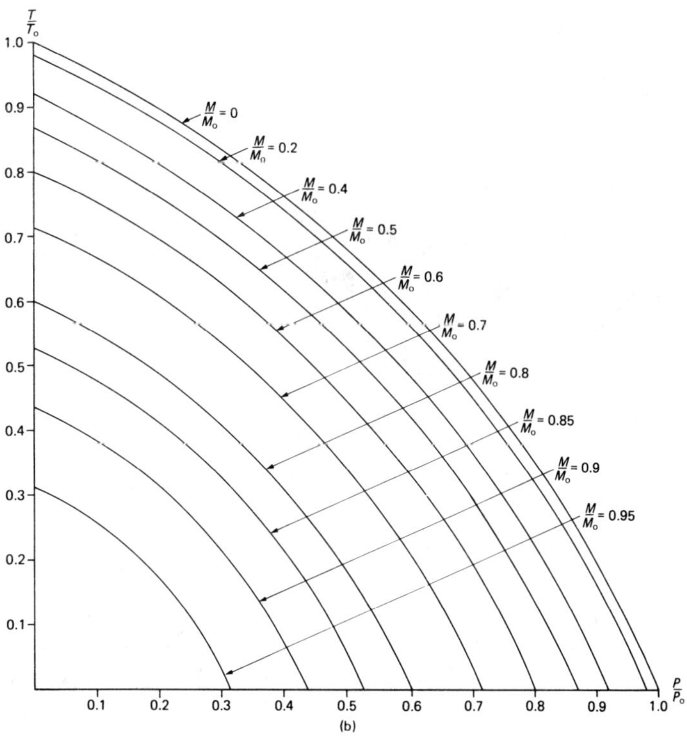

Figure 8.15 Interaction diagrams for weld groups. (a) $\varphi = 90°$; (b) $\varphi = 65°$;

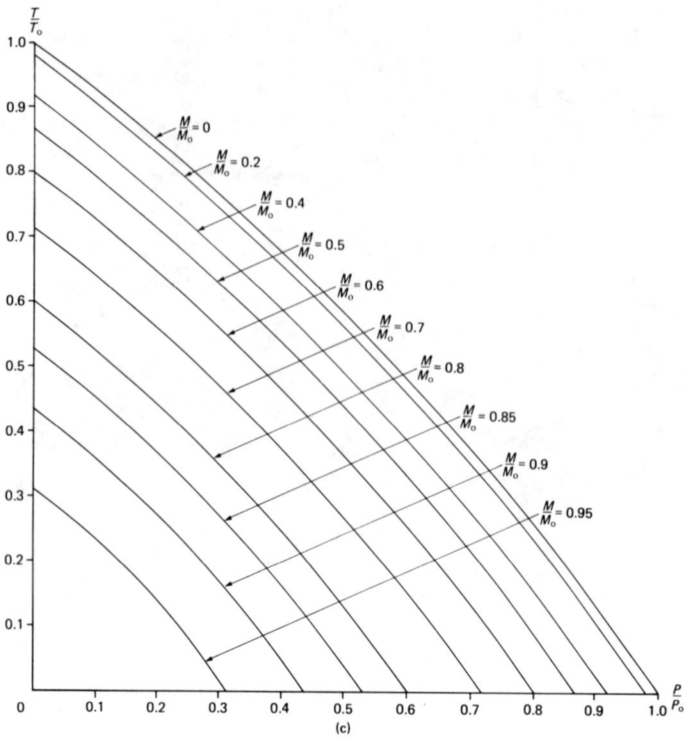

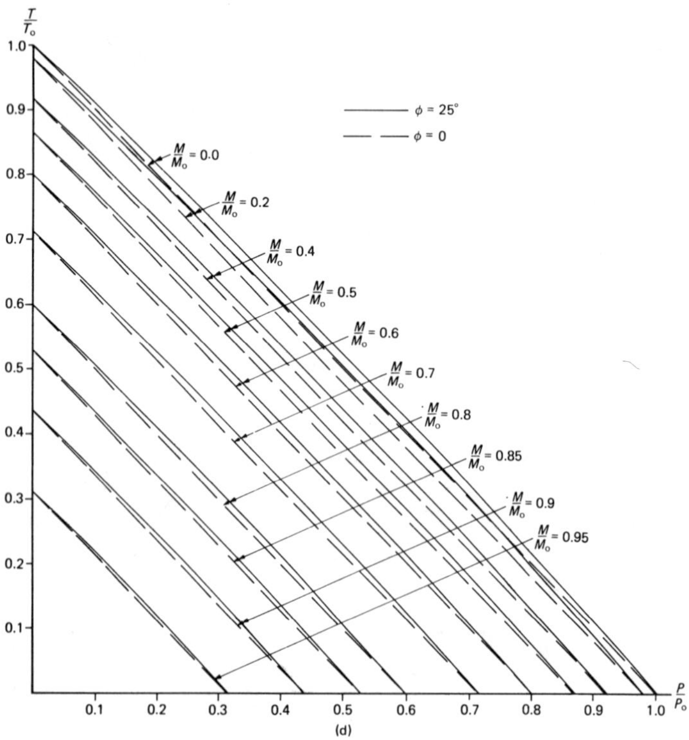

Figure 8.15 (*cont.*) (c) φ = 45°; (d) φ = 25° and 0°

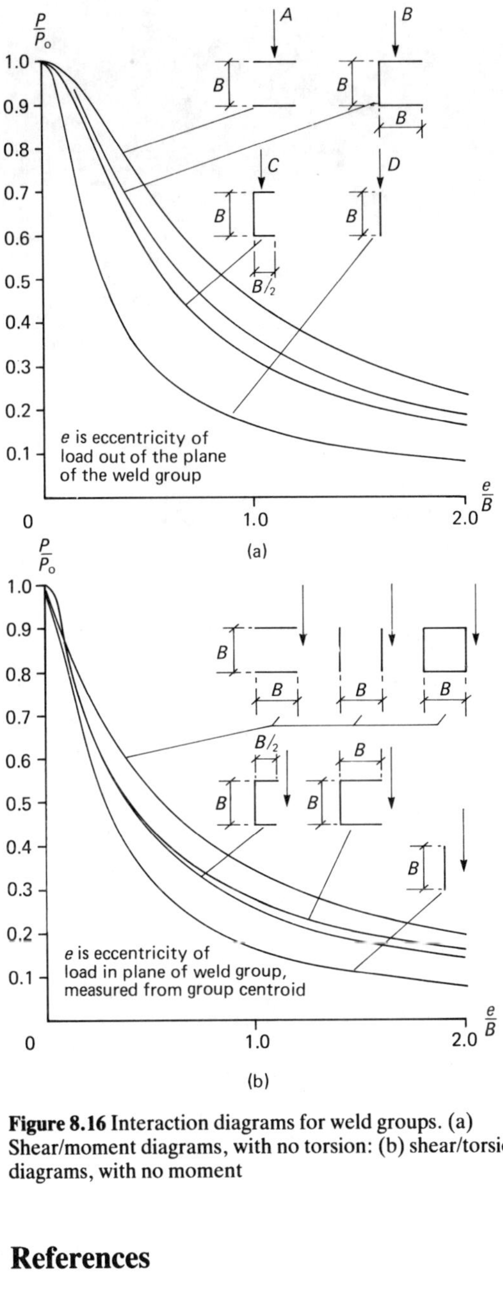

Figure 8.16 Interaction diagrams for weld groups. (a) Shear/moment diagrams, with no torsion: (b) shear/torsion diagrams, with no moment

References

1. Laidlaw, I., 'A practical method for analysing the load capacity of high strength friction grip bolt group using the instantaneous centre of rotation method and plastic analysis', Private communication.
2. Surtes, J. O., Gildersleeve, C. P. and Watts, C. J., 'A general tabular method for elastic and plastic analyses of eccentrically loaded fastener groups', *The Structural Engineer*, **59A**, No. 6 (1981).
3. Abolitz, A. L., 'Plastic design of eccentrically loaded fasteners', *AISC Engineering Journal*, July (1966).

Commentary: 1/1

The example illustrates the design of a bolt group with an inclined load eccentric to the centroid of the bolt group. The eccentricity should be kept to a minimum. In this example if the inclined load had passed through the centroid of the bolt group the maximum bolt load would have been reduced by 47%. HSFG bolts have been used to provide a stiffer connection.

Section 8.2.2 (8.11)

The maximum shear occurs on bolt 'X'. The calculation of the resultant shear is illustrated in the vector diagram. AB is the resultant shear in the bolt ignoring the eccentricity (i.e. 250/6). BC is the shear component due to the eccentricity moment and is perpendicular to the line from the centroid to bolt 'X'.

Note: section references are in each case to sections in the preceding chapter (e.g. Section 8.2.2 here refers to pages 107–108 and (8.11) to equation 8.11 on page 107.

Structural Steelwork Connections	Subject: Eccentrically loaded bolt group			Chapter Ref. 8
	Design Code: BS 5950 Part 1			Calc. Sheet No. Example 1/1
	Calc. by: B.D.C.	Date: Aug '87	Check by: G.L.O.	Date: Nov '87

Code Ref.	Calculations	Output

<u>Eccentrically loaded bolt group.</u>

Design a bolt group to carry a load of 250 kN (due to factored loads) at an inclination of 30° to the vertical and at an eccentricity of 90 mm from the centroid of the bolt group.

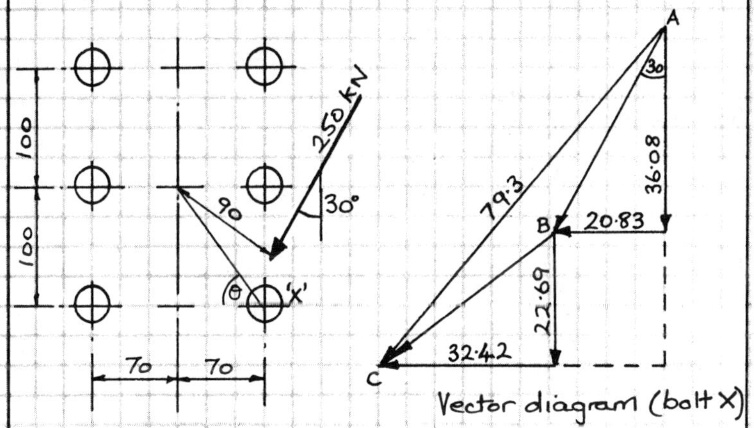

Try the bolt arrangement in sketch

Vertical component of load = P_x = 250 cos 30
= 216·5 kN

Horizontal component of load = P_y = 250 sin 30
= 125 kN

Moment = M = 250×90 = 22500 Nm

Number of bolts = n = 6

$\Sigma(x^2+y^2) = 6\times70^2 + 4\times100^2 = 69400 \text{ mm}^2$

Maximum resultant shear on bolt

$= \sqrt{\left[\frac{216\cdot5}{6} + \frac{22500\times70}{69400}\right]^2 + \left[\frac{125}{6} + \frac{22500\times100}{69400}\right]^2}$

$= \sqrt{[36\cdot08 + 22\cdot69]^2 + [20\cdot83 + 32\cdot42]^2}$

= 79·3 kN

Commentary: 1/2

BS 5950: Part 1. Clause 6.2.3 requires a minimum end distance

$= 1.40 \times$ diameter of hole
$= 1.40 \times 24 = 33.6\,\text{mm}.$

In this example the minimum edge distance (36 mm) required to provide adequate bearing resistance governs.

Before carrying out the design calculations an initial rough check could be carried out to see if the proposed bolt arrangement is likely to be suitable. Although for the relatively small bolt group in the example an initial rough check is probably unnecessary, for larger bolt groups it could well be worthwhile.

Section 8.2.6

For a small bolt group the following procedure, using the interaction diagrams of Figure 8.7, is appropriate

$$\frac{e}{D} = \frac{90}{244} = 0.37$$

From Figure 8.7(a)(i) For three bolts in column
$P/P_0 = 0.67$
From Figure 8.7(b)(i) For two bolts in row
$P/P_0 = 0.57$

Adopt lesser value, $P/P_0 = 0.57$

$$\text{Shear per bolt without moment} = \frac{250}{6} = 41.67\,\text{kN}$$

Estimated shear per bolt with moment

$$= \frac{41.67}{0.57} = 73.1\text{kN}$$

Add 10% to allow for effect of inclined load.
Estimated shear per bolt $= 1.1 \times 73.1 = 80.4\,\text{kN}$
M22 HSFG general grade bolt would probably be adequate; therefore go ahead with design.

Structural Steelwork Connections	Subject: Eccentrically loaded bolt group			Chapter Ref. 8
	Design Code: BS 5950 Part 1			Calc. Sheet No. Example 1/2
	Calc. by B.D.C.	Date Aug. '87	Check by G.W.O.	Date Nov '87

Code Ref.	Calculations	Output
	Try M22 General grade HSFG bolts in single shear.	
6·4·2·1	Slip resistance $= 1{\cdot}1\, K_s \mu P_o$	
	$= 1{\cdot}1 \times 1{\cdot}0 \times 0{\cdot}45 \times 177$	
	$= 87{\cdot}6$ kN	
	$> 79{\cdot}3$ kN O.K. Use	M22 general grade HSFG bolts
6·4·2·2	Bearing resistance $= d t p_{bg} < \frac{1}{3} e t p_{bg}$	
	therefore using grade 43 steel	
	minimum ply thickness $= \dfrac{79{\cdot}3 \times 10^3}{22 \times 825}$	
	$= 4{\cdot}4$ mm	
	Assuming 8 mm ply	
	minimum edge distance $= \dfrac{3 \times 79{\cdot}3 \times 10^3}{8 \times 825}$	
	$= 36{\cdot}0$ mm	

Commentary: 2/1

The example illustrates the design of a weld group with an inclined load eccentric to its centroid. The eccentricity should be kept to a minimum. In this example if the inclined load had passed through the centroid of the weld group the design loading on the weld would have been reduced by 57%.

Ten millimetres are deducted from the horizontal welds to allow for end craters. Normally, the leg length of the fillet weld (in this case, 6 mm) is deducted for each end crater. The weld size is not known at this stage in the calculation; therefore, the largest likely size of weld is assumed and a slightly conservative calculation is obtained.

Section 8.4.1

$$\int r^2 ds = \int x^2 ds + \int y^2 ds$$

The terms $\int x^2 ds$ and $\int y^2 ds$ are the second moments of area of the weld group about the YY and XX axes for a weld of unit width (throat thickness).

The maximum shear on the weld occurs at point 'X', and the calculation of the resultant shear is illustrated in the vector diagram. AB is the resultant shear ignoring the eccentricity (i.e. 190/500). BC is the shear component due to the eccentricity moment and is perpendicular to the line from the centroid to the point 'X'.

Capacity of fillet weld = 0.7 × (leg length) × p_w

Normally, the strength of a fillet weld is read directly from a table of capacities.

Structural Steelwork Connections	Subject: Eccentrically loaded weld group			Chapter Ref. 8
	Design Code: BS 5950 Part 1			Calc. Sheet No. Example 2/1
	Calc. by B.D.C.	Date Aug '87	Check by G.O.	Date Nov '87

Code Ref.	Calculations	Output

Eccentrically loaded weld group.

Design a weld group to carry a load of 190 kN (due to factored loads) at an inclination of 30° to the vertical and at an eccentricity of 90 mm from the centroid of the weld group.

Try the weld arrangement in the sketch

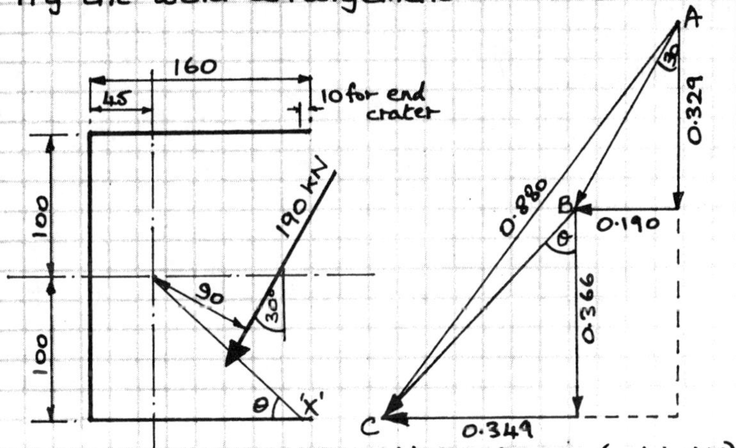

Vector diagram (weld at X)

Vertical component of load $= P_x = 190 \cos 30$
$= 164.5$ kN

Horizontal component of load $= P_y = 190 \sin 30$
$= 95.0$ kN

Moment $= M = 190 \times 90 = 17100$ Nm

6.6.5.2 Assume effective length of each horizontal weld $= 160 - 10 = 150$ mm

Total effective length of weld
$= 2 \times 150 + 200 = 500$ mm

Distance from centroid of weld group to vertical leg $= \frac{2 \times 150 \times 75}{500} = 45$ mm

$$\int r^2 ds = 2\left[\frac{150^3}{12} + 150\left(\frac{150}{2} - 45\right)^2\right] + 200 \times 45^2 + 2 \times 150 \times 100^2 + \frac{200^3}{12}$$

$$= 4.90 \times 10^6 \text{ mm}^3$$

Commentary: 2/2

Before carrying out the design calculations an initial rough check could be made to see if the proposed weld group is likely to be suitable. For vertical loads, interaction diagrams such as those in Figure 8.16 may be used.

In this example the load is inclined and the reaction diagrams do not apply. Nevertheless a rough check may be carried out as follows.

Section 8.4.4

Try weld group in sketch:

$B = 200\,\text{mm} \qquad e = 90\,\text{mm}$

$$\frac{e}{B} = \frac{90}{200} = 0.45$$

Length of horizontal weld = 150 mm = 0.75 × B

Using Figure 8.16(b) and interpolating between curves for horizontal weld lengths of $B/2$ and B:

$$\text{for } \frac{e}{B} = 0.45 \quad \frac{P}{P_0} = 0.46$$

Shear per unit length of weld without moment

$$= \frac{190}{500} = 0.38\,\text{kN/mm}$$

Estimated shear with moment

$$= \frac{0.38}{0.46} = 0.83\,\text{kN/mm}$$

Add 10% to allow for effect of inclined load.

Estimated shear on weld = 1.1 × 0.83 = 0.91 kN/mm

A 6 mm fillet weld would probably be adequate; therefore go ahead with design.

Structural Steelwork Connections	Subject: Eccentrically loaded weld group			Chapter Ref. 8
	Design Code: BS 5950 Part 1			Calc. Sheet No. Example 2/2
	Calc. by B.D.C.	Date Aug '87	Check by G.W.O.	Date Nov '87

Code Ref.	Calculations	Output
	Maximum resultant shear on weld	
	$=\sqrt{\left[\frac{164.5}{500}+\frac{17100\times105}{4.9\times10^{6}}\right]^2+\left[\frac{95.0}{500}+\frac{17100\times100}{4.9\times10^{6}}\right]^2}$	
	$=\sqrt{[0.329+0.366]^2+[0.190+0.349]^2}$	
	$=0.880$ kN/mm	
6.6.5	Capacity of 6 mm fillet weld	
	$=0.7\times6\times215\times10^{-3}$	
	$=0.903$ kN/mm	
	o.k. Use	6 mm fillet welds

9

Practical considerations for economic design

9.1 Introduction

The total cost of a steel structure is the sum of the costs of the material, design, fabrication, erection and corrosion protection. The breakdown between these components varies with type of structure, time and location. The complexity, degree of repetition within the structure and the severity of design specification all vary with structural type and all have a major influence on economic design. Figure 9.1 shows the changes in the relative cost of material and labour that have taken place in recent decades for various locations. It is clear that even for a particular location the most economical design will vary with time, with a general tendency for economic designs to become simpler and less labour intensive. Economic design will certainly vary from one country to another. Within a country the most economic details will vary from fabricator to fabricator and will depend on the equipment and skills of a particular shop. It is particularly notable that productivity has improved in recent years, with the greatest advances occurring in the more advanced economies. Improvements in cutting and welding technology have played their part, but the greatest progress has been due to the increased use of automatic plant.

With such a complex and variable situation it becomes very difficult to offer guidance on economic design. The comments in this chapter are inevitably of a general nature, and in almost all cases it will be possible to think of counter-examples where the reverse to the guidance offered will be appropriate. In most cases economic design is little more than applied common sense; none of the advice given should be applied in a context where it disagrees with that guidance.

There is one generality to which there can be no counter-examples. It is that connection design and detailing are of primary importance for the economy of the completed structure. Material costs are unlikely to be more than 40% and may well be as low as 20% of the total cost. Corrosion protection is unlikely to exceed 20% and is usually considerably less. The remaining costs of design, fabrication and erection (rarely less than 50% and frequently a greater percentage of the total cost of the steelwork) are primarily concerned with connections. A carelessly conceived welded connection which requires an unnecessary rotation of the assembled fabrication with special cranage to complete the welding could spoil an otherwise economic design.

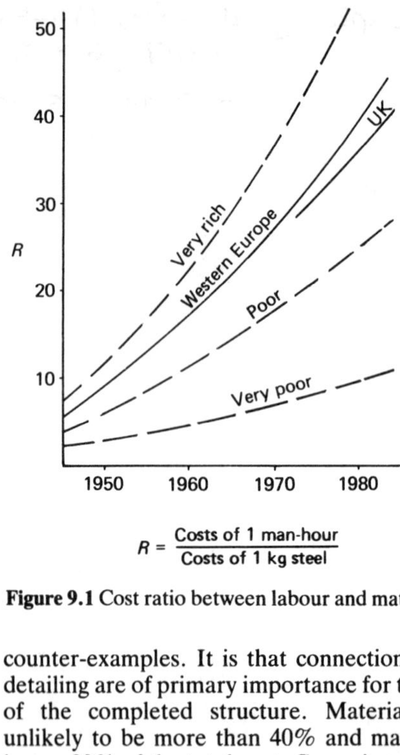

$R = \dfrac{\text{Costs of 1 man-hour}}{\text{Costs of 1 kg steel}}$

Figure 9.1 Cost ratio between labour and material

A choice of truss arrangements that requires angled cuts and precise fitting of the web members could add 10 hours per tonne to fabrication time. Lack of provision of erection bolts could double erection costs. Inadequate provision for adjustment could lead to expensive remedial measures on site to correct a lack of fit. The list could be endless. In almost all cases such mistakes simply arise from a lack of foresight, practicality and common sense.

9.2 Choice of method of connection

9.2.1 General

Selection of the method of connection is governed by two criteria that are frequently in conflict. Clearly, the method of connection must be compatible with the design requirements for the connection and the structure of which it is part; strength, stiffness and deformation capacity need to be considered. Second, the choice of method of connection must take account of economic factors. The designer is looking for the most economical method of connection that is structurally acceptable. In the following sections detailed guidance is given on the relative economies of different types of connection and general guidance on structural acceptability. For more specific guidance on this latter topic, the reader is referred to Chapters 10–16.

9.2.2 Choice between welded and bolted connections

Under controlled conditions at ground level it is generally cheaper to fabricate a welded connection to transmit a particular shear or tension load than a bolted one. Thus for connections carried out in a fabrication shop there is a clear preference for welding. It is only in quite special circumstances that bolting would be preferred (for example, if the holing lines in the shop were underutilized and the welding bays overloaded).

On site the situation is less clearcut, although, in general, bolting is favoured for work in the UK for the following reasons:

1. Better access must be provided to a welded joint than a bolted joint, both for the welders and their equipment.
2. Welders must be protected from the elements if they are to achieve satisfactory connections. Preheating is considerably more difficult on site than in a fabrication shop.
3. Traditionally, welders on site were charged with full shop overheads, and this made them extremely expensive operatives. (However, some site welding is now carried out by subcontract welders without such artificial penalties.)
4. Bolting can be carried out by semi-skilled operatives.
5. Certification of welders and procedures for site work is frequently required for each site.
6. Bolted joints may be more readily inspected.

Items 1, 2 and 5 involve considerable expenditure when site welding is set up. Thus it is most unlikely to be economic for small-scale construction where this initial overhead could not be justified. It is generally thought that site welding should only be considered for structures of over 500–1000 tons. Even above this size any difficulty in meeting the other requirements noted above will preclude its use on economic grounds. If the decision is made to opt for site welding, design of the connections should certainly recognize the greater difficulty of site work.[2]

9.2.3 Choice between fillet and butt welding

Figure 9.2 compares the cost of fillet and butt welds for varying plate thicknesses. It can clearly be seen that there are considerable economic advantages in choosing the former where static strength is the sole criterion. Note that the cost comparison is based on a 6 mm fillet weld, which is the largest single-run fillet weld that can be laid by traditional manual welding. The difference in cost becomes much less if larger, multipass, fillet welds have to be used. Table 9.1 summarizes the relationship between fillet weld size and number of passes for manual welding. Since fillet weld cost is primarily a function of number of passes it can be seen that there is a very non-linear relationship between fillet weld size and cost. It is rarely economic to use more than a 12 mm or 15 mm fillet weld, unless submerged arc welding is used.

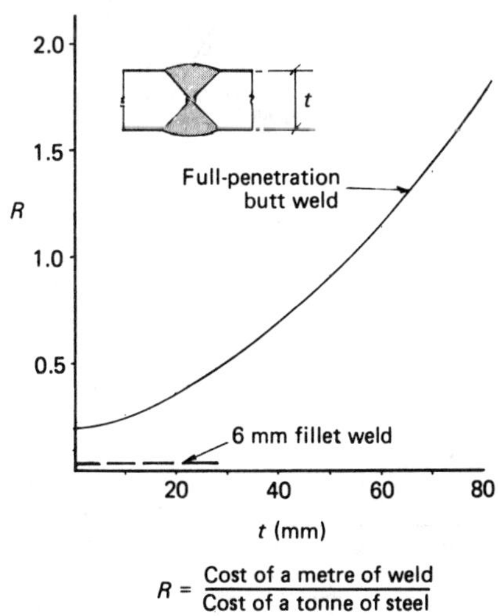

Figure 9.2 Comparative costs of butt and fillet welds

Table 9.1 Relationship between fillet weld size and number of passes for manual metal arc welding in the downhand position

Leg size	<6	8–10	11–13	14–16	19	22	25
Number of runs	1	2	3	4	5	7	10

Other reasons for choosing butt welds despite their greater costs are:

1. Greater fatigue endurance;
2. A more pleasing appearance;
3. If lap plates are necessary for fillet welds and would lead to corrosion traps.

9.2.4 Choice between dowel and HSFG bolts

Figures 9.3 and 9.4 compare the design strength of Grade 8.8 dowel or bearing bolts with HSFG bolts. It can be seen that, except for situations where tensile loads predominate, the former has a greater design strength. In addition, there is a price differential of approximately 40% in favour of Grade 8.8 bolts and their installation costs are generally substantially less. The greater installation cost of HSFG bolts primarily relates to the need for powered torquing equipment and the associated requirement for temporary staging. Grade 8.8 bolts can be installed in a spanner-tight condition without any staging. Thus the economic arguments are strongly in favour of using dowel bolts wherever possible.

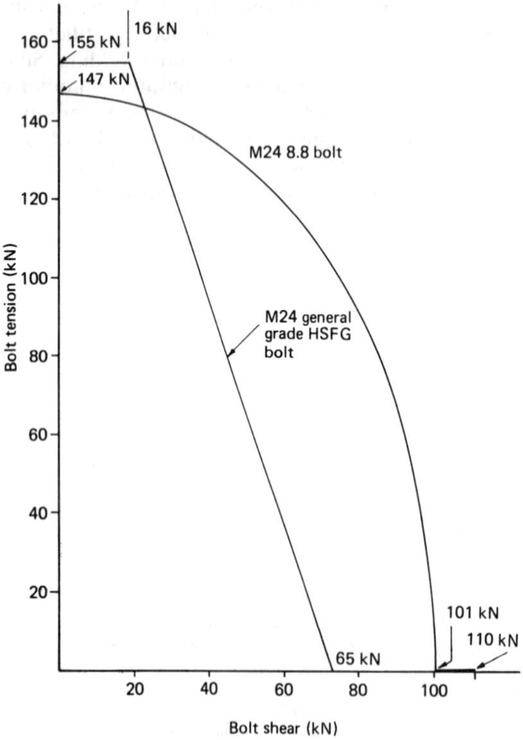

Figure 9.3 Design strengths of HSFG and Grade 8.8 bolts to BS 5400: Part 3

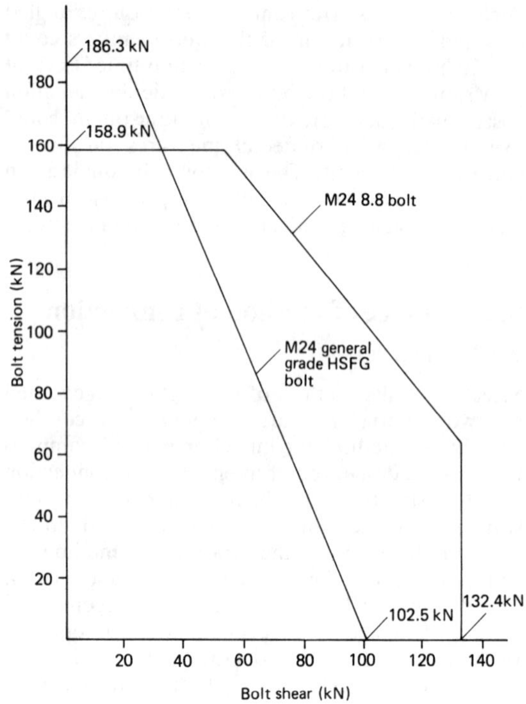

Figure 9.4 Design strengths of HSFG and Grade 8.8 bolts to BS 5950: Part 1

The primary structural difference between the two fasteners is the greater flexibility of the bearing bolt. The bolts are always used in clearance holes and, because they are untorqued and the faying surfaces of the joint are not prepared to enhance friction, they will slip into bearing under low shear load. If this potential movement can be accommodated then economics dictate that these fasteners should be used; if not, then HSFG bolts should be specified. There is some disagreement about the classification of situations where bearing bolts may or may not be used. In commonsense terms a shear displacement (for example, where a beam end plate is bolted to the face of a beam) will not generally create difficulty for the structure. (In practice, at least some bolts will be in bearing due to the self-weight of the beam.) However, a 'flexural' movement (for example, where a beam is spliced with cover plates and bolts in shear) is likely to lead to unacceptable deformations in the structure, HSFG bolts should therefore be used.

There are two other situations where spanner-tight dowel bolts would not be acceptable. Where fatigue is a consideration in design, either in shear or in tension, it is essential that these fasteners are avoided. With fluctuating shear loads HSFG bolts should be used in frictional joints to minimize fretting between the plies. With fluctuating tensile forces the bolts must be adequately torqued to reduce stress fluctuations, as discussed in Section 6.3.3. A common way of achieving this condition without special provision is to specify HSFG bolts.

In the above discussion the primary comparison has been between general grade HSFG bolts and 8.8 dowel bolts. The reason for limiting the discussion in this way is that these are the most popular strength grade for both classes of connector. Mild steel bolts (4.6) used to be very popular; however, their price per kN of capacity in both shear and tension is significantly less than that of 8.8 bolts and their use is now limited to situations where loads per bolt are necessarily low – for example, stitching bolts at maximum pitch. Higher grades of both classes of connector are not common, and their specification is likely to cause difficulties of supply, sometimes with severe economic consequences.

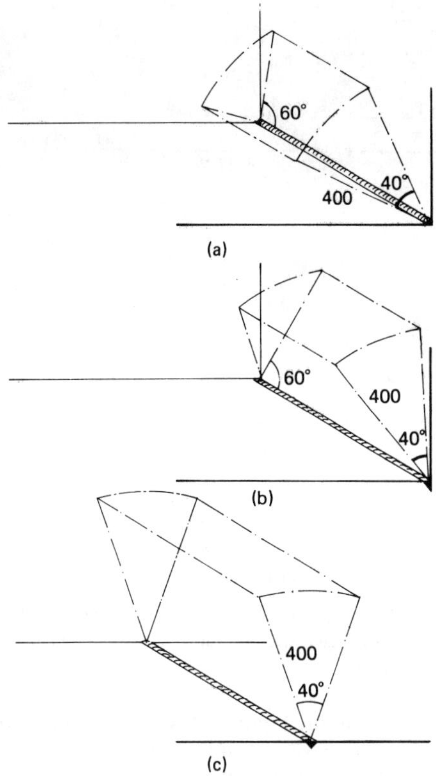

Figure 9.5 Access requirements for manual welding. (a) Fillet weld; (b) corner butt weld; (c) planar butt weld

9.3 Access for fabrication and assembly

Ready access during fabrication and erection is clearly essential for maximum economy. It is important to differentiate between ready access, i.e. a situation where the operatives may work without hindrance, and minimum access, where it is possible for them to carry out their task but only with difficulty and at increased cost.

Figure 3.14 summarizes the requirements for ready access for bolted connections. Dowel bolts are most conveniently tightened by podger spanner and their use governs the minimum clearance from upstands. Lesser dimensions may be used but will lead to a requirement for more compact and less convenient spanners. Torqued bolts are usually tightened by torque wrench and the same figure gives the spatial requirements for the direct application of this tool to one side of the fastener. Where necessary, angled drives may be employed, although these can create considerable difficulty for calibration when used with torque control. Manual torque wrenches can also be used but while they require a considerably smaller distance along the bolt axis, they need a considerable lateral sweep for the handle.

Figure 9.5 shows the ready access requirements for manual welding. The figures quoted are for manual metal arc welding and are governed by the electrode length. Semi-automatic torches will require less space. In addition, space should be provided for the welders, so that they may visually monitor the weld pool; most protective masks are bulky. It is more difficult to specify minimum access requirements; welders can show remarkable versatility in awkward circumstances. Mirrors can be used to overcome problems of visibility and bent or short electrodes those of space. However, in such circumstances productivity will suffer severely and quality may well be suspect.

It is important to consider the needs of the inspector when detailing a welded joint. Requirements will vary with the type of inspection. Visual inspection, dye penetrants and magnetic particle inspection require little more than visibility, but subsurface techniques are more difficult to accommodate. X-ray inspection generally requires access to one side of the weld for the radiation source and access to the other for the film. Shielded radiation sources for site use are bulky and awkward to handle. Their use in purpose-made inspection chambers is only acceptable for small components. Ultrasonic testing is more acceptable for site use but may still present severe practical difficulties. The probes are small and easily manipulated, but there may be difficulties (sometimes insurmountable) in finding a pulse path that can be monitored. In extreme circumstances this can lead to a complete

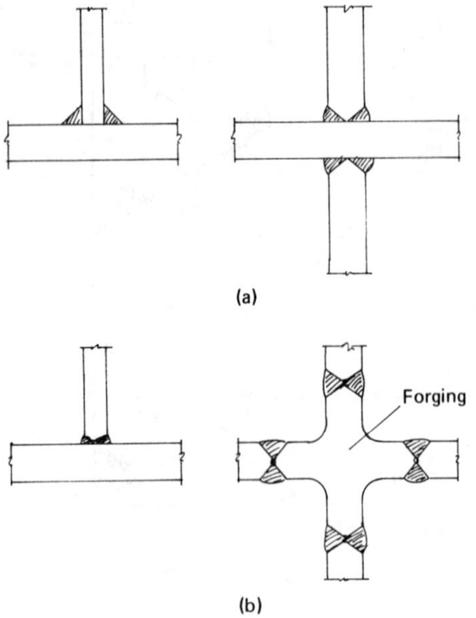

Figure 9.6 Influence of inspection requirements on weld layout. (a) Subsurface inspection difficult; (b) full weld inspection possible

redesign of the joint. Figure 9.6 shows examples of joints where inspection requirements were important factors in determining weld layout.

The possibility of repairs should also be considered when considering weld layout. Chipping out is expensive and only possible if direct thrust can be applied to the chipping hammer. Grinding out also requires considerable space. Arc air gouging requires considerably less room but may only safely be used where there is free passage for the air that removes the remelted metal. For example, cope holes must be provided at internal corners. Failure to do so would lead to metal with very high carbon content (from the arc electrode) being redeposited, resulting in severe embrittlement.

9.4 Weld preparations

Weld preparation is not generally required for fillet welds. It is only in special circumstances of the kind shown in Figure 9.7 that the need to ensure side-wall fusion requires the modest weld preparation shown.

Weld preparations for manual butt welds must satisfy the following criteria. They must:

1. Be cheap to prepare;
2. Minimize weld volume;
3. Provide a suitable root face for the root run;
4. Provide proper access for the root run;
5. Ensure that side-wall and interpass fusion can be achieved during filler runs;
6. Permit cleaning between runs.

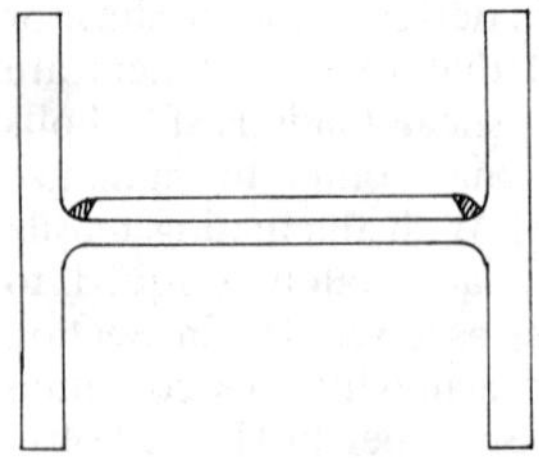

Figure 9.7 Welding detail for reinforcing plate on column web or vertical leg of wide seating angle

Guidance on choice of weld preparation to satisfy these criteria is given in BS 5135. Figure 2.5 shows examples of its recommendations.

Weld preparations for automatic butt welds may differ considerably from those for manual welding. The greater currents used achieve greater penetration; in many circumstances this can be used to reduce deposited weld volume. Narrow gap techniques are increasing in popularity because of the savings in deposited weld volume.

9.5 Holing

9.5.1 Methods of hole making

Holes may be punched or drilled; the former is cheaper but may not be an acceptable method of manufacture in many circumstances.[3] The punching operation is shown schematically in Figure 9.8. The punch is driven against the plate which is held against the die, forcing a plug of steel out through the latter. For a good-quality hole to be produced

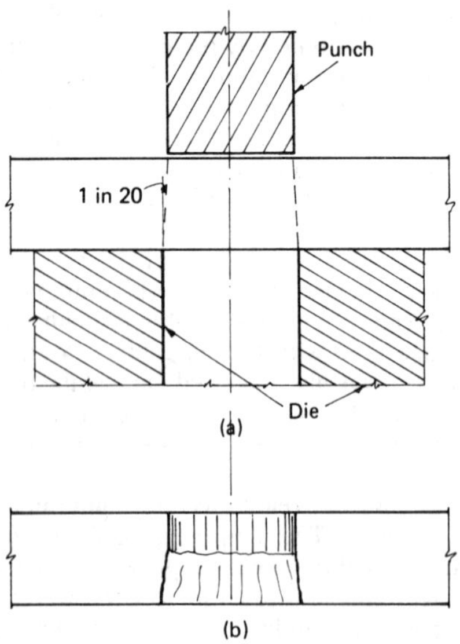

Figure 9.8 (a) Schematic diagram of hole punching; (b) cross-section through punched hole

the punch and die must be concentric and free from edge distortions. There must be a clearance (usually 5% of the plate thickness on radius) between the punch and die. If the clearance is too large the hole formed will be too great a deviation from the ideal cylinder. If it is too small there will be too much tearing of the material. If either the punch or die is worn or blunt the material will be distorted from the flat. There is always likely to be a small burr on the die side of the hole, and this should be removed to ensure that neighbouring plates can achieve proper mating.

In practice, it is not difficult to achieve a properly punched hole provided that due attention is paid to the points indicated above. However, the surface of the hole will still differ from that of a drilled hole, and it is this difference which limits the application of punching. The shearing action of the punch produces severe straining close to the surface of the hole. This cold working leads to embrittlement of the material within approximately 1 mm of the surface of the hole and leaves a jagged surface. This embrittled material is very likely to fracture at low tensile strain. For safety, these cracks have to arrest in the surrounding, ductile, material. Current restrictions on the use of punching are designed to ensure that the deformation capacity of the punched material is not less than approximately 2 mm when subject to tensile strains. While this is considerably less than that of plate with drilled holes it is comparable with the deformation capacity of a high-strength bolt in shear, and is judged to be appropriate for general design for static strength.

Drilled holes are much less sensitive to variations in workmanship, and the most common problem is that of mispositioning. This can occur even with proper setting out if the drill is incorrectly sharpened so that its tip does not lie on its longitudinal axis. Pilot holes will overcome the problem if extraordinary accuracy of hole positioning is required.

9.5.2 Oversize and slotted holes

Holes are normally specified with 2 mm or 3 mm clearance on bolt diameter. The former figure is used for bolts up to 24 mm in diameter, the latter for larger sizes. Such holes should be able to accommodate conventional tolerances of element size and hole position.

If greater provision for adjustment is required then oversize, short- or long-slotted holes may be specified for the inner plies of a connection, as listed in Table 3.10. These holes may only be used with HSFG bolts in friction connections unless there is some special requirement for movement. The simplest holes to form are oversize, circular ones.

Table 9.2 Standard backmarks, cross-centres and transverse pitches for rolled sections

Section	*Flange width or leg length*	*Backmark (S) or cross-centre (C)*	*Transverse pitch (P)*	*Maximum diameter of bolt*
Channel	102	55		24
	89	55		20
	76	45		20
	64	35		16
	51	30		10
Angle	200 (2 bolts)	75	75	30
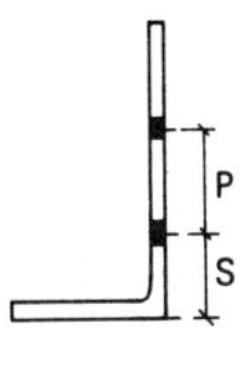	200 (3 bolts)	55	55	20
	150	55	55	20
	125, 120	45	50	20, 16
	100	55	–	24
	90	50	–	24
	80, 75	45		20
	70	40		20
	65, 60	35		20, 16
	50	28		12
Columns,	424 – 362 (2 or 4 bolts)	140	75	24
beams and	322 – 300 (2 bolts)	140	–	24
Tees	322 – 312 (4 bolts)	120	60	24
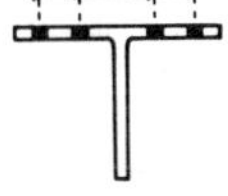	322 – 300 (4 bolts)	120	60	20
	294 – 203 (2 bolts)	140	–	24
	193 – 162 (2 bolts)	90	–	24
	162 – 150 (2 bolts)	90	–	20
	146 – 130 (2 bolts)	70	–	20
	127 – 98 (2 bolts)	54	–	12

These will provide freedom of movement in any direction; unfortunately, it is often the case that movement is only required in one direction, and some extra restraint may be necessary during bolt tightening to limit movement in other directions. In such circumstances slotted holes would be more useful during erection but are more expensive to form.

9.5.3 Hole positions

Table 9.2 summarizes the standard gauge lines or backmarks for hole positions for rolled sections.

9.6 Plate and section edge and end preparations

Steelwork edges may be prepared by the following processes.

Cropping of flats and angles and guillotining of plates. These are the fastest and therefore the cheapest methods of end preparation. The resulting surface is rough and not suitable for any bearing contact; in addition, there is likely to be some minor out-of-plane distortion close to the cut edge. The sheared edge is likely to be embrittled and should not be left in any situation where fatigue is a design consideration. The maximum that can be sheared is generally approximately 20 mm; the maximum thickness that can be guillotined is of the order of 12 mm.

Cold sawing of sections. A milling saw may be used to cut sections to length. If such a saw is properly sharpened and adjusted it should leave a flat, plane surface that may be used to transmit compression by bearing.

Flame cutting. This is used for cutting sections to length and plates to shape. Hand cutting is usually used for sections and one-off plate cutting. Automatic equipment is usually used for repetitive plate cutting; profile cutters enable complex shapes to be copied accurately, and multiple-head machines can complete a butt weld preparation at the same time as the plate is cut. The flame cut edge is jagged, particularly if hand cut, and may well be embrittled. Grinding may be used to dress the edge if required for cosmetic reasons. The cut edge will be too rough to permit any compressive load transfer by bearing, so any welds should be designed for total loads. Where the edge is not to be welded and fatigue or brittle fracture are major design considerations the hardened edge zone should be removed by machining to a depth of 3 mm.

Milling of cut edges of plates may be used either for the reasons given above or to cut a weld preparation.

Milling or 'ending' of cut section ends may be used to provide a true surface for bearing and/or to control section length accurately.

9.7 General guidance on economic fabrication

The most important general point to make is that every effort should be made to standardize connections and achieve repetition of details within a particular structure. Many connections require some particular skill for their economic completion, and they have their own learning curve. The first connection of a series may well take three times as

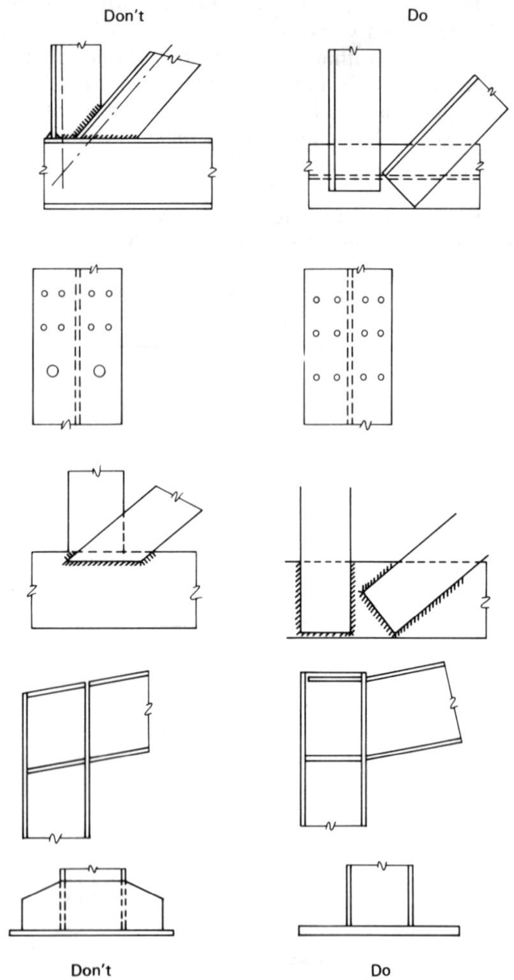

Figure 9.9 Examples of economic detailing to minimize fabrication

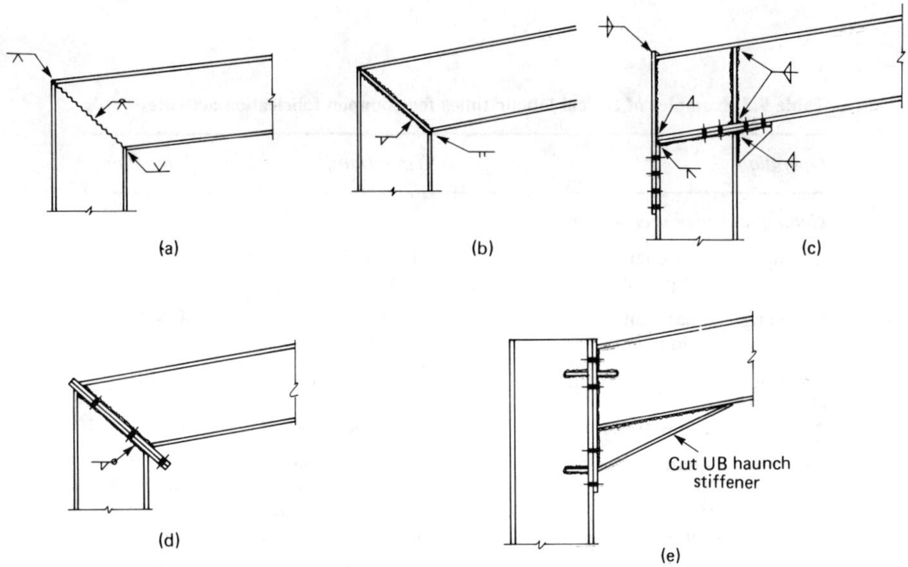

Figure 9.10 Costs comparisons for eaves connection in portal frames. (a) Welded knee joint detail where stiffeners are not required (cost ratio 1.0 (base)); (b) welded knee joint detail with division plate – square butt weld (cost ratio 1.1); (c) bolted knee joint (cost ratio 1.8); (d) diagonal bolted knee joint (cost ratio 1.3 but more difficult to erect); (e) typical knee joint with haunch stiffener (cost ratio 1.5)

long to fabricate as the tenth, and further savings will have been achieved by the fiftieth.

Second, every effort should be made to reduce labour content to a minimum.[5,6] Figure 9.9 illustrates how this may be achieved in many commonly occurring situations. Precise cutting to length and angled cuts should be avoided. Wherever possible, holes should be on standard gauge lines (see Section 9.5.3) and of constant diameter. This is particularly important where automatic drill or punch lines are used. Plated fabrications should be arranged so that small fillet welds can be used as frequently as possible (see Section 9.2.3). Doubly fitted stiffeners should be avoided unless really necessary. Thicker, unstiffened plates will often be more economic than thinner, stiffened elements.

The influence of such considerations on fabrication costs is illustrated in Figure 9.10, where various approaches to common connections have been compared economically. The precise ratios will obviously vary from one shop to another, but the general message is very clear.

A further understanding of the cost implications of fabrication can be gained by studying Table 9.3, which presents general guidance on typical times for the common operations.

9.8 General guidance on economic erection

There are several basic, commonsense principles which must be applied for economy of erection:

1. Erection bolts must generally be provided. These must be capable of being inserted quickly and of supporting the element safely. 'Hook' time is frequently the greatest single contributor to total erection cost.
2. Where possible, clearance should be provided to ensure easy positioning.
3. Where sufficient clearance cannot be provided, the connection arrangement must be such that the steel elements being connected may be eased, levered, podgered or drifted into position. As a counter-example, Figure 9.11 shows a situation that will cause considerable difficulty in erection if the columns cannot be 'sprung' apart. The cap plates will prevent the beam being lowered into place in a horizontal position; its diagonal dimension is too long for it to be 'angled' into position.
4. The connection must be able to cope with rolling tolerances and other lack of fit. Figure 9.12

Table 9.3 Examples of typical labour times for common fabrication activities

Operation		*Time (min)*	*Limitations*
Cutting and edge preparation			—
Sawing:	square cut	$A/20 + 4$	
	oblique cut	$A/10 + 8$	—
Cropping:	square cut	0.5–1.0	$A \ngtr 15$
	oblique cut	1.0–2.0	$A \ngtr 15$
Shearing		3.5 per metre	$t \ngtr 20$
Hand burning		$0.5\,t$ per metre	—
Automatic burning (square cut)		$0.2t/H$ per metre	—
Chipping:	square	t per metre	—
	30° bevel	$0.1t^2$ per metre	—
	45° bevel	$0.15t^2$ per metre	—
	J prep.	$0.5t$ per metre	—
Planing:	square or bevel	$1.5 + 0.1\,t$ per metre	—
	J prep.	$3 + 0.2\,t$ per metre	—
Automatic flame cut edge preparation		$4 + 0.25\,t$ per metre	—
Holing			
Radial drilling		$1.54 + 0.00005\,t \times D^2$	—
Punching		0.1	
Mark out and assembly			
Bolted connections		4 min per bolt	—
Mark set and tack cleat		12 min	—
Mark set and tack base or end plate (depends on size)		15–60 min	—
Fit stiffener for rolled section		45 min	—
Welding: hand			
Fillet welding:	Horizontal/vertical	17 min per metre per pass	—
	vertical	34 min per metre per pass	—
Butt welding:	flat	$0.2\,t^2$–$0.4\,t^2$ min per metre	—
	vertical	$0.4\,t^2$–$0.8\,t^2$ min per metre	—
Welding: automatic submerged arc			
Setting-up time		70 min per second	—
Run-on and run-off plate		90 min per second	—
Preheating – add 10% to welding times where $\Sigma t > 35$ mm			—
Fillet welds:	5–13 mm	10 min per metre	—
	14–19 mm	20 min per metre	—
Butt weld		$0.15\,t^2$ per metre	—

A: Cross-sectional area (cm^2).
t: Thickness (mm).
D: Hole diameter (mm).
H: Number of heads for multiple-cutting machines.

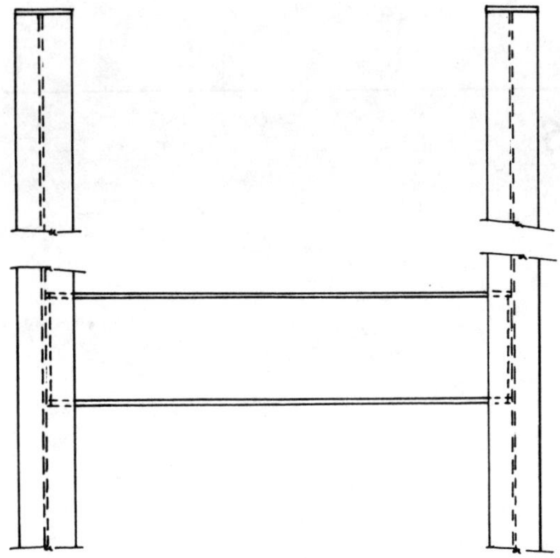

Figure 9.11 Beam framing into minor axis of columns that will create difficulties in erection

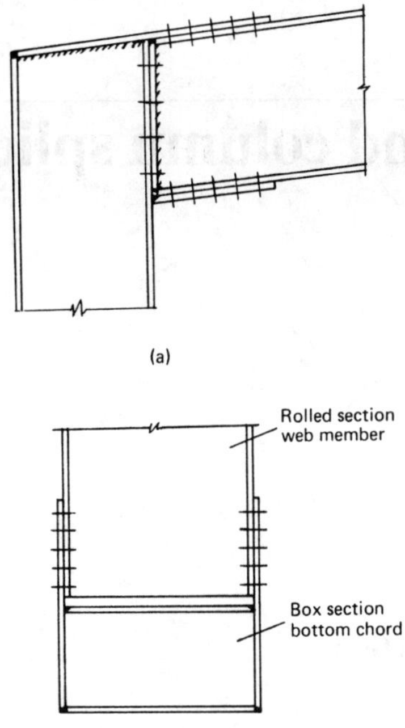

Figure 9.12 Examples of connections where difficulties may be expected in erection. (a) Portal frame eaves connection; (b) cross-section through truss connection

shows counter-examples. In both cases provision for erection clearance with suitable packs would alleviate the problems but not eliminate them entirely, since angular lack of fit would still create difficulties.

5. It must be possible to manoeuvre the element into position without undue difficulty. Deep end plates can cause particular problems in some circumstances.
6. Access for bolting and/or site welding should be provided

References

1. van Douwen, A. A., 'Welded and bolted connections', 2nd Int. ECCS Symposium on New Developments in Steel Construction, London, 1978.
2. Allen, J. S., 'Site welding of structural steel – the design of connections', *Metal Construction*, January, **13**, No. 1 (1981).
3. Driver, P. J., *Punched Holes in Structural Steelwork*, PhD thesis, University of London, 1988.
4. *National Standard Specification for Structural Steelwork*, BCSA, London, 1988.
5. Firkiss, A., and Hogan, T. J. (eds), *Economics of Steelwork Details*, Australian Inst. Steel Const., 1975.
6. *European Recommendations for Steel Construction: Commentary*, ECCS, 1978.
7. Oakes, D. L. T., 'Philosophy for economical design, fabrication and erection', *Steel Construction*, **17**, No. 4 (1983) (Aust. ISC).

10

Beam and column splices

10.1 Introduction

10.1.1 Location

Wherever possible, splices should be located away from critical sections. Of course, axial loads in columns do not vary significantly within a particular storey. However, in both beams and columns it is usually possible to find a splice location where the moment is significantly less than maximum. In beams it is also possible to find a position where the shear force is well below the section capacity. For any given section, connection cost is not a linear function of percentage of section capacity. The greatest rate of increase in connection cost occurs as the connection design strength approaches the section capacity. Thus moving a splice to a position where its design values for axial and shear forces and bending moments are even 20% less than the element capacity can lead to significant economies.

If the member being spliced is subject to instability (and this is almost always the case with columns and frequently so with beams) the splice should be located near a point of effective restraint, if possible. If this cannot be achieved then special consideration has to be given to the splice design, as discussed below.

10.1.2 Continuity requirements in the presence of instability

If, in a member that is subject to instability, a splice has to be located away from a point of lateral restraint special consideration must be given to the design of that splice. The design method for the member concerned will be based on the assumption that there is continuity of stiffness along the length of the member, and clearly that must be maintained effectively through the connection.

It is more difficult to define continuity of strength requirements effectively. Figure 10.1 shows schematically some bending moment distributions for typical columns and beam/columns. Similar phenomena would be observed in the distribution of weak axis, secondary, moments along a beam bent about its major axis. It is not always easy to define

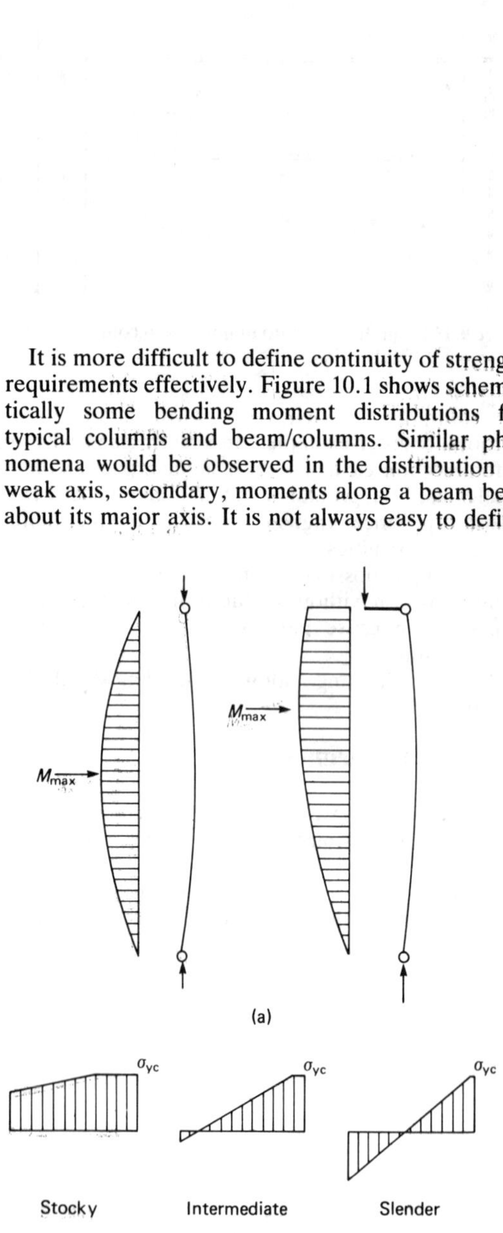

Figure 10.1 (a) Column bending moment distributions at failure; (b) stress distributions at critical cross-sections at failure

either the position of the critical cross-section or the distribution of moments away from that cross-section. In addition, the magnitudes of these moments are difficult to determine at the failure load of the element. In a slender beam or column they would probably be the dominant loading on the critical cross-section, but as the member becomes more stocky they would become less significant. In the limit, with a stocky, axially loaded column the entire critical section will remain in compression up to the point of failure, and failure would be governed by compressive yielding of the relevant extreme fibres. It is difficult to give precise design guidance in these circumstances. Splices in slender elements away from points of restraint should clearly be able to develop substantial weak axis bending strength; for simplicity, design could be based on the section weak axis capacity. The strength of the stocky, axially loaded column discussed above would not be impaired if there was only a nominal weak axis bending capacity in a splice located at the mid-height. However, since axially loaded conditions are difficult to achieve in practice some nominal percentage of the weak axis bending capacity (say, 30%) should be used as a basis for design. Any lesser percentage than this would be unlikely to give continuity of EI_y of the section. In the context of splice plate connections, this later requirement leads to full-width splice plates and not less than two rows of bolts in each half of the connection; using Grade 8.8 bolts such a connection would achieve 30% of minor axis bending strength. Sections of intermediate slenderness should be designed for an appropriate percentage between 30% and 100%.

Even if the splice is located close to a point of restraint, the designer must ensure that any local weakness or discontinuity will not invalidate the design assumptions. Clearly, any splice near the base of a cantilever must achieve full continuity and satisfactory strength. Even in a column, if, for example, the design is based on an effective length of $0.7l$ (implying full continuity at the ends), any connection near the ends of such a column must have adequate strength in order not to invalidate the member design assumptions.

10.1.3 Analysis and load partition

In many situations it is permissible to simplify the force paths through the splice, thus reducing design time and, in many cases, achieving a more economical connection. If a rolled section beam splice is located away from a point of maximum moment it is common practice to assume that the flange splices carry all the moment and the web splice the shear, i.e. $F = M/d$ and $M_{web} = 0$ in Figure 10.2. Because the splice moment is less than the section capacity there is unlikely to be any overstress in the section.

If the splice is fully loaded, resisting M_p at sections XX and YY in Figure 10.2(b), it is still possible to make the same simplifying assumptions by treating the beam as if it were a partially

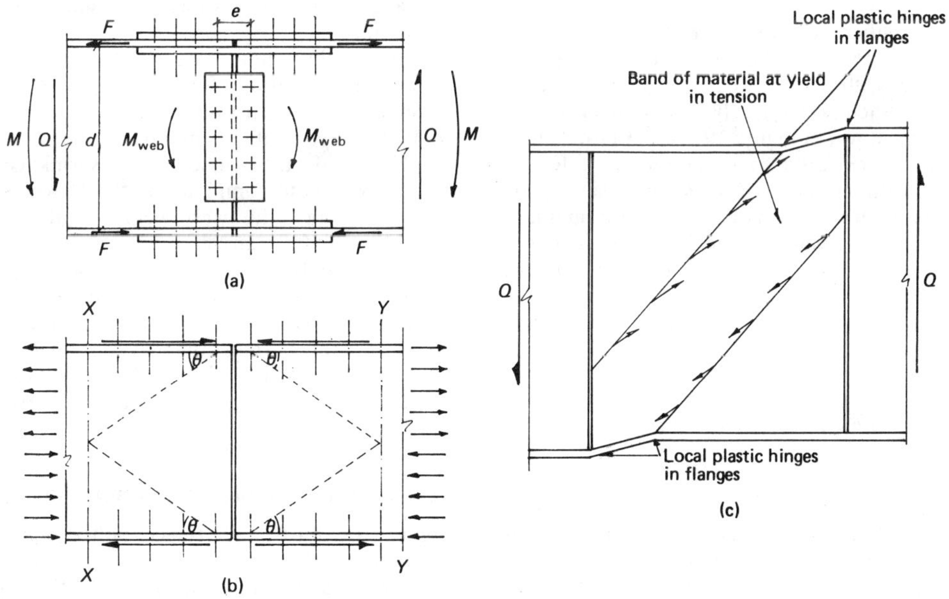

Figure 10.2 (a) Analysis of beam splice; (b) diffusion of moment into flange splice at highly loaded connection; (c) tension field action in slender web panel

connected member in accordance with Table 7.4. If the beam were axially loaded the most conservative dispersion angle would be used, as is appropriate in the presence of a free boundary, i.e. at the web. However, in bending no more than half the web force needs to be transferred to the flange, because of the greater lever arm of the latter. In these circumstances a dispersion angle (θ) of 45° should be satisfactory. The critical section is at the first line of bolts (X–X) and must be checked using the effective area of the flanges. Note that this design approach can lead to a large minimum length requirement for the flange splice, and this may reduce its economy.

Of course, it is quite acceptable to partition the total moment between the flanges and the web in accordance with the stress distribution in the beam away from the splice, i.e. $M_{web} \neq 0$ and $F = (M - M_{web}) \div d$. In this case the web splice has to be designed to resist this moment (M_{web}) in addition to the moment arising from the local eccentricity (e) between the centroids of the two bolt groups. This eccentricity moment $Q.e$ is divided equally between the two bolt groups. When the two moments are summed algebraically it is found that on one side they counteract each other, and the net web splice moment is $M_{web} - Q.e/_2$. On the other half of the splice they accumulate to give a total web splice moment of $M_{web} + Q.e/_2$. This governs the web splice design which becomes equivalent to design under a shear Q at an eccentricity of:

$$\left(\frac{M_{web}}{Q} + e/_2\right)$$

The simplified partition of resisting all the moment on the flange splices should not be used for plate girders with slender webs (i.e. those requiring stiffening). Such webs partly resist shear by the *tension field* shown in Figure 10.2(c). A more robust web splice is required to resist the local tension forces, and this may be achieved by designing the web for both shear and its portion of the moment. Some traditional design authorities recommend that a double row of bolts be used on each side of such web splices, and this would seem to be appropriate for more slender webs where tension field action is predominant.

10.2 Types of beam splice

10.2.1 Butt-welded connections

The simplest form of beam splice is one where all three elements are connected by full-strength butt welds. No special provisions are necessary in design. Unfortunately, this is one of the most expensive forms of connection; its use will almost certainly be limited to situations where aesthetics require unobtrusive connections.

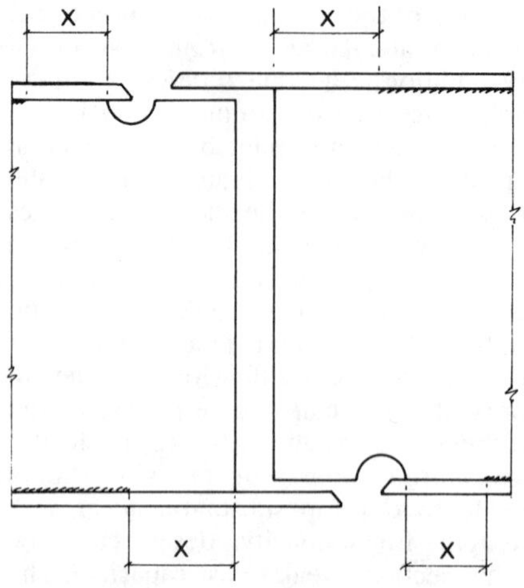

Figure 10.3 Butt-welded beam splice

Certain practical matters have to be considered:

1. Proper provision must be made for temporary support and location during welding. The staggered connection shown in Figure 10.3 is helpful in this respect though somewhat more expensive to prepare. Alignment will be helped if the flange/web welds of the sections to be joined are omitted for a short distance X on either side of the connection prior to assembly. It is essential that they are completed after the main welding has been carried out. Failure to do so would remove locally the interactive support between flanges and web, which is essential for their stability.
2. Cope holes should be used to improve the welding conditions for the flanges. No attempt should be made to fill them.
3. It is important to minimize the distortional effects of transverse weld shrinkage. One approach is to complete the flange welds before carrying out the web welds. If the order is reversed, the transverse weld shrinkage of the flange welds will cause severe buckling of the already welded web because of the slenderness of the latter. Alternatively, individual runs of weld on the three elements may be carried out in sequence, starting with the flanges. This has the effect of balancing the shrinkages between the elements.

10.2.2 Bolted splice plate connection

Figure 10.4 illustrates a typical bolted beam splice. In almost all circumstances the deformations associated with slip into bearing with this type of connection would be unacceptable, so HSFG bolts should be used. Details of the analysis of this type of connection have been discussed in Section 10.1.3.

A designer has a choice of single- or double-splice plates. Major connections, and any that are required to develop full section strength, should have double-splice plates to all elements. For most connections double-splice plates are, in any case, likely to lead to greater economy because of the reduction in the number of bolts. Single-splice plates may, however, be used for the lighter rolled sections (for example, with flange widths less than 200 mm).

Several practical matters merit special consideration:

1. Matching between the two halves of the connection. Either because of rolling tolerances, or because of differences in fabricated shape, problems can arise because of differences in size between the beams being spliced, as illustrated in Figure 10.4(b). In extreme circumstances where HSFG bolts are being used this can seriously reduce the efficiency of the line of bolts nearest the centreline of the connection. The problem can be alleviated by ensuring that the distance between bolt-lines X is the maximum allowable. Alternatively, the first line of bolts may be ignored when calculating the slip resistance of the connection.
2. Although it is possible to use faced beams and rely on bearing to transmit some of the compressive flange forces, this is not recommended in practice because of the difficulty of achieving adequate bearing contact.
3. Where the splice coincides with a change in flange or web thickness (either in a plate girder or by changing rolling weight within a serial size in a rolled beam) shims or packs may be used. However, it is essential that these comply with the general requirements for faying surfaces.

10.2.3 Welded splice plate connection

It is possible to use splice plates for welded connections in minor beams and thus overcome the very costly requirements of good fit that are necessary for butt-welded connections, as shown in Figure 10.5(a). However, it is not easy to use double-splice plates for this type of connection, and this limits the scale of beam for which this is applicable. By attaching top and bottom splice plates to alternate halves of the connection in the fabrication shop it is possible to make the connection self-supporting during erection.

10.2.4 Part-welded/part-bolted splices

As shown in Figure 10.5(b), it is possible to use hybrid or composite connections with splice plates where these plates are attached to one half of the splice by welding in the fabrication shop; the site connections are completed by bolting. This reduces the number of bolts to a minimum and eliminates site welding. HSFG bolts should be used for this type of connection to provide adequate stiffness at

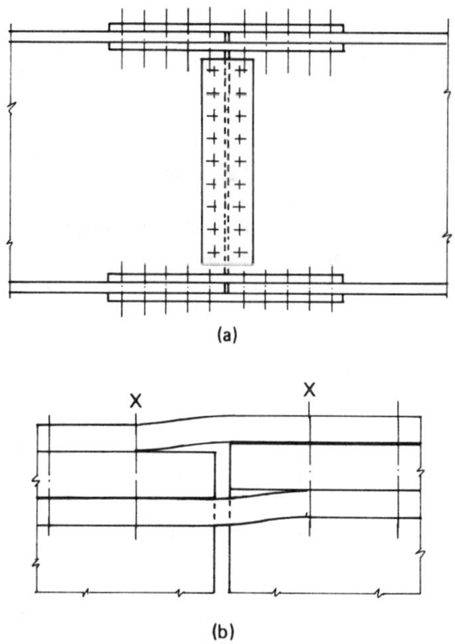

Figure 10.4 Bolted beam splice. (a) General arrangement; (b) detail at flange junction, showing effect of mismatched sections

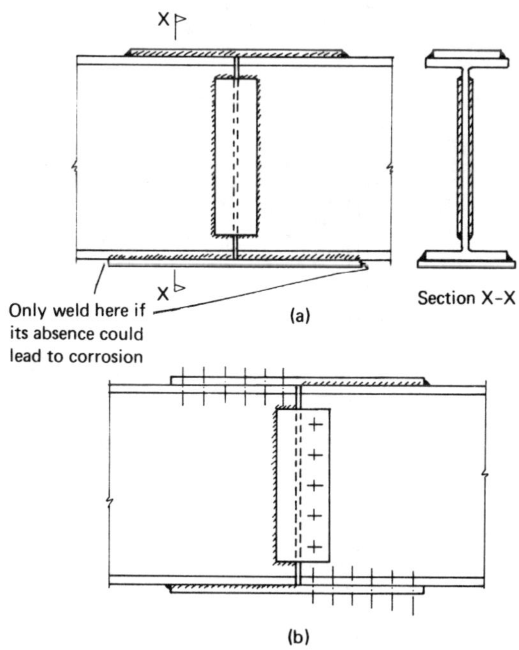

Figure 10.5 Light beam splices. (a) All welded; (b) bolted and welded

working load. Once again, it is only practicable to use single-splice plates, and this limits the scale of application. The provision for adjustment is halved when compared with a fully bolted connection, but this should not create difficulties provided that reasonable fabrication standards are maintained. This connection is not popular; costs are increased because all elements require both welding and drilling in the fabrication shop. In addition, the welded splice plates are prone to damage during transportation.

10.2.5 End-plate splices

Figure 10.6 shows the three forms of end-plate connection for beam splices. Short end-plate connections can be used when the beam is subject only to modest moments. Singly extended end plates can be used when high moments of one sign only have to be resisted and doubly extended end plates for high moments subject to full reversal.

This form of connection is becoming increasingly popular. By putting the bolts in tension it produces a bolted connection with high stiffness without the costs of HSFG bolting and frictional surfaces. However, it has less inherent robustness than the traditional splice plate connection and it is essential that proper attention be paid to the tensile flange connection if the design is to be satisfactory and to ensuring that lamellar tearing does not occur in the end plate. (The reader's attention is drawn to the discussion of this problem in Section 2.7.8.) In making the necessary subjective judgements about precautions against lamellar tearing designers must assess both the likelihood and consequence of this defect occurring. They can probably afford to be sanguine about an end-plate splice under modest moment where the end plate is less than 25 mm thick and the structure has sufficient redundancy for a tear not to lead to a major collapse. However, proper precautions should be taken under high moment, with an end-plate thickness greater than 25 mm and hence greater susceptibility to tearing, and in an isostatic or near-isostatic situation, where a tear could cause a major failure.

The tension flange connection may either be analysed as a Tee stub or the nominal bolt forces may be determined by assuming a centre of rotation for the whole connection at the centroid of the compression flanges. Whichever method is used, it is important to take prying action into account in evaluating the actual bolt forces.

It is also possible to modify some of the specialized methods of analysis for end plate beam-to-column connections for such splices. However, note that such methods are based on yield line analysis of the end plates. These imply that significant plastic deformation will have developed within the connection when the design strength is attained. The implications of this rotation on the overall structure should be considered before such methods are used in practice.

Three practical points merit special mention for end-plate splices:

1. Even if the connection is not subject to moment reversal, a pair of bolts should be placed close to the compression flange, to ensure that the compressive bearing surfaces are pulled into close contact; without them the joint may be very 'springy'.
2. Reasonable care should be given to the flatness of the end plate which will tend to distort in the presence of so much welding on one face. As a minimum, it should be clamped to a strong back during welding. Equally, inspectors should not insist on an unduly high standard of fit, particularly for connections where corrosion is unlikely to occur. Provided that contact is achieved in the vicinity of both flanges, small gaps near the mid-length of the end plate should be tolerated.
3. Such connections have no allowance for length corrections. Beams should therefore be fabricated to +0, −5 mm tolerances and provision made for packs in long beams where several connections are in line.

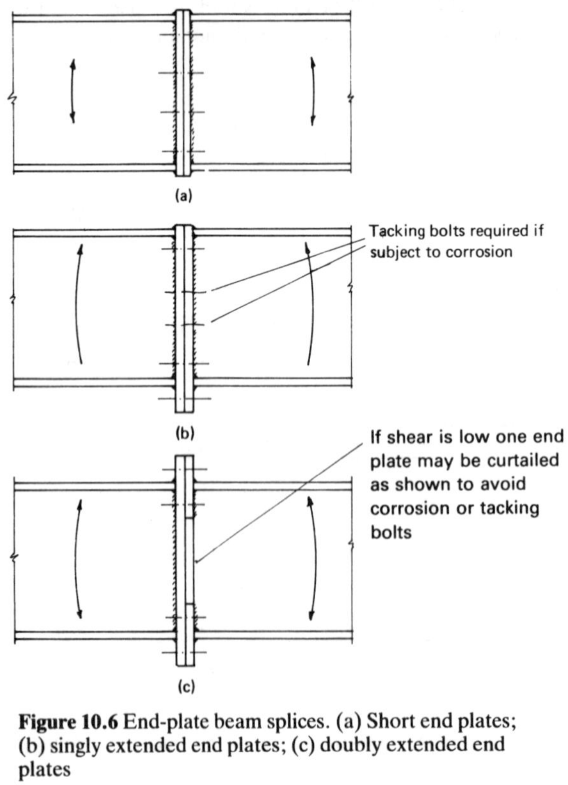

Figure 10.6 End-plate beam splices. (a) Short end plates; (b) singly extended end plates; (c) doubly extended end plates

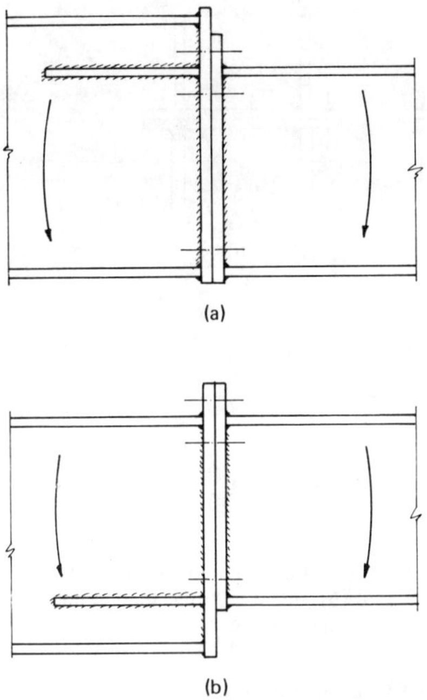

Figure 10.7 End-plate beam connections between elements of different serial size. (a) Coplanarity of compression flange; (b) coplanarity of tension flange

Most end-plate connections are between beams of the same overall size. Note that they can accommodate changes in rolling weight without difficulty. If a change in size occurs at an end-plate splice, it is usually possible to accommodate this by introducing longitudinal stiffeners to the larger beam, as shown in Figure 10.7.

10.3 Column splices

Generally, the same forms of connection may be used for column splices as for beam splices. Design conditions are less onerous because the high tensile forces that caused the greatest difficulty with beam splices are usually absent.

10.3.1 Butt-welded splices

Figure 10.8 shows the various types of butt welded splice. Full-penetration welds (a) can be used but it is usually cheaper to face the column ends for bearing and use partial penetration welds (b). If the ends are faced, much of the load may be transferred by direct bearing. If the column sections are concentric, changes in rolling weight may be accommodated in a straightforward way (c). If one face has to be maintained flush (d) or there is a change in serial size (e) a division plate will generally be required. In the latter case stiffeners may be necessary for the larger section.

The lifting provisions should be noted; in the absence of a division plate, the lifting lugs can be used conveniently to locate the upper column.

10.3.2 Bolted splice plate connection

Figure 10.9 shows traditional column splices with splice plates. These are still the most common form of column splice, and they are particularly economic if automatic saw and punch or drill lines are used in the fabrication shop. Generally, the column will be faced (by the milling saw) and the compressive load is transmitted by bearing. Provided that neither flange goes into net tension and that continuity is not required as in Section 10.1.2, it is then possible to use nominal bolted connections to maintain alignment of the column elements; these may be made with bearing bolts. If one of the flanges is subject to significant tension, the column ends are not faced for bearing, or full continuity is required, then HSFG bolts would have to be used. These should be avoided where possible for economy. Changes in rolling weight can generally be accommodated with this type of connection by using packs (Figure 10.9(b)). Where these are used with

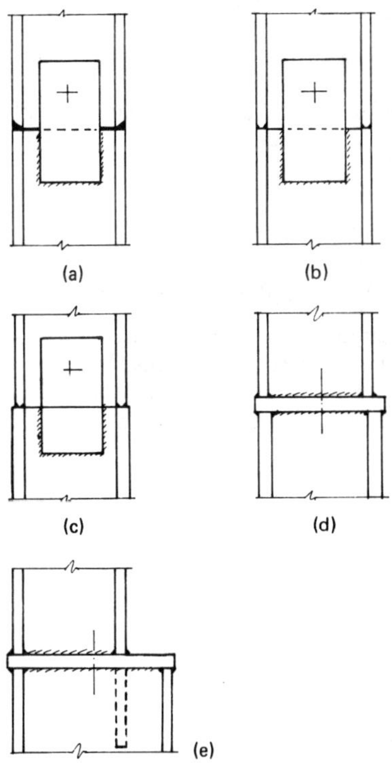

Figure 10.8 Butt-welded column splices. (a) and (b) Same serial size; (c) and (d) different rolling weight, same serial size; (e) different serial size

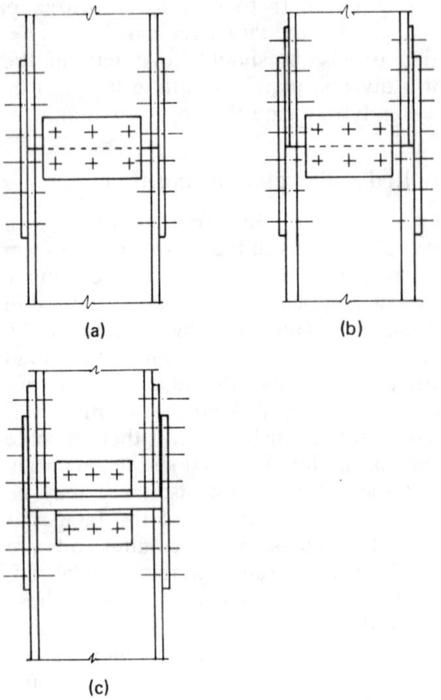

Figure 10.9 Bolted, splice plate column splices. (a) Similar columns; (b) different rolling weights, same serial size; (c) different serial size

HSFG bolts it is essential that their surfaces comply with the general requirements for faying surfaces. Changes in serial size require the additional use of division plates (Figure 10.9(c)), and this leads to very cumbersome connections.

10.3.3 Welded splice plate connection

It is possible, although relatively unusual, to use welded splice plates for column splices. If they are used, they should be in conjunction with column ends which have been faced for bearing.

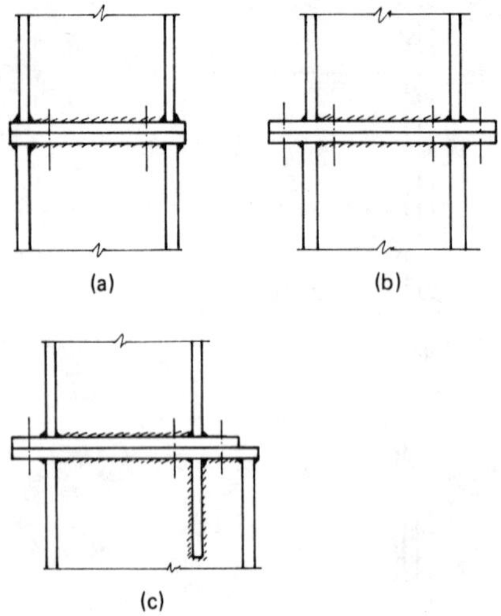

Figure 10.10 Column splices using end plates. (a) Similar sections, low moment; (b) similar sections, high moment; (c) different serial sizes, high moment

10.3.4 End-plate connections

It is possible to achieve very neat column splices using short end plates, as shown in Figure 10.10(a), if the column is subjected only to low moments and the splice is near a point of lateral restraint. If significant moment capacity is required, extended end plates (Figure 10.10(b)) must be used and the connection becomes much larger. However, in practice it is often possible to locate this connection within the floor slab, in which case it should not cause difficulty with architectural detailing. In design these connections can be treated in a similar way to end-plate beam splices. They may readily be used with changes in serial size with appropriate stiffening, as can be seen in Figure 10.10(c).

Reference

1. *Engineering for Steel Construction*, AISC, 1984.

Worked examples

Commentary: 1/1

Load paths

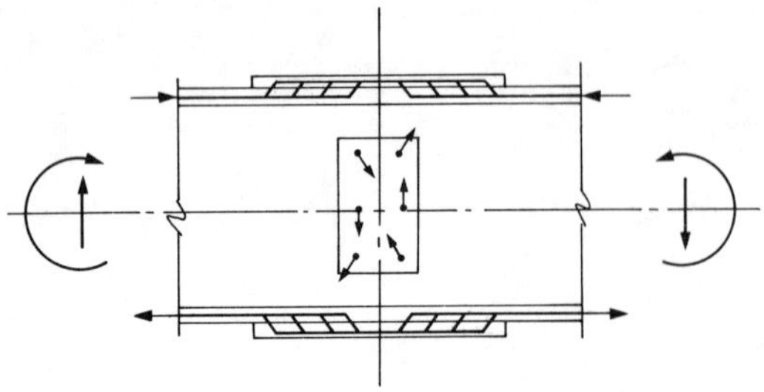

Forces in tension and compression are shown as load paths. In the web splice, the forces applied by the splice plates to the bolts (and beam web) are shown. The moment is carried by compression (and tension) in the beam flanges; shear in the flange bolts and compression (and tension) in the splice plates. Shear is carried by shear in the web bolts and by the web splice plates

Section 10.2.2

Single-cover (splice) plates are used for the flange splices. This is satisfactory for a beam of the proportions being considered. If double-cover plates were used the number of bolts and the length of the plates could be reduced, but this would be offset by the introduction of four, rather narrow, additional cover plates on the inside of the flanges.

In this example it is considered that, provided reasonable care is taken in checking the beam sizes and the fit-up of the splice, it is not necessary to add an extra row of bolts to allow for a possible mismatch of the sections. There is, in fact, some reserve in that eight bolts have been used against the 7.2 required.

It is assumed in the example that the splice is part of a medium-sized structure where M20 general grade HSFG bolts are being used.

Structural Steelwork Connections	Subject: Bolted cover plate splice for Universal Beam			Chapter Ref. 10
	Design Code: BS 5950 Part 1.			Calc. Sheet No. Example 1/1
	Calc. by B.D.C.	Date Aug '87	Check by G.W.O.	Date Nov' 87

Code Ref.	Calculations	Output

Bolted cover plate splice for Universal Beam

Design a bolted cover plate splice for a 457 × 152 × 60 UB in grade 50 steel.
The splice is to carry the following loading (due to factored loads):

Bending moment 227 kNm
Shear 113 kN

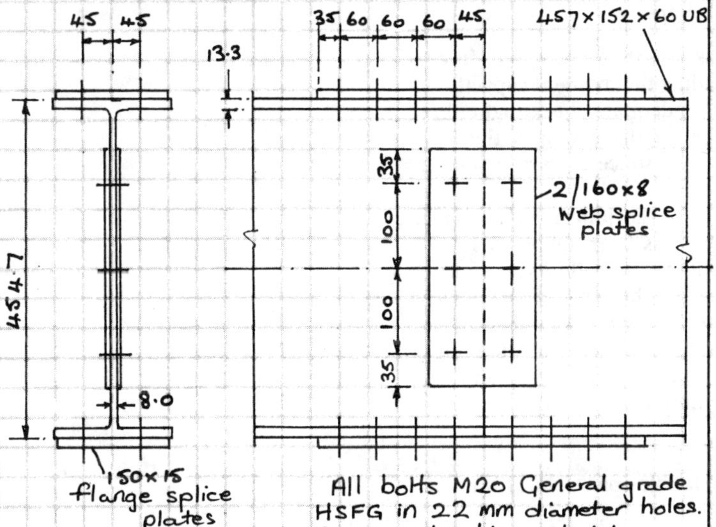

All bolts M20 General grade HSFG in 22 mm diameter holes.
Beam and splice material — Grade 50.

Assume that all of the moment is carried by the flange splices and that the web splice only carries the shear.

Flange splices

$$\text{Flange force} = \frac{227 \times 10^3}{454{\cdot}7 - 13{\cdot}3} = 514{\cdot}3 \text{ kN}$$

Commentary: 1/2

Section 10.2.2

HSFG bolts are used to increase the stiffness of the splice.

Slip resistance = P_{sl} = $1.1K_s\mu P_0$

K_s = 1.0 for clearance holes

Bearing resistance = P_{bg} = $dtp_{bg} \leqslant etp_{bg}/3$

If the end distance (e) is less than $2d$, the second check for bearing resistance is required. However, in this example it can be seen that the slip resistance will govern, even if the end distance is reduced to the minimum.

Section 7.3.1

Effective area of flange = K_e × net area ≯ gross area

Section 10.1.3: Figure 10.2(b)

If the flange capacity is less than the flange force, check the section, including the moment capacity of the web. If the section is adequate check that either the joint is long enough for dispersion of the web moment or design the web splice to carry some of the moment.

It is assumed that slip of the joint is undesirable and, as a precaution, K_e has been taken as 1.0 to avoid undue reduction of the preload (and therefore the slip resistance) due to thinning of the plates under tensile loading (see Section 5.5.5). The splice plate is thicker than the flange; therefore the flange governs the bearing check for the bolts and no further calculation is required.

Web splice

It is customary to use double-splice plates for web splices.

Slip resistance (two interfaces) = P_{sl}

= $2 \times 1.1K_s\mu P_0$

K_s = 1.0

Bearing resistance = P_{bg} = $dtp_{bg} \leqslant etp_{bg}/3$

In this instance, with the bolt in double 'shear', the end distance bearing check could govern.

Minimum end distance for slip resistance to govern

$$= \frac{3P_{sl}}{tp_{bg}} = \frac{3 \times 142.6 \times 10^3}{8 \times 1065} = 50.2\,\text{mm}$$

However, in this example the resultant shear on the bolts is well below the slip resistance and the minimum end distance of $1.4D$ could be used.

Section 8.2.2

Horizontal shear force on bolt due to moment due to eccentricity

$$= \frac{P_x e_x r_i}{\sum_n r_i^2}$$

There are two bolt groups (one each side of the centre line of the splice) each having one column of three bolts.

Equal eccentricity (e = 45 mm) is taken on each bolt group.

Structural Steelwork Connections	Subject: Bolted cover plate splice for Universal Beam			Chapter Ref.: 10
	Design Code: BS 5950 Part 1.			Calc. Sheet No.: Example 1/2
	Calc. by: B.D.C.	Date: Aug '87	Check by: G.G.O.	Date: Nov '87.

Code Ref.	Calculations	Output
	For M20 General grade HSFG bolts in single shear	
6·4·2·1	Slip resistance (per bolt) = 1·1 × 0·45 × 144 = 71·3 kN	
6·4·2·2	Bearing resistance of flange (per bolt) = 20 × 13·3 × 1065 × 10^{-3} = 283·3 kN	
	No of bolts required = $\frac{514 \cdot 3}{71 \cdot 3}$ = 7·2	
	Use 4 rows of 2 bolts.	8/M20 General grade HSFG bolts in each flange (each side of joint)
3·3·3	Effective area of flange = 1·0 × (152·9 − 2×22) × 13·3 = 1448 mm^2	
	Flange capacity = 355 × 1448 × 10^{-3} = 514·2 kN = Flange force O.K	
	Try 150 mm wide splice plate	
	Thickness of splice plate required = $\frac{514 \cdot 3 \times 10^3}{1 \cdot 0 \times 355 (150 - 2 \times 22)}$ = 13·7 mm	
	Use 15 mm	150 × 15 flange splice plates (Grade 50)
	<u>Web splice</u>	
	For M20 General grade HSFG bolts in double shear	
6·4·2·1	Slip resistance (per bolt) = 2 × 1·1 × 0·45 × 144 = 142·6 kN	
6·4·2·2	Bearing resistance of web (per bolt) = 20 × 8 × 1065 × 10^{-3} = 170·4 kN	
	Try 3 bolts at 100 mm vertical pitch 45 mm from the centre of the joint	
	Horizontal shear force on bolt due to moment due to eccentricity = $\frac{113 \times 45 \times 100}{2 \times 100^2}$ = 25·4 kN	

Structural Steelwork Connections	Subject: Bolted cover plate splice for Universal Beam			Chapter Ref. 10
	Design Code: BS 5950 Part 1.			Calc. Sheet No. Example 1/3
	Calc. by B.D.C.	Date Aug '87	Check by G.W.O.	Date Nov '87.

Code Ref.	Calculations	Output
	Vertical shear force per bolt $= \frac{113}{3} = 37.7$ kN	
	Resultant shear force $= \sqrt{25.4^2 + 37.7^2}$	
	$= 45.5$ kN	3/M20 General grade HSFG bolts at 100 mm pitch in web (each side of joint)
	< bolt capacity o.k.	
	Use 2 No. 8 mm thick splice plates	2 No. 8 mm web splice plates.

Commentary: 2/1

Section 10.2.2

Double-cover (splice) plates are used for the flange splices. The plate girder flanges are wide enough to accommodate three lines of bolts each side of the web. Using double-cover plates provides a more efficient splice than single-cover ones as the bolts are in double 'shear'. The centroid of the 'splice' is also closer to the centroid of the flange.

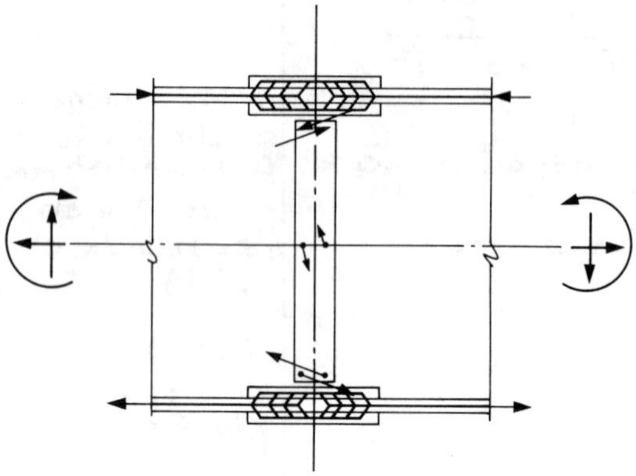

Forces in tension and compression are shown as load paths. In the web splice, the forces applied by the splice plates to the bolts (and beam web) are shown. For clarity the forces are only indicated on the end and central bolts.

The shear (and portion of axial load and moment that stresses the web adjacent to the web splice) is carried by shear in the web bolts and by the web splice plates.

The balance of the moment and axial load is carried by compression (and tension) in the beam flanges; shear in the flange bolts and compression (and tension) in the splice plates.

Structural Steelwork Connections	Subject: Bolted cover plate splice for Plate Girder			Chapter Ref.: 10
	Design Code: BS 5950 Part 1.			Calc. Sheet No.: Example 2/1
	Calc. by: B.D.C.	Date: Aug 87	Check by: G.O.O.	Date: Nov '87

Code Ref.	Calculations	Output

Bolted cover plate splice for Plate Girder

Design a bolted cover plate splice for a 1500 mm × 600 mm plate girder, with 40 mm thick flanges and a 15 mm thick web in grade 43 steel.

The splice is to carry the following loading (due to factored loads):

Bending moment	4456	kNm
Shear	620	kN
Axial tensile load	106	kN

Properties of plate girder $I = 2{\cdot}916 \times 10^6\ cm^4$

Area $= 693\ cm^2$

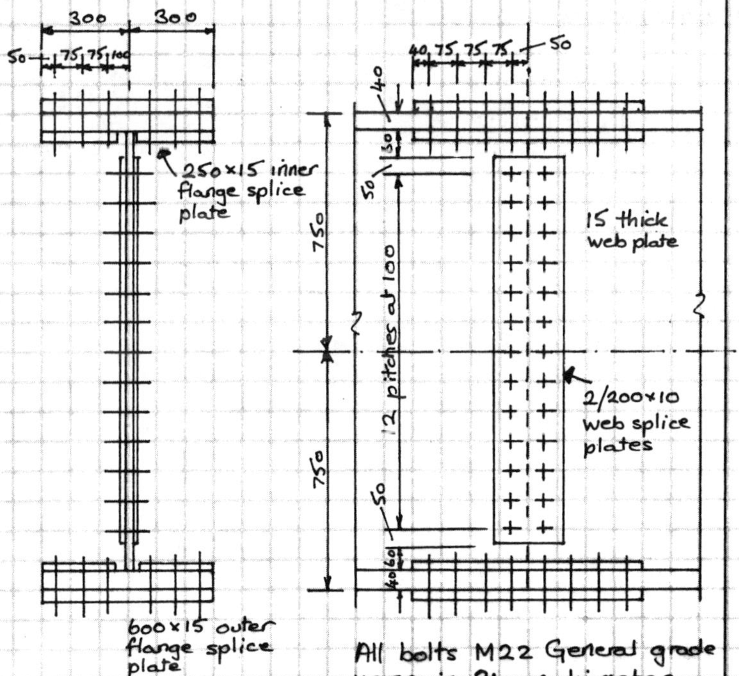

All bolts M22 General grade HSFG in 24 mm diameter holes.

Plate girder and splice material grade 43.

Commentary: 2/2

Section 10.1.3

In this example the web splice is designed to resist part of the applied moment.

At the level of loading being considered for the design of the splice, the plate girder will behave elastically; therefore an elastic stress distribution is assumed for the design of the splice:

$$f_b = \frac{M_y}{I}$$

Horizontal force on bolt due to moment = f_b × bolt pitch × t

$$\text{Average stress due to axial load} = \frac{\text{Axial load}}{\text{Area of girder}}$$

Horizontal force on bolt due to axial load = Av.stress × bolt pitch × t

Note that the resultant shear forces on the web bolts vary. The forces due to eccentricity are added to those due to the design moment on one side of the joint and are subtracted on the other. The forces due to the axial tension are added in the lower half of the joint and subtracted in the upper half.

Section 8.2.2

Horizontal shear force on bolt due to moment due to eccentricity:

$$\frac{P_x e_x r_i}{\sum_n r_1^2}$$

It is assumed in the example that the splice is part of a large industrial structure where M22 general grade HSFG bolts are being used.

Section 10.2.2

HSFG bolts are used to increase the stiffness of the splice.
(*Note*: The capacity of bolts could be taken from published tables of bolt capacities.)

Slip resistance = $P_{sl} = 1.1 K_s \mu P_0$
$K_s = 1.0$ for clearance holes
Bearing resistance = $P_{bg} = dtp_{bg} \leqslant \frac{1}{3}etp_{bg}$

Slip resistance governs, provided that end distance

$$\not< 3d \times \frac{P_{sl}}{P_{bg}} = 3 \times 22 \times \frac{175.2}{272.2}$$

$$= 42.5\,\text{mm}$$

Note that the 'end distance' is virtually the horizontal distance to the end of the web, because the resultant force on the bolts is almost horizontal. The end distance required is inversely proportional to the thickness.

Therefore, for splice plates,

$$\text{end distances} \not< 42.5 \times \frac{15}{2 \times 8} = 40\,\text{mm}$$

Structural Steelwork Connections	Subject: Bolted cover plate splice for plate girder			Chapter Ref. 10
	Design Code: BS 5950 Part 1			Calc. Sheet No. Example 2/2
	Calc. by B.D.C	Date Aug, '87	Check by G.G.O.	Date Nov '87

Code Ref.	Calculations	Output
	<u>Web splice</u>	
	Try a single row of 13 bolts at 100mm vertical pitch, 50 mm from the centre of the joint.	
	Bending stress in web of girder at level of top and bottom bolts due to moment $= \dfrac{4456 \times 10^6 \times 600}{2.916 \times 10^6 \times 10^4} = 91.7\ \text{N/mm}^2$	
	Horizontal shear force on bolt due to moment $= 91.7 \times 100 \times 15 \times 10^{-3} = 137.5\ \text{kN}$	
	Average stress on girder due to axial load $= \dfrac{106 \times 10^3}{693 \times 10^2} = 1.53\ \text{N/mm}^2$	
	Horizontal shear force on bolt due to axial load $= 1.53 \times 100 \times 15 \times 10^{-3} = 2.3\ \text{kN}$	
	Vertical shear force on bolt due to shear $= \dfrac{620}{13} = 47.7\ \text{kN}$	
	Horizontal shear force on bolt due to moment due to eccentricity $= \dfrac{620 \times 50 \times 600}{2(100^2 + 200^2 + 300^2 + 400^2 + 500^2 + 600^2)} = 10.2\ \text{kN}$	
	Resultant shear force $= \sqrt{(47.7)^2 + (137.5 + 2.3 + 10.2)^2} = 157.4\ \text{kN}$	
	For M22 General grade HSFG bolts in double shear	
6·4·2·1	Slip resistance (per bolt) $= 2 \times 1.1 \times 0.45 \times 177 = 175.2\ \text{kN}$	
6·4·2·2	Bearing resistance of web (per bolt) $= 22 \times 15 \times 825 \times 10^{-3} = 272.2\ \text{kN}$	13/M22 General grade HSFG bolts at 100 mm pitch in web (each side of joint)
	O.k. Use	
	Use 2 No 10 mm thick splice plates.	2 No 10 mm web splice plates

Commentary: 2/3

Section 10.1.3

Area of web associated with the web splice is assumed to extend half a bolt pitch beyond the top and bottom bolts.

The slip resistance of the bolts is the same as in the design of the web splice. From comparison of the flange, web and cover plate thicknesses it can be seen that bearing or end distance will not govern the design.

Section 10.2.2

The splice is assumed to be an important detail where slip at working load is unacceptable. The extra row of bolts is an allowance for lack of fit. In addition, care must be taken to see that the bolts are preloaded correctly and that the condition of the faying surfaces is adequate. In particular, the presence of mill scale is not acceptable if the slip factor is taken as 0.45 (Figure 5.15).

Section 7.3.1

Effective area of flange and splice plates = $K_e \times$ net area $\not>$ gross area

Consideration should be given to the effect of any loss of preload (and therefore in slip resistance) of the bolts that could occur due to thinning of the plates under tensile loading. In the example, the number of bolts used is 50% greater than the theoretical requirement, and this should cover any loss of preload as well as lack of fit. If the number of bolts is nearer to the theoretical it would be a wise precaution to take $K_e = 1.0$ in the design of the cover plates (see Section 5.5.5).

Structural Steelwork Connections	Subject: Bolted cover plate splice for plate girder			Chapter Ref. 10
	Design Code: BS 5950 Part 1			Calc. Sheet No. Example 2/3
	Calc. by B.D.C	Date Aug, '87	Check by G.W.O	Date Nov, '87

Code Ref.	Calculations	Output
	<u>Flange splices.</u> Second moment of area of portion of web which has its share of the applied moment and axial load carried by the web splice $= I_w' = \frac{15 \times 1300^3}{12 \times 10^4} = 0.275 \times 10^6 \text{ cm}^4$	
	Area of the same portion of web $= A_w' = 15 \times 1300 \times 10^{-2} = 195 \text{ cm}^2$	
	Proportion of applied moment carried by web splice $= \frac{I_w'}{I_{girder}} = \frac{0.275 \times 10^6}{2.916 \times 10^6} = 0.094$	
	Proportion of axial load carried by web splice $= \frac{A_w'}{A_{girder}} = \frac{195}{693} = 0.281$	
	Force to be carried by flange splice $= \frac{4456 \times 10^3}{1500 - 40} \times (1 - 0.094) + \frac{106}{2}(1 - 0.281)$ $= 2765.2 + 38.1 = 2803.3 \text{ kN}$	
	No of M22 General grade HSFG bolts required in double shear $= \frac{2803.3}{175.2} = 16.0$	24/M22 General grade HSFG bolts in each flange (4 rows of 6 bolts each side of joint)
	Assuming that slip is critical use 4 rows of 6 bolts (which includes an extra row to cover any lack of fit)	
3.3.3	Effective area of flange $= 1.2 \times (600 - 24 \times 6) \times 40 = 21{,}888 \text{ mm}^2$ (Gross area $= 600 \times 40 = 24{,}000 \text{ mm}^2$)	
	Flange capacity $= 265 \times 21888 \times 10^{-3}$ $= 5800 \text{ kN}$ > flange force O.K	
	Effective area of splice plates required $= \frac{2803.3 \times 10^3}{275} = 10194 \text{ mm}^2$	

Structural Steelwork Connections	Subject: Bolted cover plate splice for plate girder			Chapter Ref. 10
	Design Code: BS 5950 Part 1			Calc. Sheet No. Example 2/4
	Calc. by B.D.C	Date Aug, '87	Check by G.G.O.	Date Nov, '87.

Code Ref.	Calculations	Output
	Outer splice plate, try 600 mm × 15 mm flat	
	Effective area = $1{\cdot}2 \times (600 - 6\times 24) \times 15$	
	$= 8208 \text{ mm}^2$	
	$> \frac{1}{2} \times 10{,}194 \text{ mm}^2$ O.k.	600×15 outer flange splice plates
	Inside splice plates, try 2 No 250 mm × 15 mm flats.	
	Effective area = $1{\cdot}2 \times 2 \times (250 - 3\times 24) \times 15$	
	$= 6408 \text{ mm}^2$	
	$> \frac{1}{2} \times 10{,}194 \text{ mm}^2$ O.K	250×15 inner flange splice plates

Commentary: 3/1

Section 10.3.2

The design examples are for columns which are not prepared for full contact in bearing. Normally the columns are cut square and the axial loads are transmitted in direct bearing. In such cases, where the splice is located near a point of lateral restraint, (a) a nominal bolted splice would be used for erection and continuity. This would be similar to the splice designed in this example, but M20 ordinary 4.6 bolts could be used instead of the HSFG bolts. For Universal Column sections, when the shear force is small and could be transmitted by friction, the web connection could be omitted for the nominal connection.

The part of the axial load which is distributed in the web of the column is indicated, for the sake of clarity, by two load paths.

The axial load is carried by compression in the column; shear in the bolts and compression in the splice plates.

Load paths

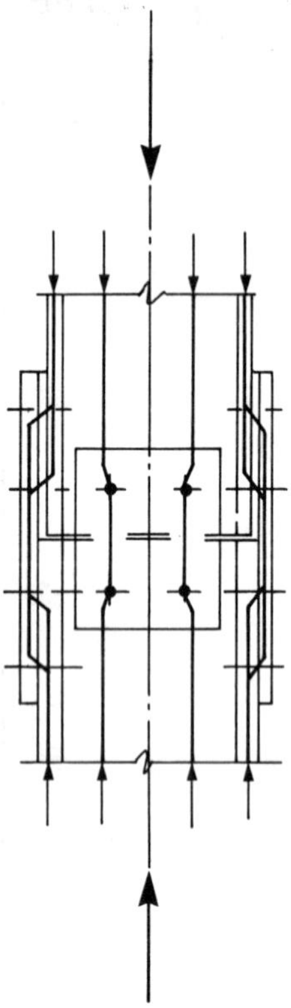

Structural Steelwork Connections	Subject: Bolted cover plate splice for Universal Column			Chapter Ref. 10
	Design Code: BS 5950 Part 1			Calc. Sheet No. Example 3/1
	Calc. by: B.D.C	Date: Aug, '87	Check by: G.W.O.	Date: Nov, '87

Code Ref.	Calculations	Output

Bolted cover plate splice for Universal Column

Design a bolted cover plate splice for a 203 × 203 × 60 UC connected to a 203 × 203 × 86 UC. Both columns are grade 43 steel. The splice is to carry an axial load of 700 kN (due to factored loads.)

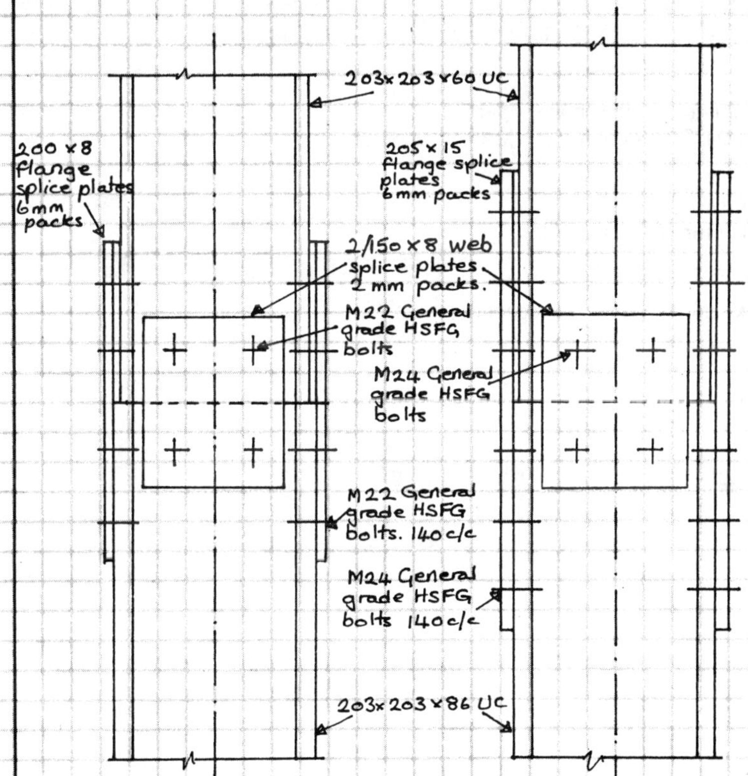

Commentary: 3/2

The axial load is shared between the web and flange elements in proportion to their areas. If shear continuity is not required it would be possible to leave out the web connection and share the load between the flanges (in which case the local capacity of the flanges should also be checked).

HSFG bolts are used to provide adequate stiffness. It is important that the surfaces of packs and shims (when used) should comply with the requirements for faying surfaces.

It is assumed in the example that the splice is part of a large structure where M22 general grade HSFG bolts are being used.

Slip resistance = $P_{sl} = 1.1 K_s \mu P_0$
$K_s = 1.0$ for clearance holes

Bearing resistance = $P_{bg} = dtp_{bg} \leqslant etp_{bg}/3$

The end distance check comes from the second requirement.

Structural Steelwork Connections	Subject: Bolted cover plate splice for Universal Column			Chapter Ref. 10
	Design Code: BS 5950 Part 1			Calc. Sheet No. Example 3/2
	Calc. by B.D.C	Date Aug, '87	Check by G.W.O.	Date Nov, '87

Code Ref.	Calculations	Output
6·1·7·2	<u>Case (a)</u> The splice is near a point of lateral restraint. The ends are not prepared for full contact in bearing	
	Area of 203 × 203 × 60 UC = 75·8 cm^2	
	Area of web = (209·6 − 2 × 14·2) × 9·3 × 10^{-2} = 16·9 cm^2	
	<u>Web splice</u>	
	Portion of load carried by web = $700 \times \frac{16{\cdot}9}{75{\cdot}8}$ = 156·1 kN	
	For 2 No. M22 General grade HSFG bolts in double shear	
6·4·2·1	Total slip resistance = 2 × 2 × 1·1 × 0·45 × 177 = 350·5 kN O.K	
6·4·2·2	Total bearing resistance = 2 × 22 × 9·3 × 825 × 10^{-3} = 337·6 kN O.K	
6·4·2·2	End distance ≮ $\frac{156{\cdot}1 \times 10^3}{2 \times \frac{1}{3} \times 9{\cdot}3 \times 825}$ = 30·5 mm	
Table 31	Also end distance ≮ 1·4(22+2) = 33·6 mm	2/M22 General grade HSFG bolts in web
	Use 2 No 8 mm thick splice plates	2 No 8 mm web splice plates.
	<u>Flange splices</u>	
	Portion of load carried by each flange = ½ (700 − 156·1) = 272·0 kN	
	For 4 No M22 General grade HSFG bolts in single shear	
6·4·2·1	Total slip resistance = 4 × 1·1 × 0·45 × 177 = 350·5 kN O.K.	
6·4·2·2	Total bearing resistance = 4 × 22 × 14·2 × 825 × 10^{-3} = 1031 kN O.K.	
6·4·2·2	End distance ≮ $\frac{272{\cdot}0 \times 10^3}{4 \times \frac{1}{3} \times 14{\cdot}2 \times 825}$ = 17·4 mm	4/M22 General grade HSFG bolts in each flange
	Also ≮ 33·6 mm	

Commentary: 3/3

The bolts are placed at 140 mm cross-centres (which is almost the maximum that can be used to comply with the edge distance requirements of 1.25*D*) to provide stiffness on the y-y axis: 140 mm is the standard cross-centre distance for a 203 mm wide flange.

Section 16.4

The effective width of plate is taken as the edge distance to the outside and a dispersion of 30° to the inside.

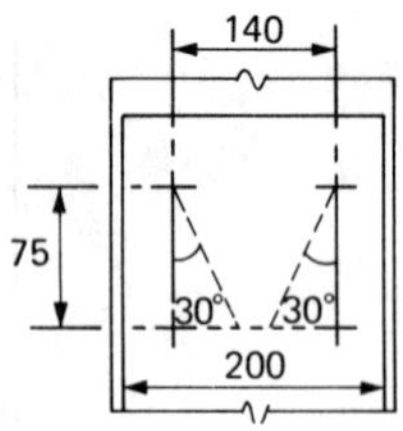

If the detail is exposed the flange splice plates should be increased to 10 mm, to comply with BS 5950, Clause 6.2.2.

Clause 6.2.2.
Case (b)
For 'braced' frames, of simple beam and column construction, it is normal practice to place the column splices immediately above a floor level. If this is near to a point of inflexion, as far as column buckling is concerned, the moment due to strut action is negligible and a procedure similar to that in (a) above can be used. In this example, the splice is at mid-point and the moment induced by strut action is taken into account.

The moment of 38.7 kNm has been calculated using the procedure in Appendix C3 of BS 5950.

Structural Steelwork Connections	Subject: Bolted cover plate splice for Universal Column		Chapter Ref. 10
	Design Code: BS 5950 Part 1		Calc. Sheet No. Example 3/3
	Calc. by B.D.C	Date Aug, '87	Check by G.L.O.	Date Nov' 87

Code Ref.	Calculations	Output
	200 mm wide splice plate Bolts at 140 mm cross centres 75 mm pitch Thickness of splice plate required $= \dfrac{272 \times 10^3}{275(200-140+2\times 75 \tan 30^\circ)}$ $= 6{\cdot}7$ mm Use 8 mm plates	200 × 8 flange splice plates.
6·4·2·2	Bearing resistance on bolt $= 4\times 22\times 8\times 825\times 10^{-3} = 581$ kN O.K.	
	Case (b) The splice is at the midpoint of the effective length. The ends are not prepared for full contact in bearing.	
6·1·7·2	As the splice is not near to the end of the member, account is taken of the moment induced by strut action. For an effective length of 6m, the moment is equal to 38·7 kNm about the y-y axis. Web splice. As in (a) above, but use M24 bolts to match flange splice.	2/M24 General grade HSFG bolts in web 2 No. 8 mm web splice plates
	Flange splices Portion of axial load carried by each flange $= \frac{1}{2}(700 - 156{\cdot}1) = 272$ kN Assume 2 lines of bolts at 140 mm centres in each flange Axial load per line of bolts $= \dfrac{272}{2} = 136$ kN Load per line of bolts due to moment $= \dfrac{38{\cdot}7 \times 10^3}{2\times 140} = 138{\cdot}2$ kN Total load per line $= 136 + 138{\cdot}2 = 274{\cdot}2$ kN	

Commentary: 3/4

Note that if a standard 200 mm wide flat is used for the cover plates, the cross-centre distance must be reduced to allow the minimum edge distance for M24 bolts. This would decrease the lever arm and increase the bolt loads. Also note that bending about the y-y axis of the column is being considered and as far as the moment is concerned it is the modulus, not the area, of the splice plate that should be not less than that of the associated flange.

If the columns had been cut square and were in bearing contact the loads that need to be carried by the bolted splice would be less.

Where the flange cover plates are in one piece, as in this example, the bolt group could be designed using a procedure similar to that for an eccentrically loaded bolt group. The procedure used here is conservative, but it has the advantage that it is simpler and that the number of bolts required is determined directly.

Structural Steelwork Connections	Subject: Bolted cover plate splice for Universal Column			Chapter Ref. 10
	Design Code: BS 5950 Part 1.			Calc. Sheet No. Example 3/4
	Calc. by B.D.C	Date Aug, '87	Check by G.W.O.	Date Nov, '87

Code Ref.	Calculations	Output
	For M 24 General grade HSFG bolts in single shear	
6·4·2·1	Slip resistance per bolt = $1{\cdot}1 \times 0{\cdot}45 \times 207$ = 102·5 kN	
	Bearing resistance O.k by inspection	
	Number of bolts required in each line = $\frac{274{\cdot}2}{102{\cdot}5} = 2{\cdot}68$	
	Use 3×2/M 24 bolts	6/M 24 General grade HSFG bolts in each flange
	The size of the flange splice plates should be not less than that of the associated flanges (of the lighter section when there is a change of section.)	
	Use 205×15 flange splice plates	205×15 flange splice plates.

11

Column bases

11.1 Introduction

This is a neglected topic in structural engineering. A literature search reveals few relevant papers and the subject is given little attention in most design texts. The most helpful text concentrates on practical aspects of design.[1]

This dearth of research information undoubtedly arises partly because of a general tendency for research workers to specialize in one or other of main structural materials. There is a strong temptation not to tackle problems at their interface. However, one of the major motivations for research is the occurrence of failures or at least mishaps in practice. A shortage of research effort would therefore seem to indicate that these connections are not troublesome in service.

Nonetheless, it is still most important that proper attention be given to the practicalities of design if economy is to be combined with durability. Therefore in addition to a discussion of traditional and alternative approaches to design, this chapter also discusses such matters as the control of bolt position during concreting, practical tolerances and corrosion prevention.

11.2 Pinned bases under axial load

11.2.1 Traditional design

In most traditional codes (and indeed in BS 5950: Part 1) the minimum thickness of a slab base is given by an empirical formula such as:

$$t = \sqrt{\left[\frac{2.5w}{p_{yp}}(a^2 - 0.3b^2)\right]} \qquad (11.1)$$

where t = slab thickness (mm),
a = the greater projection of the plate beyond the stanchion (mm),
b = the lesser projection of the plate beyond the stanchion (mm),
w = the pressure on the underside of the base assuming a uniform distribution (N/mm^2),
p_{yp} = the design strength of the plate but not greater than 270 N/mm^2.

This formula is based on the use of a modulus of 1.2Z. The design strength is limited to 270 N/mm^2, regardless of the grade of steel, in order to limit slab flexibility at working loads (see BS 5950: Part 1, Clause 4.13.2.3).

The adequacy of this approach for traditional construction is amply demonstrated by experience. However, it is most important to appreciate that it has limited applicability. It was developed, and may continue to be used, for stanchions of concentrated cross-section, such as UC columns as shown in Figure 11.1(a) and RHS. In these cases the base plate flexure is primarily governed by its projections beyond the rectangular envelope of the stanchion. However, with a more widely dispersed column cross-section, such as the UB column shown in Figure 11.1(b), flexure within the rectangular envelope to the column cannot be ignored. If the traditional formula is used it will lead to unsafe base plate thicknesses. It is also worth noting that the empirical formula can offer no guidance on the base plate thickness for the stiffened base shown in Figure 11.1(c).

11.2.2 Effective area design

Where a base plate of limited stiffness is used in conjunction with a column of dispersed cross-section

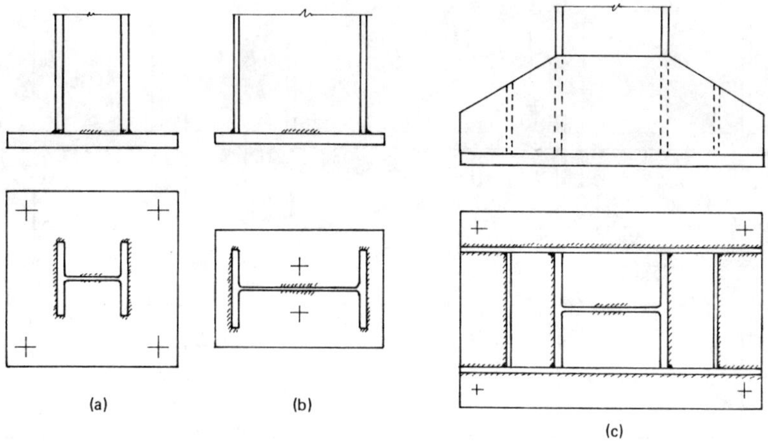

Figure 11.1 Column bases for UB and UC column sections. (a) and (b) Unstiffened slab bases; (c) stiffened base

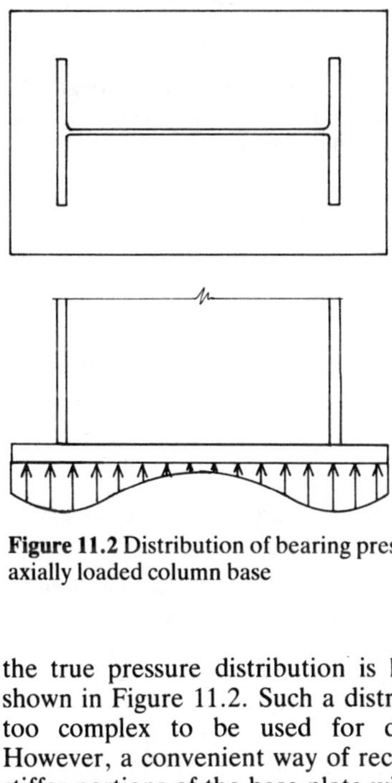

Figure 11.2 Distribution of bearing pressure beneath axially loaded column base

the true pressure distribution is likely to be that shown in Figure 11.2. Such a distribution is clearly too complex to be used for design purposes. However, a convenient way of recognizing that the stiffer portions of the base plate will attract most of the load is to introduce the concept of effective areas. These are areas which are arranged around the elements of the column cross-section, and examples are shown in Figure 11.3.

It is a well-established principle of limit design that, apart from certain situations where instability influences behaviour or there is limited ductility, it is not possible to reduce strength by adding material. Thus if a base plate design, including the determination of plate thickness, is based on the 'effective' portions of the base plate, then the addition of the 'ineffective areas' cannot detract from that strength.

The effective areas are usually arranged symmetrically about the elements of the column. The base plate thickness may thus be determined from consideration of simple cantilever bending of the effective areas.

Note that this use of effective areas relies on a definite partition of load among the elements of the column cross-section. It is necessary to ensure that each element can resist its portion of the total load; it is unlikely for a serious overstress to occur in practice but the check should be carried out for completeness.

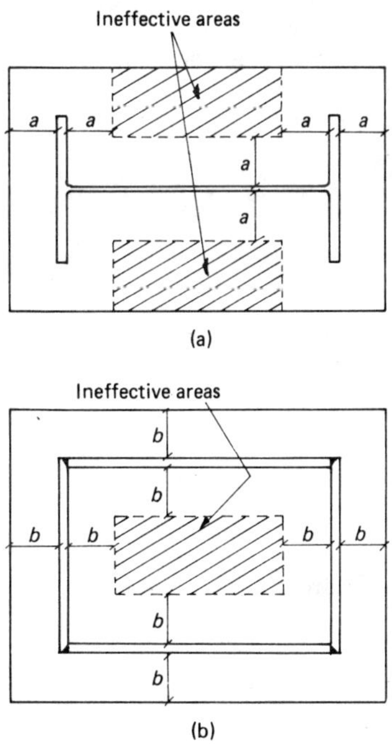

Figure 11.3 Effective area concept for axially loaded column bases

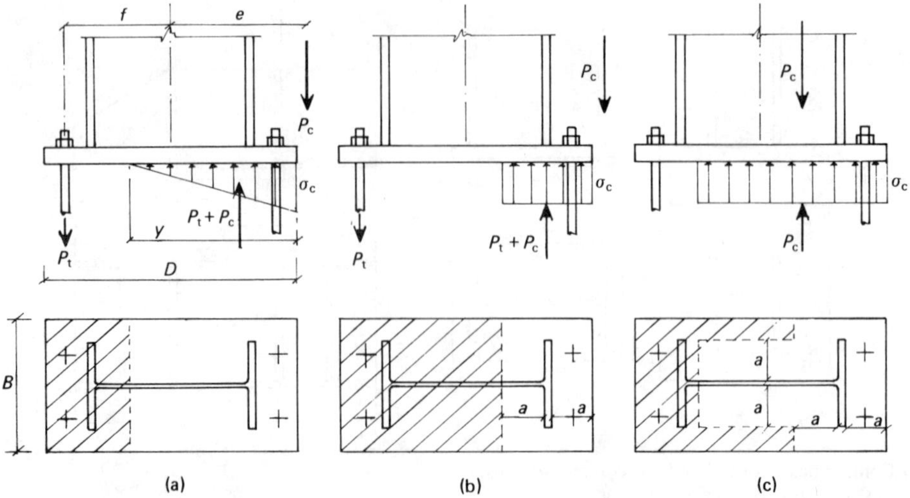

Figure 11.4 Stress distributions for eccentrically loaded column bases. (a) Elastic analysis; (b) plastic analysis and effective area concept (large eccentricity); (c) plastic analysis and effective area concept (small eccentricity)

11.3 Fixed bases under axial load and moment

11.3.1 Traditional elastic design

It is possible to carry out an elastic analysis of a fixed base.[2] This is on the questionable assumption that plane sections remain plane. With the notation of Figure 11.4(a) and where

A_s = area of holding-down bolts in tension,
σ_s = stress in steel bolts,
E_s = modulus of elasticity of steel bolt,
σ_c = bearing stress in concrete,
E_c = modulus of elasticity of concrete,
n = modular ratio Es/Ec,

there are three unknowns: P_t, y and σ_c. Vertical equilibrium gives

$$0.5y\sigma_c B - P_t - P_c = 0 \qquad (11.2)$$

Moments about column centroid gives

$$P_t.f + (P_t + P_c)\left(\frac{D}{2} - \frac{y}{3}\right) - P_c.e = 0 \qquad (11.3)$$

Plane sections condition gives

$$\frac{\sigma_s/E_s}{\sigma_c/E_c} = \frac{D/2 - y + f}{y} \qquad (11.4)$$

Eliminating P_t and σ_c from these equations, noting that $P_t = \sigma_s A_s$, gives

$$y^3 + 3\left(e - \frac{D}{2}\right)y^2 + \frac{6nA_s}{B}(f + e)y - \frac{6nA_s}{B}\left(\frac{D}{2} + f\right)(f + e) = 0 \qquad (11.5)$$

Equation (11.5) may be solved, numerically, to determine y, which may then be substituted into equation (11.3), which may be rewritten as:

$$P_t = -P_c \frac{\dfrac{D}{2} - \dfrac{y}{3} - e}{\dfrac{D}{2} - \dfrac{y}{3} + f} \qquad (11.6)$$

to determine P_t. P_t and y may then be substituted into equation (11.2) to determine σ_c.

This seems a most cumbersome procedure, and the analytical effort cannot really be justified when it is based on such a questionable assumption as that of plane sections remaining plane. It should also be considered that:

1. The effective extensional stiffness of the bolts is a function of their embedment length and preload.
2. The compressive stiffness of the concrete is a function of the bedding material behaviour and the manner of the compressive dispersion into the main foundation.
3. The analysis ignores the influence of the most flexible element, the base plate in flexure.

Thus while this method (or indeed any comparable elastic analysis) will undoubtedly give a satisfactory design, a realistic alternative should be sought which is more amenable to design office use.

11.3.2 Effective areas concept

The effective area concept may be extended to fixed bases by the means shown in Figures 11.4(b) and (c). With an eccentricity of load of more than half the column depth, an effective compressive area is postulated that is arranged symmetrically around the compression flange. Assuming uniform stress over this area, the line of action of the stress resultant is located at the centroid of the column flange. Taking moments about this point gives the magnitude of the bolt forces directly. The required effective area of concrete is given by:

$$\frac{P_t + P_c}{\sigma_c}$$

and dimension a and plate thickness t may be determined as in Section 11.2.2. Detailed design of the holding-down bolts and the associated flexure of the base plate is discussed in the following section.

If the eccentricity of load is less than half the column depth, the calculation is more awkward. The effective area is required to be extended along the web, as shown in Figure 11.4(c) until its centroid coincides with the line of action of the applied load. The calculation may then proceed as before, although it will be noted that the holding-down bolts are not subject to any external tension.

11.3.3 Design of base plate to resist tensions in holding-down bolts

The flexure of the base plate may be based on the effective-width approach outlined in Section 7.7, together with an assumption of single-curvature cantilever bending. It is not realistic to consider any prying action and associated double-curvature bending, as in an end plate, because of the relative compressive weakness of the bedding material.

11.4 Holding-down bolt design

11.4.1 Pinned bases under axial load only

In the absence of any shear loading only nominal holding-down bolts need be provided. They should be capable of resisting any construction loading. Unless some nominal moment capacity is required for boundary fire conditions the bolts should be positioned reasonably close to the column centre-line. If, in addition, they are placed near the plate boundary, grouting of the bolt pockets will be considerably eased. Where boundary fire conditions exist it will usually be necessary to use four bolts. These can usually be positioned within the column profile but will need to be reasonably close to its corners.

11.4.2 Fixed bases under axial load and moment

It is necessary to pay proper attention to the detailed design of holding-down bolt systems for fixed bases. The simplest solution, which is appropriate for UB columns up to around 600 mm in depth with bolts of up to M30 Grade 8.8, is to use an unstiffened base plate. This plate will be of no more than 50 mm in thickness and the potential savings in base plate thickness with stiffened systems will be more than offset by the increased cost of fabrication.

Above that scale of construction it is probably economic to adopt one of the stiffener arrangements shown in Figure 11.5. Here it is important to take account of the inherent eccentricity between the bolt centreline and the centroid of the tension elements in the column if bolt or local element bending is to

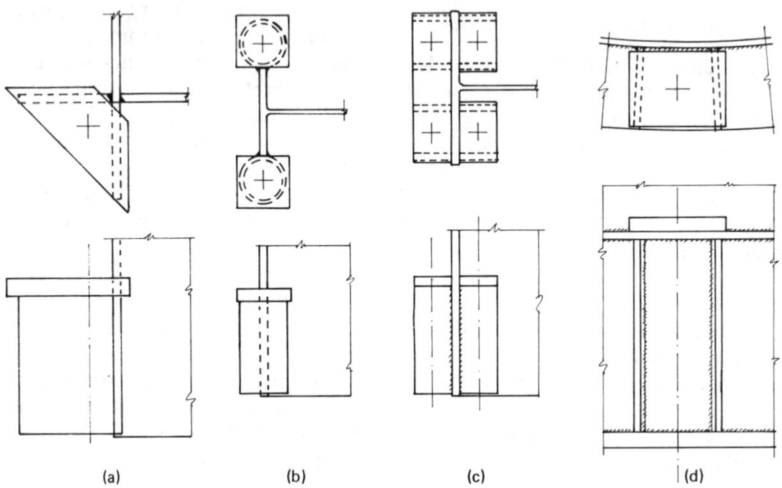

Figure 11.5 Holding-down details for column bases. (a) Box column; (b) UB section (moderate tension); (c) UB or UC section (high tension); (d) cylindrical column

be avoided. With box columns (Figure 11.5(a)) one solution is to extend the stiffeners so that the washer plate has symmetric support about the bolt centre-line. If bolt boxes are used, internal diaphragms or stiffeners should be provided to resist the associated moments. With two-bolt systems (Figure 11.5(b)) the individual bolts should be arranged so that local symmetry is achieved about the tension flange. If four bolts are used they can be arranged as shown in (c). With cylindrical bases (d) it is difficult to overcome the eccentricity directly. However, the external ring stiffener and annular ring base plate form efficient hoop structures when welded to the main cylinder. These may therefore be designed to resist, *inter alia*, the horizontal, radial couples which arise from the eccentricity moment.

With the exception of the final example where the stiffeners have to be welded directly to the base plate it is prudent to ensure that all the compressive load, including that arising from bolt preload, is contained within the main column section and excluded from the stiffeners. This is achieved simply by ensuring that the stiffeners stop short of the end of the main column cross-section. Without this precaution, flexure of the base plate could lead to a possible overstress in the stiffeners.

11.5 Resistance to shear forces

There is considerable disagreement on detailing practice for resisting base shears. More rigorous design guides preclude the use of holding-down bolts in shear. By contrast, it is common and successful industrial practice to use the holding-down bolts of pinned base portals to resist the substantial shear forces that exist in this structural form. What follows is offered as commonsense advice in an uncertain situation:

1. If shear is small compared to axial load (say, less than 20%) then no special provision is required. Provided that bedding procedures are acceptable for resisting axial load, friction should be able to accommodate this modest non-axial loading.
2. For higher shears (up to, say, 70% of the axial load) cognisance has to be taken of successful industrial practice. However, it is *essential* that proper attention be paid to the following:
   (a) The cleanliness of the holding-down pockets prior to steel erection. Any foreign matter or imperfectly removed polystyrene former is going to have a severe effect on the ability of the grout to support the bolt.
   (b) The positions of the holding-down bolts in the base plate should be arranged so that there is good access to the bolt pocket to ensure that it can be properly filled with compacted fine concrete.
   (c) The pockets for the holding-down bolts should be so arranged that there is good access for subsequent filling.
   (d) The concrete composition should be specified to minimize shrinkage. A suitable mix for general application would be, cement, sharp sand, fine aggregate and water, in proportions of 1:1¼:2:0.5. This should give a workable mortar, *not* a liquid, which will not pour but can be compacted into the bolt pocket *provided that there is good access*.
   (e) Excessive clearance between the bolts and the holes in the base plate should be avoided.

Notwithstanding these precautions, it is unrealistic to assume that some columns in such situations will not show some horizontal deformations as they take up slack in the shear path. If this cannot be tolerated then one of the more rigorous design procedures outlined below should be adopted.

Figure 11.6 shows various means of resisting shear forces without using the holding-down bolts. In Figures 11.6(a) and (b) the shear is transmitted into the base by direct bearing of the column on infill concrete. Provided that there is no requirement to separate the column base from the floor slab, the infill pour can be incorporated into this latter pour. Alternatively, a pocket base can be used, as shown in (c). This has the added advantage of providing moment fixity.

In this latter case special design procedures are necessary. Some design texts suggest that the axial load can be taken by bond between the column and the concrete, the surface area utilized being the outer faces of the flanges and both faces of the web. The inside faces of the flanges are not used because of the tendency of the concrete to shrink away from these faces. However, a more certain approach is to take the axial load in bearing on a small base plate or end cleats. The moment and shear are taken in direct horizontal bearing between the column and the concrete. An appropriate bearing width of both flanges may be utilized. Bearing stresses between steel and concrete and bending strength of the flanges are similar to those used in base plate design.

There is one special practical point about pocket bases which should be considered before they are used. They are a natural trap for debris and mud if low-lying in relation to site levels. Cleaning out is a straightforward operation prior to erection but is difficult once the steel has been erected. Cleanliness prior to concreting the pocket is clearly essential for satisfactory structural performance. They are probably not suitable on sites which are liable to flooding or in situations where there is likely to be a protracted interval between steelwork erection and concreting.

If the ground is poor it may be inadvisable to transmit the shear to the column base. In this case

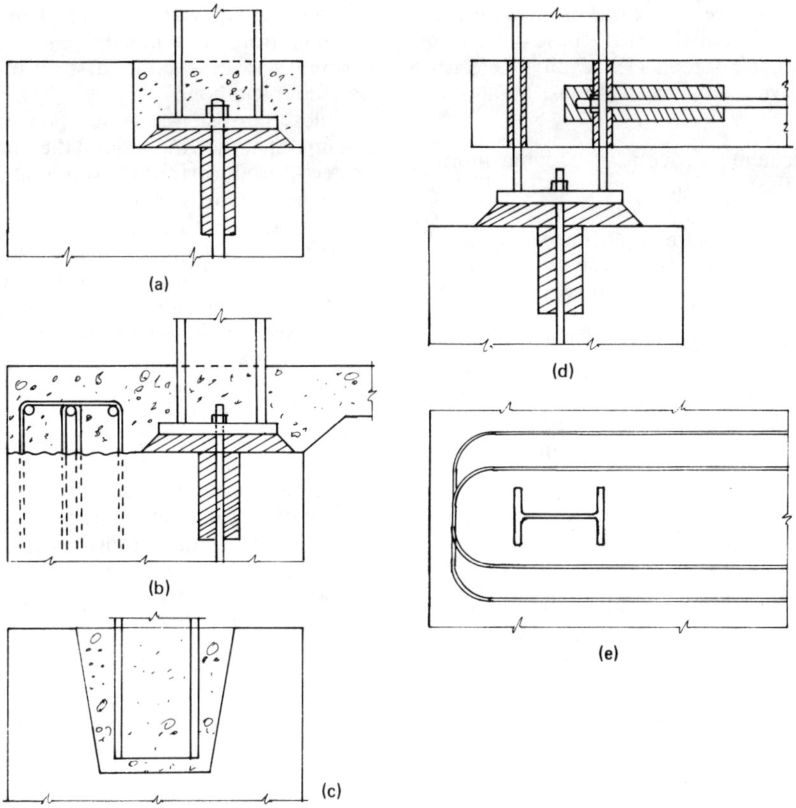

Figure 11.6 Arrangements for resisting column shear. (a)–(c) Shear transmitted into column base; (d) shear carried by tie between columns; (e) slab reinforcement to resist shear

the best solution is to provide ties between the columns. In Figure 11.6(d) a separate tie is provided. This is sometimes placed beneath the floor slab but then special protection against corrosion is required. If the ties can be accommodated within the slab depth, as shown, then the concrete will provide very effective protection. In poor ground there may be a possibility of differential settlement between the slab and column base. The detailed design can be arranged to accommodate this by means of flexible material both around the column and extending a suitable length along the tie, permitting the latter to flex without cracking the slab. This portion of tie should be wrapped as protection against corrosion. The remainder of the tie, within the slab, may be left unwrapped, in which case the tensile strains imposed locally on the slab may lead to minor cracking. If this is unacceptable the bar should be wrapped along its entire length to break its bond with the concrete. In the latter case the tie extension should be calculated and provision made for this potential movement. Consideration should also be given to this deformation in the former case, though the stiffening effect of the concrete is likely to reduce its magnitude in practice.

Alternatively, if there is no concern about differential settlement and if the slab extends at least 200 mm beyond the external column flange, horizontal shears may be resisted directly by the slab. As shown in Figure 11.6(e), reinforcement is hooked around the column and suitably lapped with the slab mesh. It is necessary to ensure that the slab mesh is continuous, or suitably lapped, across the full width of the slab.

11.6 Provision for adjustment

It is most important to appreciate the variations in practical tolerances that exist between structural steelwork and *in-situ* reinforced concrete. The former can be fabricated to tolerances of 2–3 mm; the dimensional tolerances that can be obtained with the latter, without expending excessive effort on geometric control, are 2–3 cm. For economy of construction, it is essential that the base details can accommodate these discrepancies readily. Referring to Figure 11.7, variations in level must be allowed for by:

1. Specifying a nominal depth of bedding mortar such that, if the top of the concrete is 2 cm high, it will still be possible to pack the reduced gap. A nominal depth of 40 mm is satisfactory for small bases – a greater figure should be used for larger areas. If the concrete is low, extra bedding can be provided.
2. Holding-down bolts should be detailed with a greater extension than that nominally required. Thus, even if the whole base is low, it will still be possible to use the holding-down bolts directly, without all the difficulty of collars and extension studs.

Discrepancies in plan position may be readily accommodated if:

1. The concrete base is sufficiently large to support the base plate in the presence of a realistic misalignment. In critical situations it might have to be designed for the possible eccentricity of loading.
2. There is provision for adjustment of the bolt position. The bolts may be cranked if they are no more than 24 mm in diameter and a pocket 150–200 mm deep is provided at the top of the bolt. Alternatively, a loose bolt and washer plate arrangement, as shown in Figure 11.7(b), may be used. For bolts over 24 mm diameter cranking is inappropriate and the arrangement shown in Figure 11.7(c) should be adopted.

11.7 Holding-down bolt details

11.7.1 Static loading

Above 24 mm diameter, J-bolts are most awkward to support during concreting because of their weight and lack of symmetry; a sleeved holding-down bolt is much more suitable. Even below that diameter, J-bolts are being superseded by light washer plate bolts for the same reason.

It is a common mistake to underestimate the force that wet concrete can exert on holding-down bolts. Simply clamping the bolts to a flexible plywood template is only satisfactory for very small bolts. Wiring them to reinforcement may also not work because the latter may also be disturbed. Larger bolts should be securely held in place by top and bottom templates; the latter may incorporate the washer plates.

Where washer plates are used it is important to ensure that the bolt is free for adjustment after concreting. This may be achieved either by the use of a nut cap below the washer plate or by releasing the bolt from its template immediately after the concrete has achieved its first set by waggling it to break the end bond.

Unless corrosion-resistant bolts are used it is essential that the void around the bolt be grouted to protect it from corrosion. Note that mild steel bolts are more able to sustain corrosion than high-yield bolts because of their greater diameter for a given capacity. A similar corrosive waisting on bolts of similar strength but different quality will lead to a greater reduction in strength with high-yield bolts. In corrosive conditions either a corrosion allowance should be made or corrosion-resistant bolts should be used.

Detailed design of holding-down bolts to resist static uplift forces does not present any particular difficulty. J-bolt anchorage lengths may be calculated from conventional criteria for plain reinforcing bars. With washer plate bolts no account is taken of any bond with the grout because of its likely shrinkage. All the tension is therefore resisted by the washer plate, which may be designed as a simple, upturned, base plate.

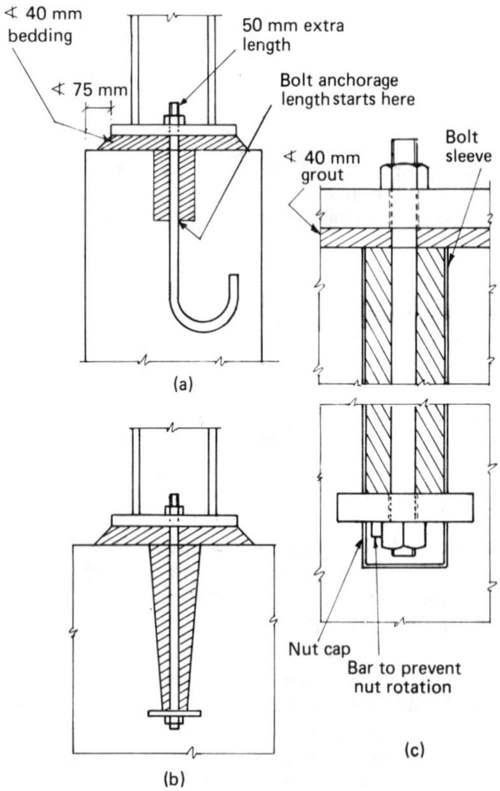

Figure 11.7 Provision for adjustment in holding-down systems. (a) J-bolt; (b) light washer plate system; (c) heavy-duty holding-down bolt

11.7.2 Fatigue loading

As in other bolt situations where fluctuations in applied bolt tensile load occur, the only way to achieve a satisfactory fatigue life is to protect the bolt itself from fluctuations in axial load. This is achieved by preloading the bolt so that fluctuations in external load produce fluctuations in the precompression between the concrete and the base plate with negligible fluctuations in bolt load. The bolts should be:

1. Of high-strength steel, corrosion resistant if necessary;
2. Properly tensioned, to a load suitably in excess of the applied load per bolt;
3. Of sufficient length so that loss in preload due to concrete creep is minimized. Most of the creep will occur in the regions of high compressive stress immediately adjacent to the washer plate and base plate, at the ends of the bolt. Ideally, the bolt should have a stressed length of at least twenty times its diameter;
4. Wrapped along their length so that they are debonded from the concrete.

In critical situations, it would be appropriate to require retightening of the bolts after most of the creep losses had occurred (say, six or twelve months after original tensioning).

11.8 Foundation bolts

Section 3.5 describes the various types of foundation bolt that are available. These are usually inserted into drilled holes, and their main advantage is that they thereby overcome all the problems of position tolerance that occur with traditional holding-down bolts. In addition, they can be used for attaching steelwork to existing concrete structures.

Most of these bolts may be used for both shear or tensile loadings. However, the following points should be carefully considered before using them to resist tensile loads in the completed structure:

1. Thermal or other movement may well cause the concrete to crack in the immediate vicinity of the hole. Such cracks would reduce the capacity of frictional fasteners severely.
2. Resinous bonding agents only adhere to the surface of the concrete. Any dirt or dust coating will reduce the bond.
3. The high quality of modern concrete drilling means that much less reliance can be placed on mechanical interlock to develop tensile capacity in the bolts (for example, a diamond-drilled hole without being roughened is unsuitable for a resin anchor).
4. It is difficult in practice to achieve satisfactory workmanship standards on small-volume mixes of either cementitious or resinous bonding agents.
5. Resinous bonding agents will creep under sustained load.

References

1. *Holding Down Systems for Steel Structures*, The Concrete Society *et al.*, 1980.
2. Blodgett, O. W., *Design of Welded Structures*, James F. Lincoln Arc Welding Foundation, 1972.

Commentary: 1/1

Section 11.2.1

The base plate design uses the semi-empirical formula in BS 5950: Part 1.

Structural Steelwork Connections	Subject: Base plate for Universal column with concentric axial load			Chapter Ref. 11
	Design Code: BS 5950 Part I			Calc. Sheet No. Example 1/1
	Calc. by B.D.C	Date Aug, '87	Check by G.O.	Date Nov' 87

Code Ref.	Calculations	Output

Design a base plate for a 254×254×89 UC with a factored axial load of 1720 kN

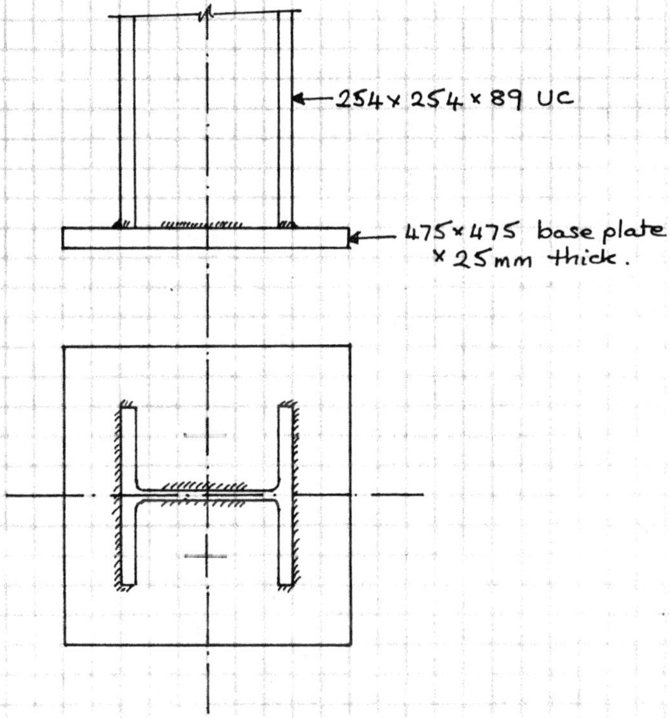

Base plate

Bearing strength of grout (and concrete) under base plate $= \omega = 8\ N/mm^2$

Area of base plate required $= \dfrac{1720 \times 10^3}{8}$

$= 215000\ mm^2$

Minimum size of square base plate

$= \sqrt{215000} = 464\ mm$

Commentary: 1/2

Although the minimum size of base plate is rounded up to 475 × 475, the extra plate is ignored and the minimum dimensions are used in determining the minimum thickness (see BS 5950: Part 1, Clause 4.13.2.1). If w, a and b for the larger plate were used in the formula, the minimum thickness would also be increased:

$p_{yp} = 265\,\text{N/mm}^2 = p_y$ for Grade 43 plate, $16 < t < 40$

In this example the welds and holding-down bolts are nominal, since only axial compressive forces have to be resisted.

Structural Steelwork Connections	Subject Base plate for Universal column with concentric axial load			Chapter Ref. 11
	Design Code BS 5950 Part 1			Calc. Sheet No. Example 1/2
	Calc. by B.D.C	Date Aug, '87	Check by G.W.O.	Date Nov, '87

Code Ref.	Calculations	Output
	Outstand from edge of flange = a	
	$= \frac{1}{2}(464 - 255\cdot9) = 104$ mm	
	Outstand from face of flange = b	
	$= \frac{1}{2}(464 - 260\cdot4) = 102$ mm	
4·13·2·2	Minimum thickness $= \sqrt{\frac{2\cdot5w}{p_{yp}}(a^2 - 0\cdot3b^2)}$	
	$= \sqrt{\frac{2\cdot5 \times 8}{265}(104^2 - 0\cdot3 \times 102^2)}$	
	= 24·1 mm	
	Use 475 × 475 base plate 25 mm thick	475 × 475 square base plate, × 25 mm thick Grade 43.
4·13·2·2	25 mm > Flange thickness (= 17·3 mm) O.K.	

Commentary: 2/1

Section 11.3.2

The applied load and moment are equivalent to an axial load of 1147 kN applied at an eccentricity of $536 \times 10^3/1147 = 467$ mm from the centreline of the column. This is outside the compression flange; therefore the axial load and moment will be assumed to be resisted by tension in the holding-down bolts and an area of concrete in compression balanced about the compression flange.

Load paths

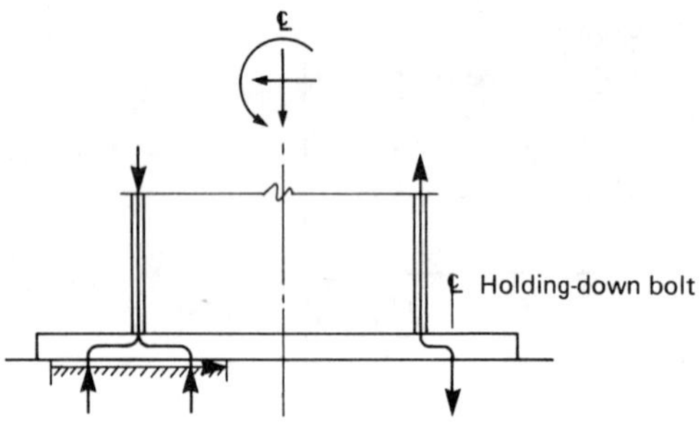

The overturning moment (in conjunction with the axial load) is resisted by uplift on the holding-down bolts adjacent to the tension flange and bearing on an area of concrete balanced about the compression flange. The eccentricity of the holding-down bolts from the tension flange and the spread of the load from the compression flange are carried by the base plate in bending.

Four hundred millimetres is chosen as the distance of the bolt line from the centreline of the column (i.e. 95.2 mm offset from the face of the column). If the bolts are too close to the column, there would be high stress concentrations in the fillet welds and adjacent to the bolts, and the effective length of the weld used in the design would need to be limited to ensure that the weld is not overstressed. (Remember, a fillet weld has limited ductility.)

Two hundred millimetres is chosen as the bolt spacing parallel to the flange. This spacing, combined with the offset chosen, places the bolts in a reasonable relationship to the welds. As a guide, make:

Bolt spacing $\not>$ 2(offset + bolt diameter)

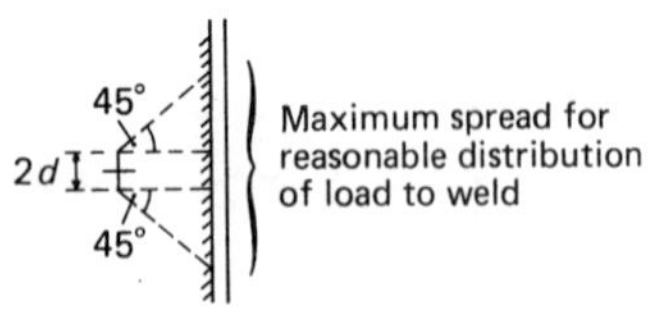

Structural Steelwork Connections	Subject: Base plate for column with axial load and moment			Chapter Ref. 11
	Design Code: BS 5950 Part 1			Calc. Sheet No. Example 2/1
	Calc. by B.D.C	Date Aug, '87	Check by G.W.O.	Date Nov' 87.

Code Ref.	Calculations	Output

Design a base plate for a 610×305×149 UB with an axial load of 1147 kN, a moment (about the XX axis) of 536 kNm and a horizontal shear of 179 kN. (All due to factored loads.)

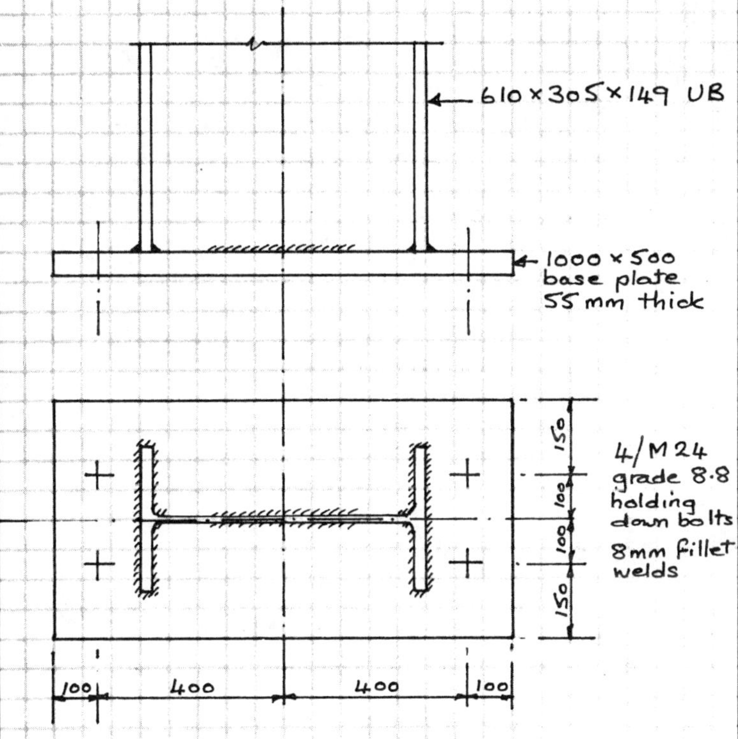

Assume the applied loading is resisted by uplift on the holding down bolts adjacent to the tension flange and a balanced area of bearing on the concrete under the compression flange.

By taking moments about centre of compression flange:

$$\text{Uplift on bolts} = \frac{536\times10^3 - 1147\times\frac{1}{2}(609{\cdot}6-19{\cdot}7)}{400+\frac{1}{2}(609{\cdot}6-19{\cdot}7)}$$

$$= 284{\cdot}5 \text{ kN}$$

Commentary: 2/2

The bending moment in the base plate is calculated at the centreline of the compression flange. If equal loads are assumed for the welds at the inside and outside of the flange, the maximum bending moment occurs at the inside of the flange. Using the lever arm to the centre of the flange to calculate the bending moment is slightly conservative. (Taking the bending moment at the face of the column would not be conservative.) If the distance from the bolts to the edge of the plate is excessive (i.e. the top and bottom edges in the figure), the whole width of the base plate may not be effective in bending. The maximum effective width can be taken as $2s + 2a + 2.8b$, as in Figure 7.17(g) (although here the plate is in single rather than double curvature and a different yield line pattern would apply). By resolving the vertical loads

Compressive load = (Uplift on bolts)
+ (applied axial load)

Bending moment at support of (358/2) mm cantilever, 500 mm wide, with a distributed load of 8 N/mm^2:

$p_{yp} = 245\,\text{N/mm}^2 = p_y$ for Grade 43 plate
$40 < t < 100\,\text{mm}$

51.1 > 33.6 mm. Compressive loading governs thickness of base plate.

Structural Steelwork Connections	Subject: Base plate for column with axial load and moment		Chapter Ref. 11
	Design Code: BS 5950 Part 1		Calc. Sheet No. Example 2/2
	Calc. by B.D.C	Date Aug, '87	Check by G.W.O.	Date Nov, '87

Code Ref.	Calculations	Output
	Try 2/M24 grade 8·8 bolts	
6·3·6·1 Table 32	Tension capacity of 2 bolts $= 2 \times 450 \times 353 \times 10^{-3}$ $= 317·7$ kN o.k.	4/M24 Grade 8·8 bolts at 800 × 200 cross centres.
	Base plate	
	Bending moment at centre line of tension flange $= 284·5 \times \{400 - \frac{1}{2}(609·6 - 19·7)\}$ $= 29887$ Nm	
4·13·2·3	Moment capacity $= 1·2\, p_{yp} Z$	
	$Z = \dfrac{wt^2}{6}$	
	If width of plate = 500 mm	
	Minimum thickness $= t = \sqrt{\dfrac{29887 \times 10^3 \times 6}{1·2 \times 265 \times 500}}$ $= 33·6$ mm	
	Compressive load on concrete under compression flange $= 284·5 + 1147$ $= 1431·5$ kN	
	Bearing strength of grout (and concrete) under base plate $= \omega = 8$ N/mm²	
	Effective area of base plate required $= \dfrac{1431·5 \times 10^3}{8} = 178938$ mm²	
	Width of effective strip $= \dfrac{178938}{500}$ $= 358$ mm	
	Bending moment at centre line of flange $= \dfrac{8 \times 500 \times (\frac{358}{2})^2 \times 10^{-3}}{2}$ $= 64082$ Nm	
	Minimum thickness $= \sqrt{\dfrac{64082 \times 10^3 \times 6}{1·2 \times 245 \times 500}} = 51·1$ mm	

Commentary: 2/3

The 8 mm fillet weld is specified to avoid hydrogen cracking, bearing in mind the thickness of the base plate. A smaller weld could be specified if welding consumables are used with a controlled hydrogen content. (See BS 5135: 1984.)

The end of the column and the base plate are assumed to be in tight bearing contact, and the compression is transmitted in direct bearing.

Section 11.5

The shear can be transmitted to the concrete base by:

1. Provision of shear keys;
2. Concreting in of the base plate;
3. By shear on the holding-down bolts (but the effect of clearances and/or play in the bolts must be considered);
4. Friction between the base plate and concrete (grout).

In this example the shear force can be transmitted by friction:

$$\mu \text{ required} = \frac{179}{1431.5} = 0.125$$

Structural Steelwork Connections	Subject: Base plate for column with axial load and moment			Chapter Ref. 11
	Design Code: BS 5950 Part 1			Calc. Sheet No. Example 2/3
	Calc. by B.D.C	Date Aug, '87	Check by G.W.O.	Date Nov, '87

Code Ref.	Calculations	Output
	Minimum length of base plate $= 609.6 - 19.7 + 358 = 948$ mm	
	Use 1000 × 500 base plate, 55 mm thick.	1000 × 500 base plate 55 mm thick Grade 43
	By taking moments about centre of compression flange (ignoring the column web)	
	Force in tension flange of column $= \dfrac{536 \times 10^3}{(609.6 - 19.7)} - \dfrac{1147}{2} = 335$ kN	
	Load on fillet welds $= \dfrac{335}{2 \times 304.8} = 0.55$ kN/mm	
	Use 8mm fillet welds	
	Force in compression flange of column $= 1147 + 335 = 1482$ kN	
	Average stress $= \dfrac{1482 \times 10^3}{304.8 \times 19.7} = 246.8$ N/mm² $< p_y$ O.k	
	Shear force = 179 kN	
	Load on fillet welds connecting web to base plate $= \dfrac{179}{2 \times 537.2} = 0.17$ kN/mm	
	Use 8mm fillet welds (over length of, say, 250 mm)	8mm fillet welds

12

Beam-to-column connections

12.1 Introduction

Building frames are classified conveniently in two ways. First, it is necessary to differentiate between *sway frames*, where horizontal forces are resisted by flexure and shear in the beams and columns, and *no-sway frames*, where horizontal forces are resisted by other means. These other means might be a system of bracing or infilled panels between neighbouring columns or may involve the use of diaphragm action of the floors, which are themselves tied into a concrete services core.

Second, as shown in Figure 12.1, it is possible to differentiate between the assumed modes of behaviour of the frame connections under gravity loading. *Simple connections* are only designed to transmit shear from the beam to the column, at some nominal eccentricity from the face of the column. *Semi-rigid connections* are those where it is recognized that the connection has a finite rotational stiffness and strength and will therefore transmit some moments from the beams to the columns. Of course, these beam end moments will be less than would occur in a similar frame with *rigid connections* between the beams and columns, where, as the name implies, it is assumed that the connection does not deform significantly under applied moment.

Rationally, simple connections can only be used for no-sway frames. However, in the past, 'simple' design, or Type II construction, as it is known in Australia and the USA, was widely used for sway frames. The same connections were assumed to act as pins under gravity loading and to be moment-resisting under sway loading! The principal justification for this anomalous design method was that it worked in practice; indeed, some very large buildings (including the United Nations Building in New York) have been proportioned by this design method. One of the significant failures of structural engineering research has been its inability to demonstrate why such a design method could produce safe structures. A principal reason for using this method in the past was the great difficulty of

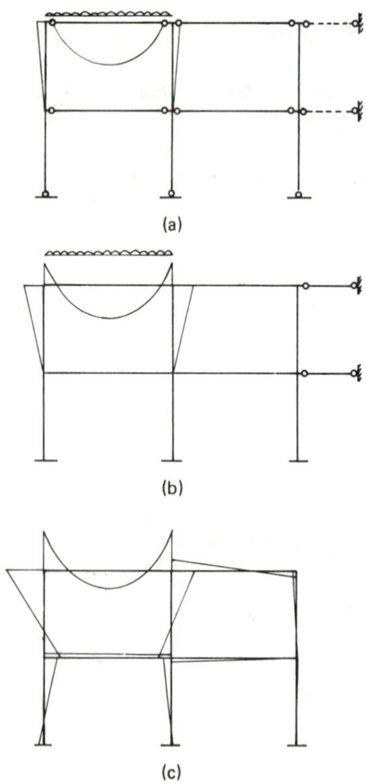

Figure 12.1 Methods of frame design. (a) 'Simple method'; (b) semi-rigid method; (c) rigidly jointed frame

analysing a rigid frame of practical size by manual techniques. It is now very much easier to analyse a practical rigid frame by computer, and it therefore seems likely that 'simple' sway frames will be less frequently used in the future. There is therefore no discussion of such connections (often called wind moment connections) in this book.

Semi-rigid design methods were originally introduced in recognition of the practical moment resistance that exists in many 'simple' connections. They were formulated for no-sway frames. Early studies showed that they offer little or no economy over simple design for frames of conventional proportions, and these design methods have found very little application in practice. The design method has never been extended to sway frames. However, there has recently been a resurgence of research interest in semi-rigid frame behaviour. At the time of writing (1989), this work has not led to any developments of direct value to the designer. In the future this situation may change and designers would be well advised to maintain a watching brief.

While the use of *rigid connections* in no-sway frames will undoubtedly reduce beam sizes compared to frames with simple connections, much of this weight saving is lost because column sizes in rigid frames will increase if they are designed by traditional elastic methods. In addition, rigid connections are significantly more expensive than their simple counterparts. Therefore they can rarely be justified in no-sway frames of conventional proportions. It is only if beam depth is limited and stiffness rather than strength governs design that rigid connections, which considerably increase frame stiffness, may be justified.

Thus the designer is currently faced with a limited practical choice of frame design methods and connection types (if beams and columns are in the same plane). Probably 90% or more of all frames may be classified as no-sway. In this context:

1. Simple connections should be used if strength rather than stiffness governs design. This is generally the case if the span:beam depth ratio is no more than 20:1.
2. Rigid connections may be justified for slender construction because of their enhanced stiffness.

Sway frames will be rigidly jointed. Up to about five storeys in height there may be some economy to be gained by using plastic methods of analysis. Above that, such a frame is likely to be too flexible and elastic analysis should be used.

The connections presented in this chapter are consistent with the above analysis of current methods of frame design.

Note that a building frame is a two-way structure, and different structural systems may be used in the two directions. For example, it is quite common for a rectangular building to be braced in its long direction but be a sway structure in the other. However, for convenience of presentation in this text, where major and minor axis connections to a particular column are illustrated or discussed, they are always of the same type.

The above entirely relates to the conventional situation where the beams and columns of a particular frame are in the same plane. However, a recent development in structural form uses pairs of beams that pass either side of the column. It is thus possible to retain the simplicity of simple construction and still utilize the benefits of continuity in beam design.

12.2 Simple connections[1]

12.2.1 General

Figure 12.2 shows the common types of simple major and minor axis beam-to-column connections. Selection of connection type is generally dictated by economics; any particular decision is primarily influenced by the fabricator's equipment but is also affected by secondary considerations such as speed of erection. However, it should be appreciated that the detailed choice of connection can influence element design, because different connections offer varying degrees of in-plane and out-of-plane restraint between members. For example, connection type (a) would justify effective length factors of less than 1.0 for minor axis column failure and lateral–torsional buckling of the major beam; effective length factors of 1.0 should be used for both these elements with connection type (b).

12.2.2 Clip and seating angle connections (Figure 12.2(a))

This form of connection would be most economic for fabricators equipped with automatic, numerically controlled, saw and drill lines or saw and punch lines. All the bolt holes in the main elements can thus be produced economically and no welding is required to the elements, except possibly for a load-bearing stiffener to the beam. The connection does not present any particular difficulty in design but attention should be paid to the following points:

1. In checking the end-bearing capacity of the beam on the seating angle, account should be taken of the possible adverse combination of tolerances in determining the minimum overlap of the beam and the vertical angle leg.
2. If the seating angle is unstiffened it is acceptable to design the seating cleat/column bolts to resist shear only. However, if the angle is stiffened the horizontal leg of the angle can no longer flex, and these bolts should be designed to resist the corresponding moment as well as the shear. The

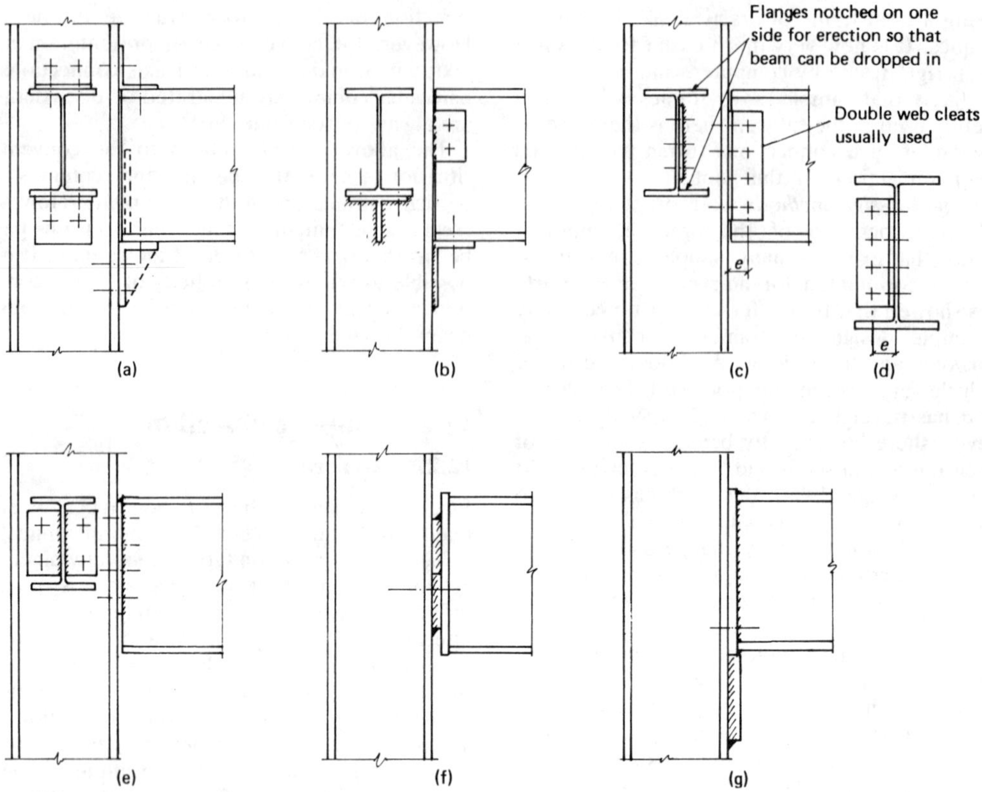

Figure 12.2 Simple beam-to-column connections. (a) Clip and seating angles; (b) shop-welded cleats; (c) web cleats; (d) single-web cleat; (e) curtailed end plates; (f) and (g) shear plates

eccentricity should be taken as the outstand of the horizontal leg.

3. The clip angle at the top of the beam can only resist shear by flexure of the legs. This is a much more flexible mode of behaviour than the bearing action of the seating angle. The clip angle should therefore not be assumed to contribute to the shear resistance of the connection.

12.2.3 Shop-welded cleats (Figure 12.2(b))

The cost of drilling holes in rolled sections increases considerably in the absence of automatic plant. Setting-out costs become significant and manipulation of the section with a radial drill or of the mobile drilling equipment if the section is held stationary also increase the fabrication cost. In these circumstances it becomes economic to minimize the number of bolts and attach the cleats to the column by welding, thus eliminating all holes in the column. Holes still have to be drilled in the beam end but this is unavoidable.

In general, angles will be used to form the cleats; they are speedily punched and cropped to length and their right-angled form helps to hold the connection square. However, with some combinations of column and minor axis beam sizes the seating cleats that would be required to be welded to the column would be a high proportion of its web width. In such circumstances it would not be possible to achieve sound welds to the vertical ends of the cleat because of access difficulties, and an alternative seating bracket should be used. In the example shown, a Tee section bracket is used. Such a bracket is very good structurally but has to be sawn rather than cropped, which increases its cost significantly.

If UB major axis beams are used it is customary to provide a small web cleat to the top of the web as shown, to stabilize the beam against torsional instability. If UC beams are employed (probably because structural floor depth is limited) there is a greater resistance to lateral–torsional buckling and these cleats may be omitted in many circumstances.

Note that the projecting cleats that have been welded to the column are usually of only modest strength. Care should be taken during transportation and erection to avoid damaging them. Irrespective of design load, they should be attached with welds of minimum size for ductility (page 49). If they are knocked they will then deform plastically rather than suffer any cracking of the welds. Minor bending damage can be corrected at site.

12.2.4 **Web cleats** (Figure 12.2(c))

As an alternative to seating angles, web cleats may be used to transmit the beam shear into the column and angle cleats may be used which can be either bolted or welded to it. Alternatively, plate cleats or fin plates can be welded to it. If they are used for attachments to the minor axis of small columns, as shown, they are protected from damage and avoid welding access problems. However, the beam flanges will usually have to be notched, so that the beam can be dropped into position.

Design texts vary in their advice about an appropriate treatment for the inherent eccentricities when angle cleats are used. The most suitable analytical model is probably to postulate zero moment at the heel of the angle. The eccentricity of load on the bolt group in the beam web is thus the angle backmark, labelled e in Figure 12.2(d). Because of symmetry, no account need usually be taken of eccentricity of load on the individual bolt groups in the column flange provided that a pair of web cleats is used and they are not less than, say, 60% of the web depth. (This second limit is to give a reasonable lever arm to resist the eccentricity moments.) Very small, or short, cleats should be designed to resist both eccentricities on the bolt groups. It is particularly important that proper consideration be given to eccentricity if a single web cleat is used, as shown in Figure 12.2(d). Attention should be paid to both deformation and strength if a single-bolted cleat is used. If the cleat is short, bolt clearance in the holes is likely to lead to significant rotation of the beam about its longitudinal axis. The bolt group should be designed for the eccentricity e shown in Figure 12.2(d).

Where fin plates are used the designer has to take care to ensure that the connection has sufficient rotation capacity.[2] This can be achieved by:

1. Using ordinary (not HSFG) bolts;
2. Ensuring that significant bearing deformation of the connected plies will precede bolt shear failure (i.e. that plate thickness is not greater than $0.5d$ for Grade 43 and $0.4d$ for Grade 50 steels);
3. Detailing edge distances that are not less than twice the bolt diameter;
4. Ensuring that resistance of the welds attaching the cleat to its support is greater than the moment that can be supplied by the bolts. For simplicity in practical design, this condition will usually be readily achieved by using welds that have the minimum capacity to ensure ductility in the connection (i.e. that $\Sigma a = 1.2t$ for Grade 43 and $\Sigma a = 1.4t$ for Grade 50, where t is the thickness of the fin plate). Designers need not concern themselves with the rare situations where such welds will not meet the criteria outlined above, because yielding of the fin plate, either adjacent to the weld or on the net section, will relieve the moment.

12.2.5 **End plates** (Figure 12.2(e))

Simple end-plate connections are now very popular. Their use can eliminate all hole drilling in the beams, and modern milling saws give sufficiently accurate dimensional control of the beam length and end squareness for the need for packs to be minimized. Design does not present any great difficulty, but some care in detailing is required if the inherent rigidity of this type of connection is not to lead to problems as the beam takes up its simply supported profile with its associated end rotations.

Some designers use full-depth end plates in this situation, welding them to both flanges; this forces the centre of the beam end rotation down to the bottom flange, increasing the deformation requirements of the top region of the connection. Other designers curtail the end plate short of both flanges, as shown in the minor axis connection in Figure 12.2(e). This produces a rather fragile end plate, prone to transportation damage. The most sensible arrangement seems to be that shown for the major axis beam connection. The end plate is attached to the top plate for robustness but does not extend unnecessarily far down the beam web, giving the least resistance to beam end rotation. It is, of course, necessary to ensure that this short end-plate design has adequate shear capacity, both within the end plate (on critical sections through the bolt holes) and in the beam web adjacent to the weld.

Ductility of the upper region under tensile deformation is partly ensured by controlling the proportions of the end plate and bolts. In *moment-resisting connections* using Grade 8.8 bolts and Grade 50 plate even plastic methods of analysis cannot justify end-plate thicknesses that are significantly less than the bolt diameter, with the bolts at the minimum eccentricity from the beam flange and web. With these proportions bolt fracture in tension is unlikely to occur before significant deformation of the end plate. In simple connections there is only a deformation requirement to be satisfied, with the provision that the bolts must be protected from excessive tension. With Grade 8.8 bolts this would seem to be achieved if the end-plate thickness is kept to no more than half the bolt diameter. Bearing in mind that Grade 4.6 bolts have half the tensile capacity of 8.8 bolts and that plate flexural strength is proportional to (thickness)^2, an appropriate upper bound on plate thickness with Grade 4.6 bolts would seem to be a third of bolt diameter. Conveniently, with these thickness limitations bolt shear rather than plate bearing will still govern bolt design strength for the common structural bolts. The cross-centres of the bolts should not be less than the 4 × bolt diameter plus web thickness that is the convenient minimum for installation.

Ductility of this region could also be impaired by premature weld rupture. This can be eliminated by providing welds joining the web to the end plate that can develop the tensile yield capacity of the web plate, i.e. double-fillet welds that comply with the requirements on page 49.

There is a scale effect with this type of connection, which does suggest that there should be an upper limit on size. Beam end rotation under factored loading is independent of beam size. It can be shown that, if the beam is fully loaded and has a span-to-depth ratio of less than 20, end rotation is unlikely to exceed 0.020 radians for Grade 43 steel or 0.026 radians for Grade 50. If a nominal deformation of, say, 5 mm to the tension region of these connections is considered acceptable at working loads (i.e. 7.5 mm under factored loads) then maximum end-plate depths should not exceed 370 mm and 280 for Grades 43 and 50 beams, respectively. (This nominal deformation may seem rather high, but in practice will be reduced partly by bedding in of the bottom of the end plate and deformations of the column.) In order to have sufficient robustness, and to ensure that the bottom flange of the beam will not bear on the face of the column, it would seem prudent to limit the minimum end-plate sizes to half the beam depth. (The local shear capacity of the beam web would be an important consideration for such proportions.) Thus the maximum beam depths for curtailed end plates would seem to be 740 mm and 550 mm for Grades 43 and 50 steel, respectively.

12.2.6 Shear plate connections (Figures 12.2(f) and (g))

There are two types of shear plate connection. These are particularly suitable where beam shears are high and/or speed of erection is an important consideration. The half plates shown in Figure 12.2(f) require more fabrication but ensure that there are no weldments to the column outside the beam profile. The combined shear and end plate in Figure 12.2(g) requires one less plate and less welding.

In either case nominal bolts should be positioned near the effective centre of rotation to ensure that the connection integrity is maintained. It must, of course, be ensured that no packs are used with this type of connection without special consideration to guarantee that adequate bearing area is maintained and that there is no possibility of the beam slipping off the shear plates. Distortion of the shear plates must also be considered, because they are only welded on three sides. Supplementary bolts may be necessary to ensure that distortion does not occur.

12.3 Semi-rigid connections

Figure 12.3, reproduced from the final report of the Steel Structures Research Committee in 1936,[3] shows the four classes of connection that were studied when semi-rigid design methods were originally developed. As can be seen from the details, they are not suitable for modern use, and are shown here for historical interest only.

Recently, attempts have been made to revive interest in semi-rigid design methods, and studies have been made of the moment/rotation characteristics of connections that would otherwise be regarded as simple.[4] As discussed in the introduction to this chapter, the use of these connections in the design of frames of conventional proportions, where strength governs design, does not appear to lead to economy. It therefore seems inappropriate to discuss them in detail in this design text.

A more appropriate use of semi-rigid design would appear to be for frames where the depth of construction is limited and where stiffness rather than strength governs design. In such situations composite beams are likely to be used, and it might be possible to use a semi-rigid composite connection.

Some initial development work has already been carried out,[5,6] and Figure 12.4 shows an example of the type of connection that might be used. The essential ductility for this design method is developed by yielding of the reinforcement in the slab. At the time of writing this development work has not been taken to a stage where a complete design method is available, and it is therefore inappropriate to present a detailed discussion at this stage. However, the initial work was encouraging, and it seems that this approach, once development work is completed, offers a possible application of semi-rigid connections in the future.

The other application of semi-rigid concepts is to recognize the actual rotational stiffness of 'simple' joints when determining the effective length of members.[7] The most likely outcome from current activity on this topic is that it will lead to refinements in element design rather than influence connection design directly.

12.4 Rigid connections

12.4.1 Conventional office buildings

Figure 12.5 shows the conventional types of rigid connections for typical office buildings. Fully site welded connections may be used, though they have not proved economic in the UK in the past. They

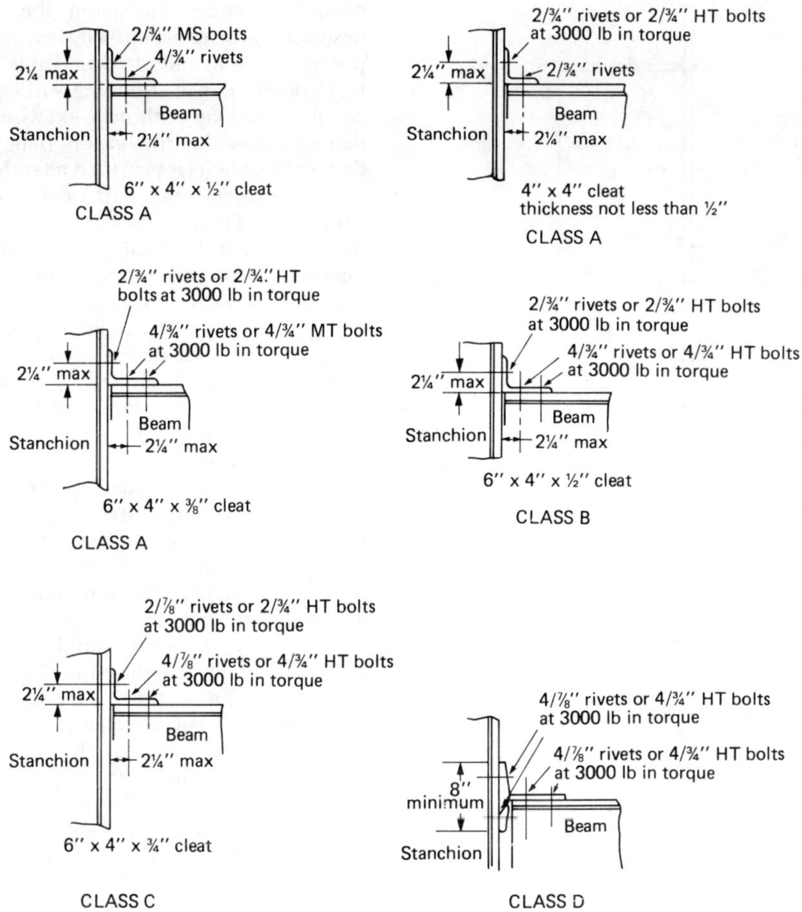

Figure 12.3 Classification of early semi-rigid connections

have found wide usage in other countries, notably in the USA. In such connections fitted column web stiffeners will generally be required for all but the heaviest column sections. Cope holes should be provided to assist in any remedial measures that may be necessary if there is poor fit of the top flange and to provide access for downhand welding of the bottom flanges.

In the early days of bolted connections Tee stubs were sometimes used to connect the flanges of the beams to the column. Because of the large number of bolts required, together with the necessity of using HSFG bolts to give the required rigidity, this form of construction has fallen into disuse. All rigid bolted connections make use of end plates, as shown in Figures 12.5(b) and (c). Short end plates are not very efficient structurally, because of the reduced lever arm for all of the bolts. They should only be used in those relatively rare situations where the connection moment is significantly less than the moment capacity of the beam. Extended end-plate connections are much more effective. They can generally be designed to develop either the yield moment or the plastic moment of the beam section. For beams of less than 500 mm depth it is usually

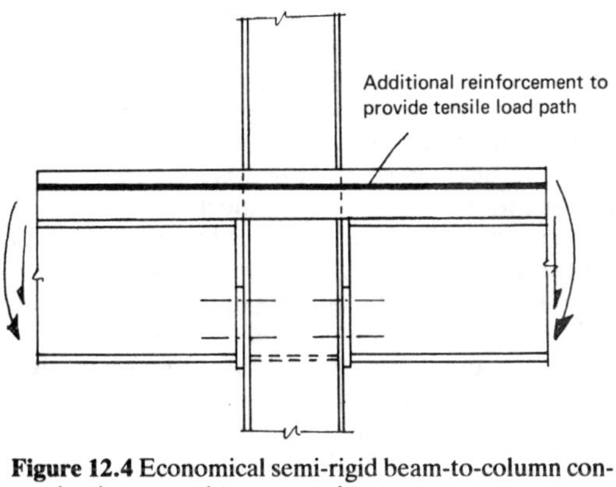

Figure 12.4 Economical semi-rigid beam-to-column connection in composite construction

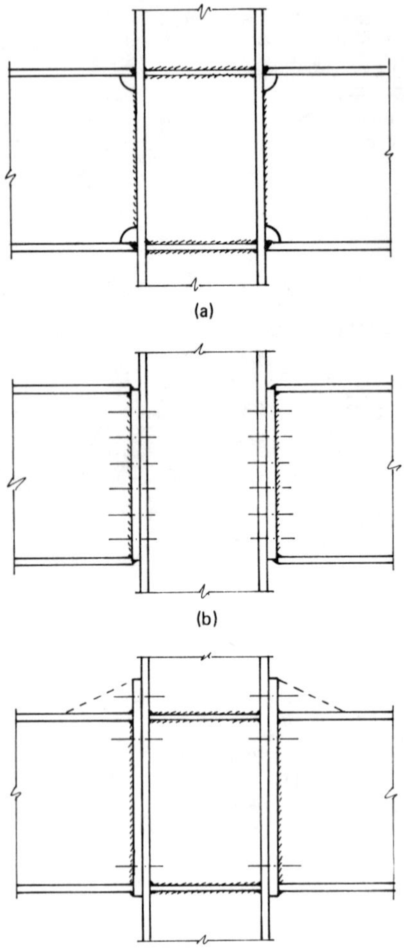

Figure 12.5 Connections for rigidly jointed frames. (a) All welded; (b) bolted with short end plate; (c) bolted with extended end plate

possible to develop the required capacity with the six bolts shown if they are of Grade 8.8 and appropriate diameter. Semi-empirical methods are available for designing this type of connection.[1,8,9] They all make use of yield line methods and are subject to the same detailed criticisms that were put forward in Section 7.7.2. However, they have been experimentally verified for connections of conventional proportions and may be used in such a context with confidence. More general methods are demonstrated in the worked examples in this chapter; they can be used in any context and are not necessarily more conservative than the semi-empirical approaches.

In extended end-plate connections that are only subjected to 70% or 80% of yield moment capacity the tension load path may be idealized as a Tee stub connection concentric about the top flange and designed in the conventional way, taking account of prying action as indicated in Table 7.5. It is likely that column flange stiffening will be required. The compressive load path may likewise be assumed to be concentrated in the bottom flange. It is essential that a row of bolts is provided near the bottom of the end plate to stop this part of the connection from 'springing'. These bolts may be utilized to transmit the shear from the beam to the column. If they are inadequate for that purpose then additional rows may be provided. As illustrated, the bolt spacing does not comply with maximum pitch criteria. For a connection that will be protected from corrosion this should not create difficulties, and indeed it is common for these criteria to be ignored in this situation. However, if there is any possibility of moisture ingress to the connection then stitching bolts should be provided. If such bolts are used then they will inevitably attract significant tension. In order to guard against premature weld rupture arising from this tension the beam web/end plate weld should be made the minimum size for ductility, at least for the tensile portion of the web. Alternatively, the welds could be designed for the load that would be induced in the bolts.

For higher proportions of the moment capacity of the section it will be necessary to mobilize the contribution that the web makes to bending capacity. The tension in the upper part of the web is transmitted to the bolts by transverse bending of the end plate, as shown in Figure 7.20(b). This will increase the load on the lower pair of bolts. In order to increase the effective capacity the triangular stiffeners shown in Figure 12.5(c) should be added.[10] Additional bolts may be provided if necessary, extending the two columns of bolts downwards. The compression in the lower part of the web is simply transmitted by direct bearing into the column.

12.4.2 Heavy construction

As the scale of construction increases it becomes more difficult to design straightforward extended end-plate connections. The major difficulty is in fitting sufficient bolts in the tension region of the connection. It is possible to provide additional bolts below the basic tensile bolt group, but they have reduced effectiveness because of the shorter lever arm and they may overload the beam web in tension. In some circumstances it may be possible to fit two lines of four bolts around the beam flange, as shown in Figure 12.6(b), but this will not work for many combinations of beams and columns. Thus in many practical instances in heavy construction it is not possible to develop the full moment capacity of the beam sections without increasing the lever arm of the connection. This may be carried out by the

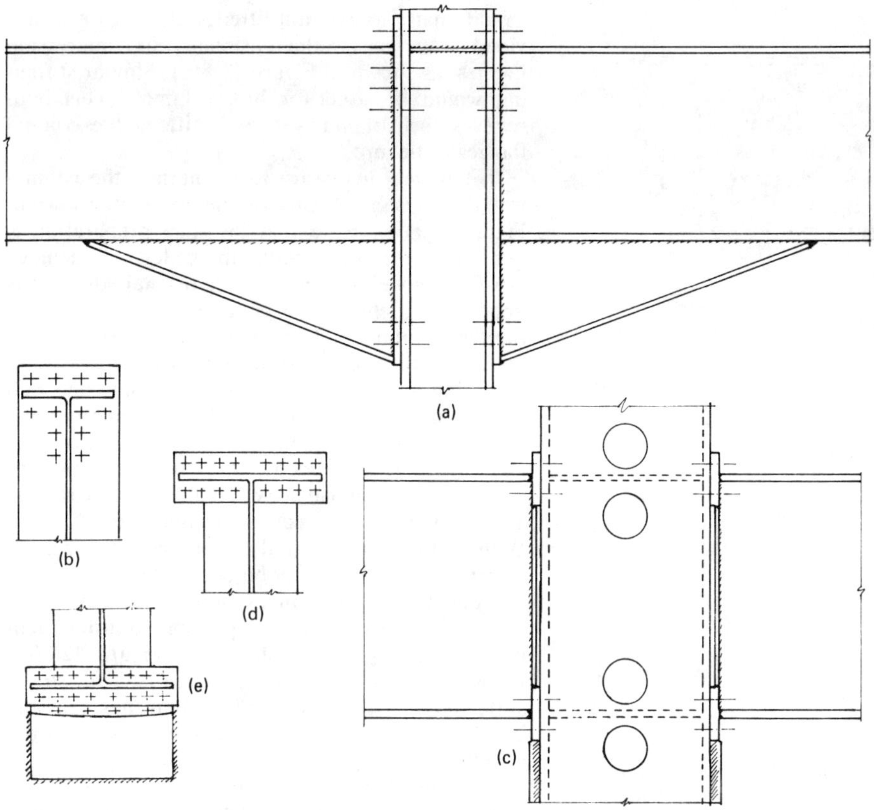

Figure 12.6 Rigid beam-to-column connections in heavy construction. (a) Using haunches; (b) bolt pattern for (a); (c) box column and plate girder construction; (d) bolt pattern for (c); (e) radiused pack for torsionally stiff girder

provision of a simple haunch, as shown in Figure 12.6(a). Once the decision to provide a haunch has been made it is sensible to provide as large a haunch as can be accommodated, up to the point where the haunch depth is similar to the beam depth. The marginal cost of increasing the size of the haunch is small, but by so doing it will be possible to reduce the number of bolts and minimize the stiffening requirements for the column. In many cases it will be most economic to cut the haunch from the same rolling as the main beam section. However, it would not generally be economic to increase the height of the building in order to accommodate a large haunch, even if this were permitted.

Figure 12.7 summarizes the two possible modes of behaviour of the haunch. In Figure 12.7(a) it is assumed that the compressive load path remains in the haunch flange. The haunch web provides the shear path for any change in flange force that may be necessary for overall equilibrium and transmits shear to the lower bolts. (Some of the shear is, of course, transmitted by the inclined flange of the haunch.) Note that this assumption of behaviour leads to a requirement for a full-strength weld at the point where the haunch meets the beam bottom flange. In Figure 12.7(b) it is assumed that the haunch web is effective in transferring the flange force C_1 into the main beam by in-plane shear and bending. In addition, it transmits the vertical shear through to the beam web. In practice, the haunch will operate in a combination of these two modes; the first will tend to predominate for short, steep haunches, the second will become more significant for longer, shallower elements. In design either mode may be assumed, or a combination of both; in any event, it would seem suitable to proportion the haunch in accordance with compact section criteria so that any necessary redistribution of forces may occur safely. The magnitude of the governing forces, C_1, C_2, T_1, T_2 are found in a straightforward way from moment equilibrium at the two ends of the haunch.

Figure 12.6(c) shows a form of rigid connection that is particularly suitable for heavy construction with box columns and plate girders. The greater width of such construction will generally enable the moment to be developed within the depth of the girder; as shown in Figure 12.6(d), it is possible to fit many more bolts across the width of the connection, and at this scale of construction it is appropriate to use large bolts (M36 or M42). The 'Tee stubs' to the flanges would typically be 40–50 mm in thickness; the beams will probably have to be machined to length once fabrication has been completed. For

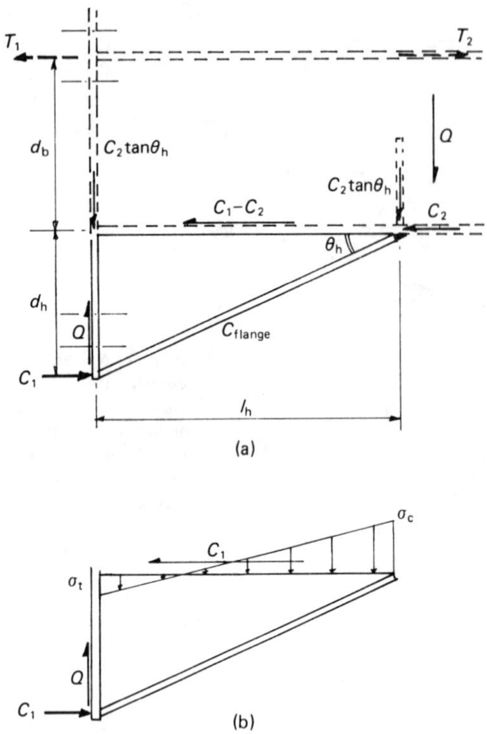

Figure 12.7 Structural behaviour of haunches. (a) Bending resistance confined to haunch flange; (b) bending and shear resisted by haunch web

these two reasons, it becomes economic to use the arrangement shown where the heavy Tee stubs are curtailed and a lighter load-bearing stiffener is positioned between them. In some instances, when the column plate would be overstressed in local flexure, it may also be necessary to use Tee stubs inside the box column. With this arrangement it is not possible to take the shear directly from the web, and a shear plate may have to be provided underneath the lower Tee stub. The box column openings for bolting access should be noted.

Simple shear plates are only satisfactory with this large-scale construction if the beams have sufficient torsional flexibility to overcome any rotational misalignment about the beam longitudinal axis. If box beams or short-span plate girders are used, problems will arise because of their torsional rigidity. If shear plates have to be used in such construction, a radiused pack plate should be incorporated to remove this potential lack of fit, as shown in Figure 12.6(e).

12.4.3 Stiffening for rigid connections

Figure 12.8 shows the various types of stiffening that are commonly employed in rigid connections. Traditionally, horizontal fitted stiffeners were provided on the lines of the tension and/or compression flanges, as shown in Figure 12.8(a). Similar stiffening would be used for box columns. This both relieves the column web and stiffens the column flanges in flexure.

If it is only necessary to strengthen the column web then it is possible to use the web plate shown in Figure 12.8(b). Note that an edge preparation is necessary for this plate in order to achieve satisfactory welds along its longitudinal edges. It is important to check web plate stability by ensuring that maximum weld spacing criteria are not exceeded. If the unsupported web plate is too slender then plug welds will be necessary. If only one web plate is used, leading to a lack of symmetry, it is customary to consider that only half its thickness is effective, unless the web is only being stiffened to increase its shear capacity. In the latter case the assymetry will not reduce strength significantly. With all these caveats and restrictions, this form of stiffening will only rarely be the best solution.

If only the column flanges on the tensile load path need stiffening in flexure it is possible to stiffen them by the backing plates shown in Figure 12.8(c). Design methods are being developed for these stiffeners, but it should be noted that they can only be effective in stiffening the column flange in the vicinity of the bolt holes. Because they are only loose plates they cannot reinforce the flanges at the critical section in transverse bending where they join the web. It follows from their mode of action that their use will increase prying forces in the bolts. Once again it is only rarely that they will be the optimum form of stiffening.

At interior beam-to-column connections most of the moment is transferred through the connection from beam to beam. At exterior beam-to-column connections the moment must be transmitted into the column, producing very high shears in the column web. This is discussed in detail in Section 14.2.1.

If the column web can accommodate these high shear forces then the most economical stiffening is the orthogonal pattern shown in Figure 12.8(d). If the column web is overstressed in shear then the traditional solution was to introduce a stiffener on the compressive diagonal, as shown in Figure 12.8(e). If this stiffener is proportioned only to improve the shear capacity of the web then the horizontal compression stiffeners must be retained. However, if the diagonal stiffener is designed to resist the diagonal resultant of the sum of the beam compressive force and the column flange upwards force at X then the lower horizontal stiffener may be omitted. The variation in behaviour implied by these different approaches to stiffening is fully discussed in Section 14.2.1.

A recent development in stiffening has been the

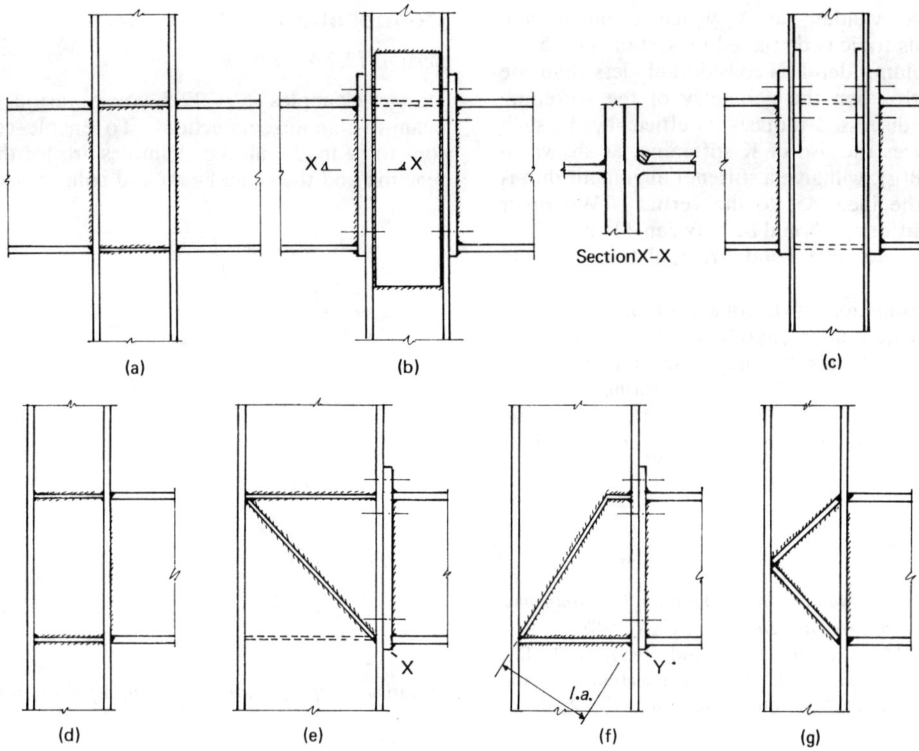

Figure 12.8 Stiffeners for rigid beam-to-column connections. (a) Traditional stiffening for interior column; (b) web plate; (c) backing plates for tension bolts; (d) exterior column; (e) compressive diagonal stiffener; (f) Morris stiffener; (g) 'K' stiffeners

introduction of diagonal tension stiffening as shown in Figure 12.8(f). There are situations where such stiffening may be more economical than that shown in Figure 12.8(e). However, it should be used with caution if the overall geometry is such that there will be a significant reduction in lever arm (l.a.) within the connection. If this occurs and is not accounted for in design there could be a premature onset of

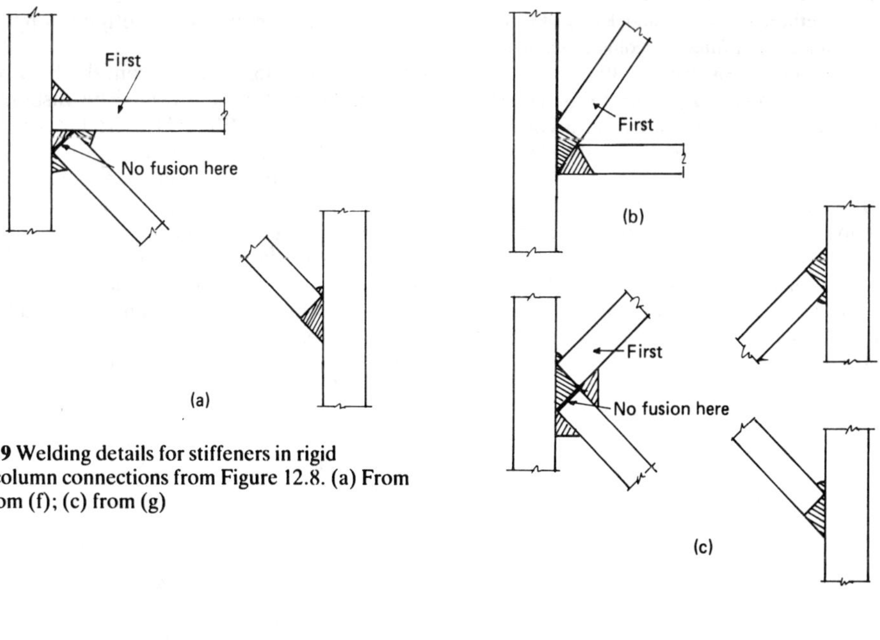

Figure 12.9 Welding details for stiffeners in rigid beam-to-column connections from Figure 12.8. (a) From (e); (b) from (f); (c) from (g)

compressive yielding at Y which could initiate failure. This topic is discussed in Section 14.2.5.

If the column depth is considerably less than the beam depth, then the geometry of the stiffening previously discussed reduces its efficiency. In such circumstances the use of K stiffening, as shown in Figure 12.8(g), will give a stiffener alignment that is closer to the ideal 45° to the vertical. (Wherever possible, stiffeners should lie between 30° and 60° to the vertical, to give good 'triangulation' of load forces.)

The introduction of diagonal stiffening elements can complicate welding, and Figure 12.9 shows some of the welding details that may arise. In general, the more critical tension stiffening is positioned first and the compression stiffening added afterwards; the weld details are chosen to suit the particular geometry.

References

1. Pask, J. W., *Manual on Connections for Beam and Column Construction*, 2nd edn, BCSA, 1988.
2. Patrick, M., Thomas, I. R. and Bennetts, I. D., 'Testing of web side plate connection', Pacific Structural Steel Conference, Auckland, New Zealand, 1986.
3. PD 3343, British Standards Institution.
4. Section 2: Semi-rigid Joints, Int. Conf. on Joints in Structural Steelwork, Teesside, 1981.
5. Johnson, R. P. and Law, C. L. C., 'Semi-rigid joints for composite frames', Int. Conf. on Joints in Structural Steelwork, Teesside, 1981.
6. Owens, G. W. and Echeta, C. B., 'A semi-rigid design method for composite frames', Int. Conf. on Joints in Structural Steelwork, Teesside, 1981.
7. Jones, S. W., Nethercot, D. A. and Kirby, P. A., 'Influence of connection stiffness on column strength', *The Structural Engineer*, **65A**, No. 11 (1987).
8. Packer, J. A. and Morris, L. J., 'A limit state design method for the tension region of bolted beam–column connections', *The Structural Engineer*, **55**, No. 10 (1977).
9. Zoetemeijer, P., 'A design method for the tension side of statically loaded, bolted beam-to-column connections', *Heron*, Vol. 20, Delft. 1974.
10. Augustyn, J., Laguna, J. and Sliwka, W., 'Charge limite des assemblages par boulons hr. et plaque d'extremité sollicités en traction ou en flexion', *Construction Metallique*, No. 2 (1982).

Commentary: 1/1

Section 12.2

Design examples 12/1–12/5 cover various simple beam-to-column connections. To enable comparisons to be made, all the examples are for the same reaction and the same beam and column sections.

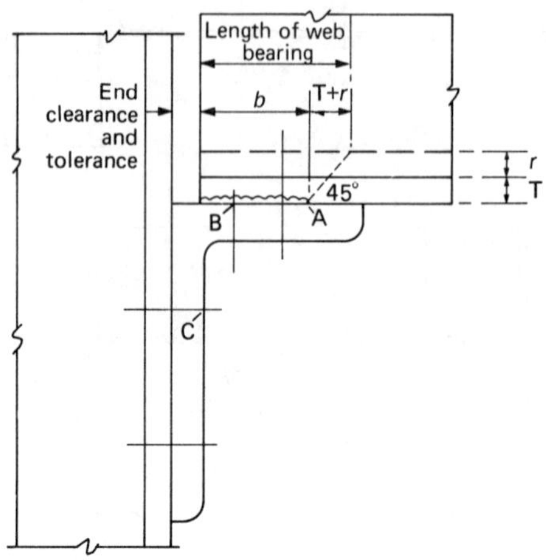

Details of dispersion through seating cleat and flange.

Section 12.2.2

Where applicable, the web buckling should also be checked. For the reaction in this example it can be seen from 'member capacity tables' that web buckling is not a problem:

Minimum length of bearing at edge of root radius =

$$\frac{\text{Reaction}}{\text{Web thickness} \times \text{design strength of web}}$$

A dispersion of 45° is taken from the bearing on the cleat to the root line. (*Note*: The dispersion at a slope of 1 in 2.5 in BS 5950: Part 1, Clause 4.5.3 is not applicable for this procedure.)

If packs are inserted between the seating angle and the bottom flange, dispersion of load should not be taken through the packs.

Check that the end of the calculated bearing length (point A) is in fact on the proposed cleat.

It is important that a tolerance is included in the design of seating cleats, since small dimensional changes can have a large effect on the capacity.

Structural Steelwork Connections	Subject Beam to column connection – Bolted seating cleat.			Chapter Ref. 12
	Design Code BS 5950 Part 1			Calc. Sheet No. Example 1/1
	Calc. by B.D.C	Date Aug, '87	Check by G.W.O.	Date Nov, '87

Code Ref.	Calculations	Output

Design a seating cleat connection for a 406 × 178 × 54 UB in Grade 43 steel, to carry a reaction of 145 kN (due to factored loads.) The connection is to the flange of a 203 × 203 × 52 UC in Grade 43 steel.

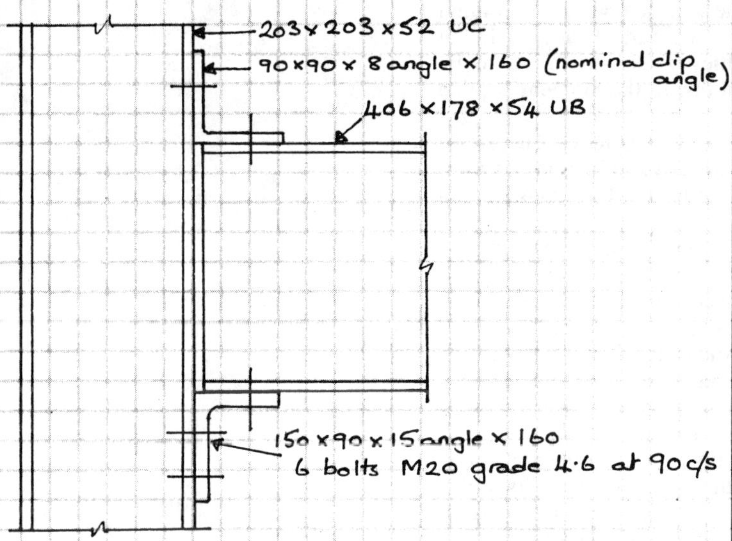

<u>Seating cleat</u>

Length of bearing required at root line of beam

4·5·3 $= \frac{R}{t p_{yw}} = \frac{145 \times 10^3}{7{\cdot}6 \times 275} = 69{\cdot}4$ mm

Commentary: 1/2

The bending resistance is checked at the tangent point of the root radius on the horizontal leg of the seating angle (point B). Theoretically, the maximum bending stress occurs to the left of this point, due to the increase in thickness (and modulus) being initially small relative to the increase in moment. This is compensated, to some extent, by taking 1.2 × $wT^2/6$ instead of the plastic modulus ($wT^2/4$). Consequently, the full plastic modulus should not be used with this procedure.

Section 12.2.2

The bolts connecting the seating angle to the column are designed only for shear (and bearing). If the angle were stiffened they would also have to be checked for the tension due to the moment applied by the beam reaction.

Reaction 0.6 × < shear capacity. Therefore, moment capacity does not have to be reduced for a high shear load (BS 5950: Part 1, Clauses 4.2.5 and 3.2.6).

The bearing capacity could be governed by the capacity of the bolt itself (on the column flange or the angle) or by the capacity of the column flange or the angle. In this example it can be seen, by inspection, that the bearing capacity of the bolts on the column flange governs.

If the seating angle were being designed as a simple bracket (and not as part of the end connection of a beam) the moment capacity would have to be checked at the upper row of bolts in the vertical leg (point c). The conventional design procedure of only checking the horizontal leg assumes that a small horizontal thrust is generated between the bottom flange of the beam and the angle. Therefore, this design procedure should only be used when the bottom flange is bolted to the seating angle and there is an additional cleat connecting the top flange (or the top of the web).

The connection is suitable for connecting a beam to a column flange (where the flange is stiffened by the column web) or to the web of a Universal Column (where the ends of the seating angle are close to the column flanges). It is not generally suitable for connecting to the web of a beam (since the web would be too flexible). It could be used if a suitable stiffener is provided on the line of the connected beam and the web is not too thin, or possibly if matching connections are made on both sides of the web and the designer is confident that the reactions will be virtually equal for all load combinations.

Structural Steelwork Connections	Subject: Beam to column connection — Bolted seating cleat			Chapter Ref. 12
	Design Code: BS 5950 Part 1			Calc. Sheet No. Example 1/2
	Calc. by B.D.C	Date Aug., '87	Check by G.G.O.	Date Nov, '87.

Code Ref.	Calculations	Output
	Length of bearing on cleat $= b = 69.4 - (T + r)$ $= 694 - (10.9 + 10.2) = 48.3$ mm	
	End clearance of beam from face of column = 5 mm Allow 5 mm tolerance	
	Try 150 × 90 × 15 angle × 160 mm long	
	Distance from end of bearing on cleat to root of angle (A to B) $= b + 5 + 5 - (t + r_1)$ of angle $= 48.3 + 5 + 5 - (15 + 12)$ $= 31.3$ mm	
	Assume uniformly distributed load over bearing length 'b'.	
	Moment at root of angle (point B) due to load to right of 'B' $= \frac{145 \times 31.3}{48.3} \times \frac{31.3}{2} = 1471$ Nm	
4.2.5	Moment capacity $= 1.2\,p_y Z$ $= 1.2 \times 275 \times \frac{160 \times 15^2}{6} \times 10^{-3}$ $= 1980$ Nm $>$ 1471 Nm o.k.	150 × 90 × 15 seating angle × 160 mm long
4.2.3	Shear capacity of outstand leg of cleat $= 0.6\,p_y \times 0.9\,wt = 0.6 \times 275 \times 0.9 \times 160 \times 15 \times 10^{-3}$ $= 356$ kN $>$ 145 kN o.k.	
	Try 4/M20 Grade 4.6 bolts for connection to column flange.	M 20 bolts grade 4.6
6.3.2	Shear capacity $= 4 \times 160 \times 245 \times 10^{-3}$ $= 156.8$ kN $>$ 145 kN o.k	
6.3.3.2	Bearing capacity (of bolts) on column flange $= 4 \times 435 \times 20 \times 12.5 \times 10^{-3} = 435$ kN $>$ 145 kN o.k	

Commentary: 2/1

Section 12.2.4

Web cleat connections are assumed to rely on the local distortion of the cleats to accommodate part of the end rotation of the beam. To obtain the flexibility required, in this example the thickness of the cleats is limited to 8 mm, and the cross-centre dimension of the bolts is 120 mm, which is close to the maximum permitted for the sections used.

In traditional design the effect of eccentricity is ignored when designing the bolts connecting the cleats to the column flange. However, with the lower safety factors adopted in current design codes it would be prudent to allow for the effect of eccentricity. This particularly applies to the smaller connections; consequently the effect is included in this example.

Shear capacity of bolt in double shear $= 2p_s A_t$
Bearing capacity of bolt $= p_{bb}dt$

Structural Steelwork Connections	Subject Beam to column connection — Bolted web cleats			Chapter Ref. 12
	Design Code BS 5950 Part 1			Calc. Sheet No. Example 2/1
	Calc. by B.D.C	Date Aug, '87	Check by G.C.O.	Date Nov, '87.

Code Ref.	Calculations	Output

Beam to column connection — Bolted web cleats.

Design a bolted web cleat connection for a 406 × 178 × 54 UB in Grade 43 steel to carry a reaction of 145 kN (due to factored loads). The connection is to the flange of a 203 × 203 × 52 UC in Grade 43 steel.

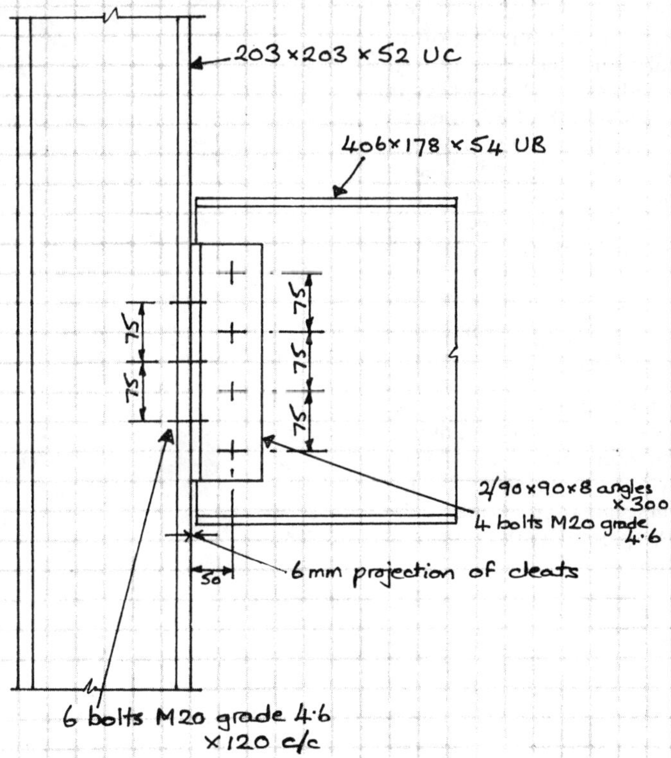

Connection to web of beam.

For M20 Grade 4·6 bolts

6·3·2 Shear capacity of bolt in double shear

$= 2 \times 160 \times 245 \times 10^{-3} = 78 \cdot 4$ kN

Commentary: 2/2

For a Grade 4.6 bolt bearing on Grade 43 steel, the bearing capacity of the bolt governs rather than the capacity of the connected ply:

Bolt capacity = lesser of the shear capacity and the bearing capacity

Section 8.2.2

Horizontal shear force on bolt due to moment due to eccentricity

$$= V_m = \frac{P_x e_x r_i}{\sum_n r_1^2}$$

Vertical shear force per bolt $= V_x = \dfrac{P_x}{n}$

Resultant shear force $= \sqrt{(V_m^2 + V_x^2)}$

The eccentricity of the load is taken as the distance from the heel of the cleats to the bolt line, i.e. the reaction is assumed to act at the face of the column. This is reasonable, since the column flange will have some relative stiffness. If the connection were to be made to a relatively flexible plate (for example, the web of a plate girder), then the eccentricity should be taken from the centreline of the web (with allowance for packs if there is a possibility that they could be used). However, the eccentricity could be taken from the face of the web if the connection is balanced by a similar connection on the other side of the web.

Shear capacity of bolt in single shear $= p_s A_t$
Bearing capacity of bolt $= p_{bb} dt$

An assessment of the maximum horizontal shear force on the bolts has been made by assuming a compressive reaction between the cleats (25 mm from the top of the cleats) and by assuming that the horizontal loads on the bolts are proportional to the vertical distances from this point. Hence, horizontal shear force on bottom bolt

$$= V_m = \frac{R}{2} \times e \times \frac{r_i}{\sum_n r_1^2}$$

Vertical shear force per bolt $= V_x$

$= \dfrac{R}{2n}$ (There are n bolts per cleat.)

Structural Steelwork Connections	Subject: Beam to column connection — Bolted web cleats			Chapter Ref.: 12
	Design Code: BS 5950 Part 1			Calc. Sheet No.: Example 2/2
	Calc. by: B.D.C	Date: Aug, '87	Check by: G.W.O.	Date: Nov, '87

Code Ref.	Calculations	Output
6·3·3	Bearing capacity (of bolt) on web of beam $= 435 \times 20 \times 7{\cdot}6 \times 10^{-3} = 66{\cdot}1$ kN	
	Try 4 bolts at 75 mm vertical pitch, 50 mm from heel of cleats	
	Horizontal shear force on bolt due to moment due to eccentricity $= \dfrac{145 \times 50 \times 112{\cdot}5}{2(37{\cdot}5^2 + 112{\cdot}5^2)} = 29{\cdot}0$ kN	
	Vertical shear force per bolt $= \dfrac{145}{4} = 36{\cdot}3$ kN	
	Resultant shear force $= \sqrt{29{\cdot}0^2 + 36{\cdot}3^2}$ $= 46{\cdot}4$ kN $<$ bolt capacity O.K	4/M20 bolts Grade 4·6
	<u>Connection to column flange.</u> For M20 Grade 4·6 bolts	
6·3·2	Shear capacity of bolt in single shear $= 160 \times 245 \times 10^{-3} = 39{\cdot}2$ kN	
6·3·3	Bearing capacity (of bolt) on 8 mm cleat $= 435 \times 20 \times 8 \times 10^{-3} = 69{\cdot}6$ kN	
	Try 6 bolts, 2 rows at 75 mm vertical pitch and 120 mm c/c.	
	Assuming centre of pressure 25 mm below top of cleat, horizontal shear force on bottom bolt due to moment due to eccentricity $= \dfrac{145 \times 0{\cdot}5(120 - 7{\cdot}6) \times 200}{2(50^2 + 125^2 + 200^2)} = 14{\cdot}0$ kN	
	Vertical shear force per bolt $= \dfrac{145}{6} = 24{\cdot}2$ kN	

Commentary: 2/3

Resultant shear force = $\sqrt{(V_m^2 + V_x^2)}$

It can be seen by inspection that the cleats and web of the beam are adequate in shear and bending.

The bolts in the column connection are staggered relative to the bolts in the web connection to give better access for tightening.

The edge and end distances should be checked (BS 5950: Part 1, Clauses 6.2 and 6.3.3.3).

Structural Steelwork Connections	Subject Beam to column connection – Bolted web cleats			Chapter Ref. 12
	Design Code BS 5950 Part 1			Calc. Sheet No. Example 2/3
	Calc. by B.D.C	Date Aug, '87	Check by G.W.O.	Date Nov, '87

Code Ref.	Calculations	Output
	Resultant shear force $= \sqrt{14 \cdot 0^2 + 24 \cdot 2^2}$	
	$= 28 \cdot 0$ kN	
	< bolt capacity o.k.	6/M20 bolts Grade 4.6
	Use 2/90×90×8 angle cleats × 300 mm	2/90×90×8 angle cleats × 300 mm

Commentary: 3/1

Section 12.2.4

A single-sided web cleat connection may be used where it is not possible to fix a pair of cleats. Otherwise, a double web cleat connection is to be preferred.

It is necessary to use friction grip fasteners for the cleat-to-column connection and to include the eccentricity of the beam reaction in their design. If ordinary bolts are used, the combined effect of the eccentric load and the clearances in the bolt holes could lead to an unacceptable twist at the end of the beam. Note that the slip resistance calculation in BS 5950: Part 1, Clause 6.4.2.1 is only a serviceability check (6.4.1), and if the top flange of the beam is not laterally restrained (for example, by a concrete floor slab) it would be prudent to keep at least 20% in hand when designing the bolts to provide a margin against slip at working load.

Structural Steelwork Connections	Subject: Beam to column connection – single bolted web cleat.			Chapter Ref. 12
	Design Code: BS 5950 Part 1.			Calc. Sheet No. Example 3/1
	Calc. by B.D.C	Date Aug, '87	Check by G.W.O.	Date Nov, '87.

Code Ref.	Calculations	Output

Beam to column connection – single bolted web cleat.

Design a bolted connection for a 406 × 178 × 54 UB in Grade 43 steel, using a single web cleat, to carry a reaction of 145 kN (due to factored loads). The connection is to the flange of a 203 × 203 × 52 UC in Grade 43 steel.

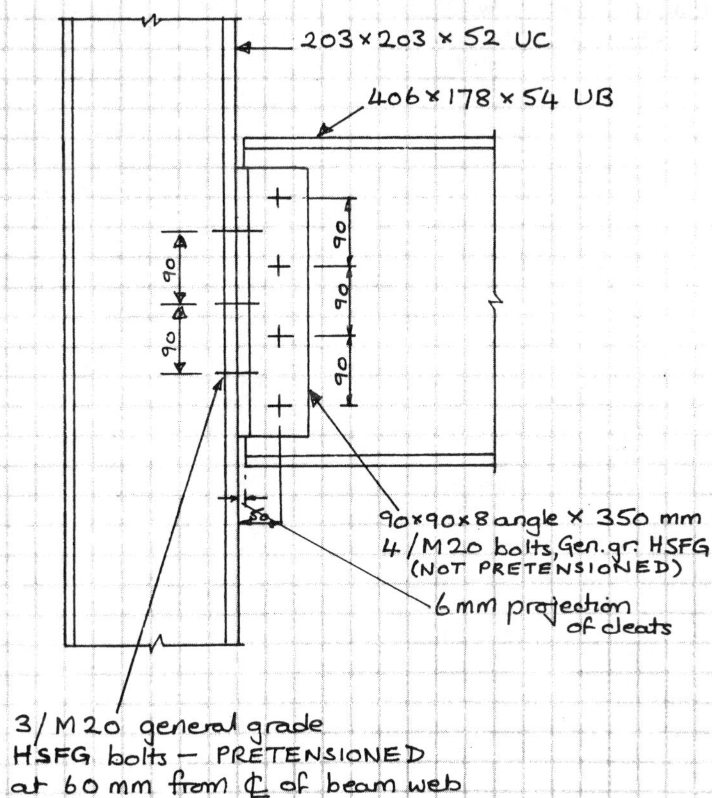

Commentary: 3/2

Shear capacity of bolt in single shear = p_sA_t

$p_s = 0.48\ U_f$ but $\leqslant 0.69\ Y_f$

Bearing capacity of connected ply = $p_{bs}dt$

Section 8.2.2

Horizontal shear force on bolt due to moment due to eccentricity

$$= V_m = \frac{P_x e_x r_i}{\sum_n r_i^2}$$

Vertical shear force per bolt = $V_x = \dfrac{P_x}{n}$

Resultant shear force = $\sqrt{(V_m^2 + V_x^2)}$

To avoid having to preload the bolts in the erected steelwork it is assumed that the cleat would be fixed to the column flange (with the preloaded bolts) before erection, and that the site connection would be made to the beam web (with no preload requirement).

Slip resistance = $P_{sL} = 1.1\ K_s \mu P_0$

$K_s = 1.0$ for clearance holes

Bearing resistance = $P_{bg} = dtp_{bg} \leqslant 1/3\ etp_{bg}$

In this example, end distance (e) > $3d$. Therefore dtp_{bg} governs for a vertical load. However, the later calculation shows that the resultant load is at 45° (e = 47.8 mm). End distance then governs bearing resistance (105.2 kN): but slip resistance still governs the bolt capacity.

Section 8.2.2

Horizontal shear force on bolt due to moment due to eccentricity

$$= V_m = \frac{P_x e_x r_i}{\sum_n r_1^2}$$

Structural Steelwork Connections	Subject: Beam to column connection – single bolted web cleat		Chapter Ref. 12
	Design Code: BS 5950 Part 1		Calc. Sheet No. Example 3/2
	Calc. by: B.D.C	Date: Aug, '87	Check by: G.b.O. — Date: Nov, '87.

Code Ref.	Calculations	Output
	<u>Connection to web of beam.</u>	
	For M20 General Grade HSFG bolts (BS 4395 part 1) used as ordinary bolts (i.e. not pretensioned).	
6·3·2	Shear capacity of bolt in single shear $= 0{\cdot}48 \times 827 \times 245 \times 10^{-3} = 97{\cdot}3$ kN	
6·3·3·3	Bearing capacity on web of beam $= 460 \times 20 \times 7{\cdot}6 \times 10^{-3} = 69{\cdot}9$ kN	
	Try 4 bolts at 90 mm vertical pitch, 50 mm from heel of cleat.	
	Horizontal shear force on bolt due to moment due to eccentricity $= \dfrac{145 \times 50 \times 135}{2(45^2 + 135^2)} = 24{\cdot}2$ kN	
	Vertical shear force per bolt $= \dfrac{145}{4} = 36{\cdot}3$ kN	
	Resultant shear force $= \sqrt{24{\cdot}2^2 + 36{\cdot}3^2}$ $= 43{\cdot}6$ kN $<$ bolt capacity O.K	4/M20 General Grade HSFG bolts (not pretensioned)
	<u>Connection to column flange</u>	
	Try M20 General Grade HSFG bolts (BS 4395, part 1) used as HSFG bolts.	
6·4·2·1	Slip resistance (per bolt) $= 1{\cdot}1 \times 0{\cdot}45 \times 144 = 71{\cdot}3$ kN	
6·4·2·2	Bearing resistance of 8 mm cleat (per bolt) $= 20 \times 8 \times 825 \times 10^{-3} = 132{\cdot}0$ kN	
	Try 3 bolts at 90 mm vertical pitch, 60 mm from centre line of beam web	
	Horizontal shear force on bolt due to moment due to eccentricity $= \dfrac{145 \times 60 \times 90}{2 \times 90^2} = 48{\cdot}3$ kN	

Commentary: 3/3

Vertical shear force per bolt $= V_x = \dfrac{P_x}{n}$

Resultant shear force $= \sqrt{(V_m^2 + V_x^2)}$

Any out-of-plane bending of the cleat required to accommodate the end slope of the beam is ignored, on the assumption that it does not significantly reduce the plastic failure load of the cleat. The important point is that the out-of-plane bending is due to an applied deflection (as distinct from an applied load).

Bending moment = Vertical load × eccentricity

The effect of the eccentricity ((7.6 + 8)/2 mm) of the beam web from the leg of the angle is not checked in the example. To limit the effect of this eccentricity, the following guidance has been observed:

Depth of cleat > 0.75 × depth of beam

Thickness of cleat $> \dfrac{5 \times (\text{end reaction})}{p_y \times (\text{depth of cleat})}$

(*Note*: The cleat must not be too thick or it may be too stiff to accommodate the end slope of the beam (see commentary on Example 2/1.))

Structural Steelwork Connections	Subject Beam to column connection – single bolted web cleat	Chapter Ref. 12
	Design Code BS 5950 Part 1	Calc. Sheet No. Example 3/3
	Calc. by B.D.C · Date Aug, '87 · Check by G.W.O.	Date Nov, '87.

Code Ref.	Calculations	Output
	Vertical shear force per bolt $= \frac{145}{3} = 48.3$ kN	
	Resultant shear force $= \sqrt{48.3^2 + 48.3^2}$	
	$= 68.3$ kN	3/M20 General Grade HSFG bolts
	< bolt capacity o.k	
	<u>Cleat angle.</u>	
	Try 90×90×8 angle × 350 mm	
	Check bending at bolt line of connection to column flange	
	Bending moment $= 145 \times 60 = 8700$ Nm	
4·2·5	Moment capacity $= p_y Z$	
	$= 275 \times \frac{8 \times 350^2}{6} \times 10^{-3}$	
	$= 44917$ Nm	90×90×8 angle cleat × 350
	> bending moment	
	o.k. Use	

Commentary: 4/1

Section 12.2.5

The strength calculations for a welded end-plate connection are relatively straightforward. It is perhaps more difficult to ensure that the connection has sufficient flexibility/ductility to accommodate the end rotation of the beam. The flexibility is achieved by limiting the thickness of the plate ($t \ngtr d/3$) and by positioning the bolts not too close to the web and flange of the beam.

Structural Steelwork Connections	Subject Beam to column connection — welded end plate			Chapter Ref. 12
	Design Code BS 5950 Part 1			Calc. Sheet No. Example 4/1
	Calc. by B.D.C	Date Aug, '87.	Check by G.W.O.	Date Nov, '87

Code Ref.	Calculations	Output
	Beam to column connection — welded end plate. Design a welded end plate connection for a 406 × 178 × 54 UB in Grade 43 steel to carry a reaction of 145 kN (due to factored loads). The connection is to the flange of a 203 × 203 × 52 UC in Grade 43 steel.	

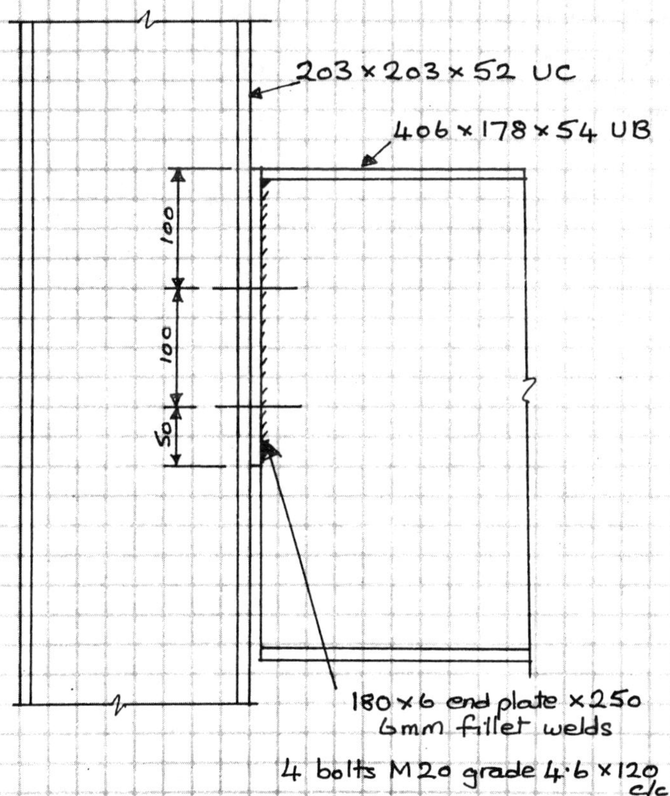

Commentary: 4/2

Section 12.2.5

Due to their lack of ductility the fillet welds must not be a 'weak link'. Generally, this would be avoided if the sum of the throat sizes of the web fillet welds is made greater than $1.2t$ for Grade 43 steel and $1.4t$ for Grade 50 (to ensure that the welds are stronger than the yield strength of the web of the beam). In this example the fillet weld size is less than this value; but the nominal capacity of the end plate in bending is so much less than the capacity of the welds that the welds cannot be considered to be the weakest link. Thus

$$\frac{Fb}{2} = \frac{p_y w T^2}{4}$$

$$\text{Nominal capacity of plate} = \frac{F}{w}$$

$$= \frac{2 \times 275 \times 6^2}{50.2 \times 4 \times 10^3} = 0.10\,\text{kN/mm}$$

which is much less than weld capacity (0.90 kN/mm)

Shear capacity of bolt in single shear = $p_s A_t$
Bearing capacity of bolt = $p_{bb} dt$
Capacity of 6 mm fillet weld = $0.7 s p_w$
= $0.7 \times 6 \times 215 \times 10^{-3}$
= 0.90 kN/mm

The stresses in the end plate are satisfactory by inspection.

Structural Steelwork Connections	Subject: Beam to column connection — welded end plate			Chapter Ref. 12
	Design Code: BS 5950 Part 1			Calc. Sheet No. Example 4/2
	Calc. by: B.D.C	Date: Aug, '87	Check by: G.W.O.	Date: Nov, '87.

Code Ref.	Calculations	Output
	<u>Bolted connection to column.</u>	
	For M 20 Grade 4·6 bolts and a 6 mm thick end plate.	
6·3·2	Shear capacity of bolt in single shear $= 160 \times 245 \times 10^{-3} = 39{\cdot}2$ kN	
6·3·3	Bearing capacity (of bolt) on 6 mm end plate $= 435 \times 20 \times 6 \times 10^{-3} = 52{\cdot}2$ kN	
	Minimum number of bolts $= \dfrac{145}{39{\cdot}2} = 3{\cdot}7$	4/M20 bolts Grade 4·6
	Use 4 bolts.	
	<u>End plate</u>	
	Use 180 mm × 6 mm plate × 250 mm deep	180×6 end plate × 250
	Length of fillet welds connecting end plate to beam web $= 250 - 10{\cdot}9 - 8$ $= 231$ mm each side	
6·6·5·5	Shear on weld $= \dfrac{145}{2 \times 231} = 0{\cdot}31$ kN/mm	
	Use 6mm fillet welds	6 mm fillet welds.
	Shear stress on web of beam $= \dfrac{145 \times 10^3}{231 \times 7{\cdot}6} = 83$ N/mm^2	
4·2·3	$\leq 0{\cdot}6\,p_y$ O.K	

Commentary: 5/1

The problem with the design of this type of connection is that the 'bending' flexibility, which is present in properly proportioned angle cleat and end-plate connections, is not available. Consequently, the movement required to accommodate the end rotation of a simply supported beam has to be achieved in some other way. In this example it is assumed that the movement is supplied by the hole clearances and where necessary by elongation of the bolt holes. To achieve this:

1. Ordinary bolts are used (not friction grip bolts);
2. The bolts are designed to fail by bearing of the connected plies and not by shear of the bolt;
3. Edge distances are made $\geqslant 2 \times$ bolt diameter;
4. The resistance of the welds attaching the cleat to the support is made greater than the moment that can be applied by the bolts. The minimum weld size relative to the web thickness to achieve ductility is also considered.

Section 12.2.5

The procedure should be suitable for beams of up to 10 m span for Grade 43 steel and 8 m span for Grade 50.

Grade 8.8 bolts are used so that failure would be in bearing and not in shear.
Shear capacity of bolt $= p_s A_t$
Bearing capacity of connected ply $= p_{bs} dt$

Edge distances are $\geqslant 2d$, so that end failure does not govern.

The beam is stopped 20 mm short of the column to give ample clearance to the fillet weld.

Structural Steelwork Connections	Subject: Beam to column connection — welded web cleat			Chapter Ref.: 12
	Design Code: BS 5950 Part 1			Calc. Sheet No.: Example 5/1
	Calc. by: B.D.C	Date: Aug, '87.	Check by: G.L.O.	Date: Nov, '87

Code Ref.	Calculations	Output

Beam to column connection — welded web cleat.

Design a web cleat connection (welded to the column and site bolted to the beam) for a 406 × 178 × 54 UB in Grade 43 steel to carry a reaction of 145 kN (due to factored loads). The connection is to the flange of a 203 × 203 × 52 UC in Grade 43 steel.

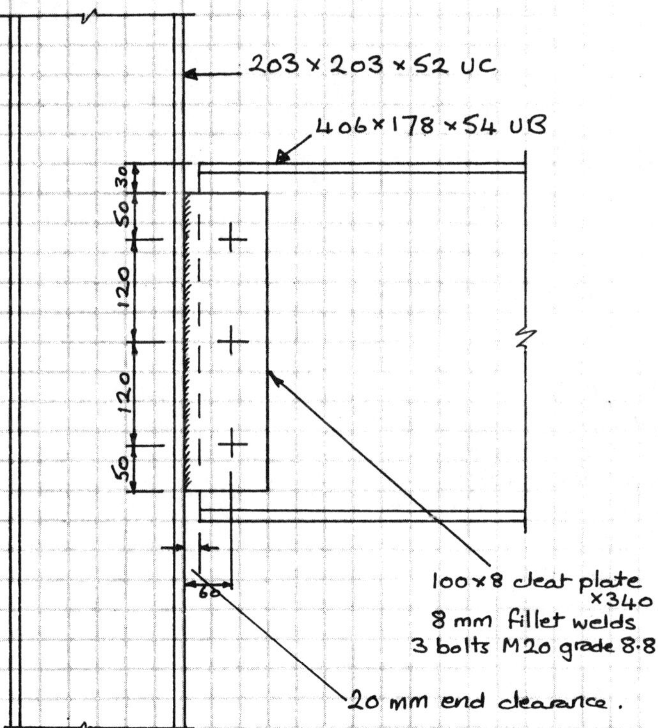

Connection to web of beam.

Try M20 Grade 8·8 bolts and a 8 mm thick cleat plate.

6·3·2 Shear capacity of bolt in single shear

$= 375 \times 245 \times 10^{-3} = 91{\cdot}9$ kN

Commentary: 5/2

The effect of the lateral eccentricity of the beam web from the cleat is not checked. To limit the effect of this eccentricity it is suggested that the following guidance is observed:

1. Depth of cleat $\nless$ 0.75 × depth of beam
2. Thickness of cleat $\nless \dfrac{5 \times (\text{end reaction})}{p_{yx}\ (\text{depth of cleat})}$

(*Note*: There is no need to check the shear capacity of the cleat since this is covered by 2.)

The moment that could be applied by the bolts if they go into bearing is checked. In particular, the welds attaching the cleat to the column are checked for this moment. (Fillet welds loaded transversely have limited ductility, and it is important that they do not become a 'weak link'.)

Section 8.2.2

Moment that could be developed by the bolts

$$= \text{capacity of a bolt} \times \frac{\sum_{n} r_i^2}{r_i}$$

The moment that could be developed by the bolts is greater than the eccentricity moment. Therefore, it is used for the design of the cleat plate and welds.

Moment capacity of plate $= p_y Z$

The leg length is deducted from each end of the weld to allow for 'end craters'. It is not considered good practice to return the fillet welds around the end of the cleat, as it may cause a severe notch in the cleat where it is most highly stressed.

$$\text{Horizontal shear} = \frac{\text{Moment}}{Z \text{ of weld group}}$$

$$\text{Horizontal shear} = \frac{\text{Reaction}}{\text{Total weld length}}$$

Section 4.4.4

Eight millimetre fillet welds are specified to ensure that the welds are not likely to be a 'weak link'.

Σ (throat size) $\nless$ 1.2 × thickness of plate (for Grade 43 steel)

or $\nless$ 1.4 × thickness of plate (for Grade 50 steel)

Structural Steelwork Connections	Subject: Beam to column connection — welded web cleat	Chapter Ref. 12
	Design Code: BS 5950 Part 1.	Calc. Sheet No. Example 5/2
	Calc. by B.D.C · Date Aug, '87. · Check by G.W.O.	Date Nov '87

Code Ref.	Calculations	Output
6.3.3.3	Bearing capacity of connected ply $= 460 \times 20 \times 7.6 \times 10^{-3} = 69.9$ kN	
	Minimum number of bolts $= \frac{145}{69.9} = 2.1$	3/M20 bolts
	Use 3 bolts	Grade 8.8
	<u>Cleat plate</u>	
	Try 100 × 8 plate × 340 mm, with bolt line 60 mm from face of column	
	<u>Check</u> that proposed cleat is acceptable for eccentricity effects due to offset of beam web.	
	Minimum thickness $= \frac{5 \times \text{Reaction}}{p_y \times \text{depth of cleat}}$	
	$= \frac{5 \times 145 \times 10^3}{275 \times 350} = 7.5$ mm	
	< 8 mm. Thickness O.k	
	Moment at face of column due to eccentricity of load $= 145 \times 60 = 8700$ Nm	
	Moment that could be developed by bolts (at their bearing capacity) $= 2 \times 69.9 \times 120$ $= 16776$ Nm	
	Design for moment of 16776 Nm.	
4.2.5	Moment capacity of plate $= 275 \times \frac{8 \times 340^2}{6} \times 10^{-3}$	100 × 8 plate
	$= 42387$ Nm > 16776 Nm. O.k.	× 340 mm
6.6.5.2	Effective length of welds $= 340 - 2 \times 6 = 328$ mm	
	Horizontal shear at end of weld due to moment $= \frac{16776 \times 6}{2 \times 328^2} = 0.468$ kN/mm	
	Vertical shear $= \frac{145}{2 \times 328} = 0.221$ kN/mm	
	Resultant shear $= \sqrt{0.468^2 + 0.221^2} = 0.52$ kN/mm	
	Minimum weld size for "ductility"	
	$= \frac{0.6t}{0.7} = \frac{0.6 \times 8}{0.7} = 6.9$ mm	
	Use 8 mm fillet welds	8 mm fillet welds

Commentary: 6/1

Load paths

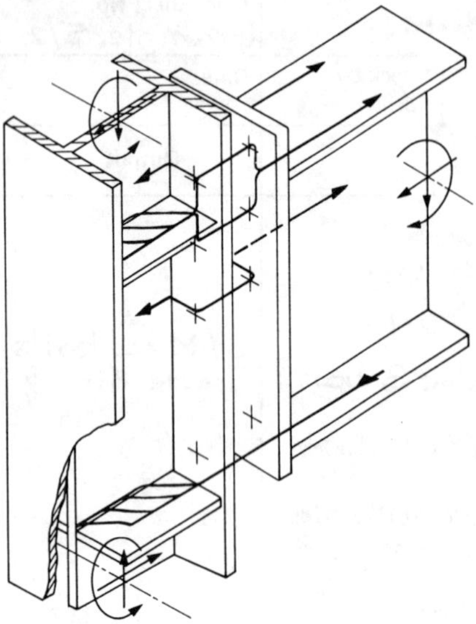

For clarity only the nearside load paths are shown. The moment in the beam is carried by tension in the top flange and upper portion of the web, bending in the end plate; tension in the bolts; bending in the column flange and tension/shear in the column web and stiffeners, together with compression in the bottom flange and compression/shear in the stiffeners and column web.

Section 12.4.1

The example illustrates the design of a bolted beam-to-column connection, where the end plate is extended above the top flange of the beam.

The bolts have been set out with a nominal minimum distance of 60 mm from the flanges and web of the beam (actually, 59.7 mm from the web). This should provide sufficient clearance for the use of power tools to tighten the bolts (at least up to and including M24 HSFG or Grade 8.8 bolts).

Section 8.3

The connection is assumed to 'pivot' about the 'hard spot' at the bottom flange and the loads in the bolts are assumed to be proportional to their distance from the centre of the bottom flange. However, some account is taken of the greater flexibility of the cantilever end plate supporting bolts F_1 compared with the portion of plate supporting bolts F_2, which is stiffened by the beam web, by assuming that the loads in bolts F_1 and F_2 are equal.

Structural Steelwork Connections	Subject: Beam to column connection — Bolted end plate			Chapter Ref. 12
	Design Code: BS 5950 Part 1			Calc. Sheet No. Example 6/1
	Calc. by B.D.C	Date Aug, '87	Check by G.W.O.	Date Nov, '87

Code Ref.	Calculations	Output

Beam to column connection — Bolted end plate.

Design a connection between a 610 × 229 × 101 UB beam in Grade 43 steel and a 305 × 305 × 158 UC column in Grade 43 steel.

The connection is to carry the following factored loads:

Bending moment	423	kNm
Axial load in beam	22	kN (tension)
Shear in beam	254	kN
Shear in column (above connection)	26	kN

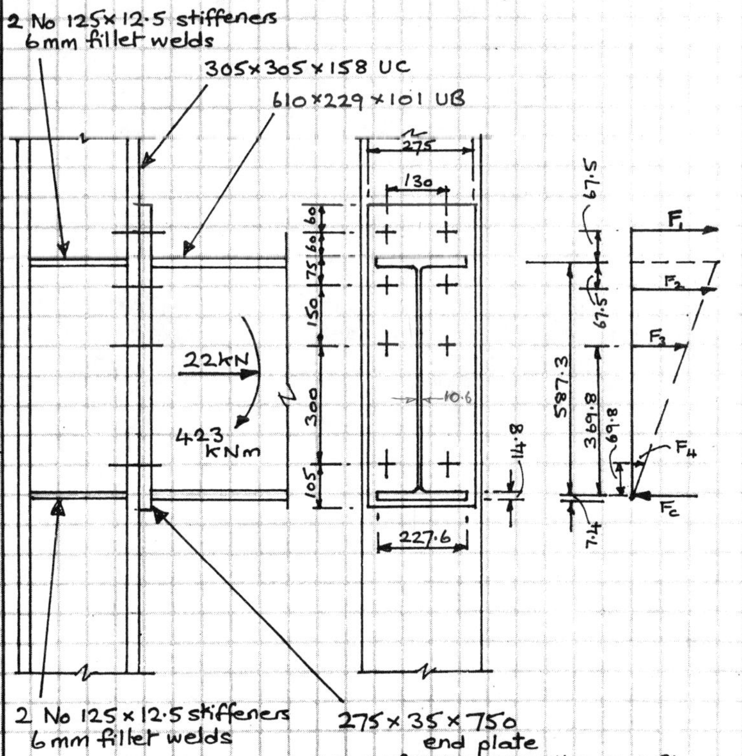

10 mm fillet welds all round flanges
6 mm fillet welds to web.
8 No. M24 General grade HSFG bolts.

Commentary: 6/2

If the end plate were finished at the top flange (i.e. all the bolts were within the depth of the beam), the compressive reaction F_c and the maximum bolt load F_1 would, in this example, both increase by about 20%. As a consequence, F_c would be greater than the flange capacity, a larger size of bolt would be required and the web of the beam would be overstressed by about 40%.

b is measured to the toe of a fillet weld but to the half radius for the root fillet of a rolled section.

Structural Steelwork Connections	Subject Beam to column connection — Bolted end plate			Chapter Ref. 12
	Design Code BS 5950 Part 1			Calc. Sheet No. Example 6/2
	Calc. by B.D.C	Date Aug, '87	Check by G.W.O.	Date Nov, '87

Code Ref.	Calculations	Output

Distribution of load.

Try bolt spacing in sketch.
By moments about bottom flange

$$423 \times 10^3 + 22 \times 293.7$$
$$= (2F_1 + 2F_2) \times 587.3 + 2F_3 \times 369.8 + 2F_4 \times 69.8$$
$$= \frac{2F_1}{587.3}\left[2 \times 587.3^2 + 369.8^2 + 69.8^2\right]$$

$F_1 = F_2 = 151.7$ kN

$F_3 = \frac{369.8}{587.3} \times F_1 = 95.5$ kN

$F_4 = \frac{69.8}{587.3} \times F_1 = 18.0$ kN

Reaction at bottom flange $= F_c$
$= 2(151.7 + 151.7 + 95.5 + 18.0) - 22$
$= 811.8$ kN

Beam flanges.

Capacity of flange $= p_y A$
$= 275 \times 227.6 \times 14.8 \times 10^{-3}$
$= 926.3$ kN

$> F_c$. Therefore bottom flange O.K
Top flange O.K. by inspection.

End plate and bolts.

Consider portion of end plate above top flange (assume 10 mm fillet welds to flange and 275 mm wide end plate)

Distance from centre line of bolt to toe of fillet weld $= b = 60 - 10 = 50$ mm

Commentary: 6/3

Section 7.7.2

$a + 1.4b = 60 + 1.4 \times 50 = 130\,\text{mm} > \frac{1}{2} \times$ pitch of 130 mm
and > edge distance of 72.5 mm

therefore $\frac{1}{2} \times 130 + 72.5 = 137.5\,\text{mm}$ governs w

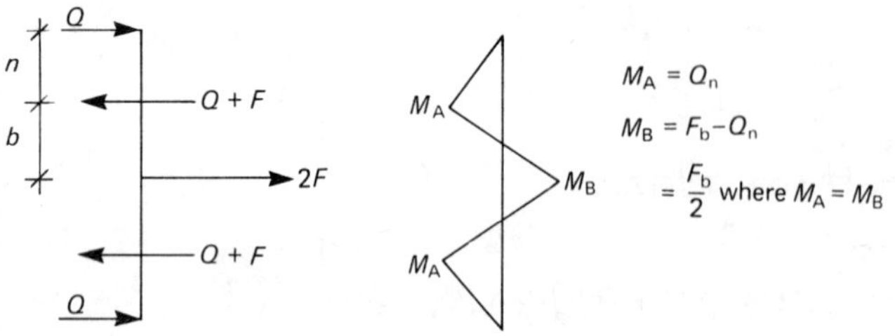

Section 5.3.2

Although the bolts have been set out to enable power tools to be used to preload the bolts, the bolts have been classified as 'non-preloaded' for the purposes of design (i.e. $\beta = 2$ is used in the formulae for n and Q). To justify this, it has been assumed that the joint is part of a structure where it is unlikely that more than a small number of joints would be overloaded and that if the overloading did occur the loss of stiffness would be acceptable. Taking $\beta = 2$ instead of $\beta = 1$ (as a preloaded bolt) is 'beneficial' as far as the design is concerned, in that it leads to a lower prying force (Q).

Where the connection is subject to dynamic loading or fatigue, or where shear is to be taken by friction, the bolts should be preloaded and $\beta = 1$ should be used in the formulae. The bolts are designed for the bolt load, including the prying force, regardless of whether the bolts are designed as ordinary bolts or 'friction grip fasteners'. The procedure in BS 5950: Part 1, Clause 6.3.6, where prying action is ignored, is not used as it could lead to an unsafe connection, when used in conjunction with double-curvature bending of the end plate.

The tension capacity for design using factored loads is taken as

$$\frac{Y_f A_t}{1.1\gamma_m} \text{ but } \not> \frac{0.7 U_f A_t}{1.1\gamma_m}$$

The above formula for tension capacity (from BS 5400: Part 3: 1982, clause 14.5.3.2) is adopted so that the design procedure is in general compliance with current codes. For bridges $\gamma_m = 1.2$, but for building structures may be taken as 1.0. For general grade HSFG bolts, $0.7\, U_f A_t/1.1$ is virtually the same as $0.9\, P_o$, the tension capacity in BS 5950: Part 1, clause 6.4.4.2.

Since the prying force is greater than 10% of the applied load (F), there is no need to check the bolts for compliance with Clause 6.3.6.

Structural Steelwork Connections	Subject Beam to column connection – Bolted end plate			Chapter Ref. 12
	Design Code BS 5950 Part 1			Calc. Sheet No. Example 6/3
	Calc. by B.D.C	Date Aug, '87	Check by y.W.O.	Date Nov, '87

Code Ref.	Calculations	Output

Effective length of end plate per bolt

$$= W = \frac{275}{2} = 137.5 \text{ mm}$$

Using "minimum thickness" design,

$$M = \frac{F_1 b}{2} = \frac{151.7 \times 50}{2} = 3793 \text{ Nm}$$

$$\text{Moment capacity} = \frac{p_y}{1.15} \times \frac{WT^2}{4}$$

$$T = \sqrt{\frac{1.15 \times 4 \times 3793 \times 10^3}{265 \times 137.5}} = 21.9 \text{ mm}$$

Try 25 mm.

Check prying force.

Classify bolts as "non-preloaded"

Distance from centre line of bolt to prying force $= n =$ least of edge distance

$$\text{or } 1.1T\sqrt{\frac{\beta P_o}{p_y}}$$

$$= 1.1 \times 25\sqrt{\frac{2 \times 512}{265}}$$

$$= 54.1 \text{ mm}$$

$$\text{Prying force} = \frac{M}{n} = \frac{3793}{54.1} = 70.1 \text{ kN}$$

Bolt load $= 151.7 + 70.1 = 221.8$ kN

Tension capacity of M30 General Grade HSFG bolt

$$= \frac{0.7 U_f A_t}{1.1}$$

$$= \frac{0.7 \times 725 \times 561 \times 10^{-3}}{1.1}$$

$$= 258.8 \text{ kN}$$

M30 bolt is required for "minimum thickness" design of end plate.

Commentary: 6/4

The redesign, using M24 instead of M30 bolts, is used to illustrate how the bolt load can be reduced by using a thicker end plate. If the M30 bolts had been retained it may have been necessary to increase the bolt distance from the flange (and web) to 70 mm to provide clearance for larger power tools.

Section 5.3.2

The formulae for the maximum distance from the bolt to the prying force (equation (5.21)):

$$n = 1.1T\sqrt{\frac{\beta p_0}{p_y}}$$

and for the minimum prying force

$$Q = \frac{b}{2n}\left[F - \frac{\beta \gamma p_0 w T^4}{27nb^2}\right]$$

are aimed at limiting the strain of the bolts to avoid loosening or breakage.

In the formula for n, p_y = design strength for the plate material. It is worth noting that a lower value can be used in the formula provided that the same value is used for the design of the plate. This means that if the plate is 'overstrong' p_y could be reduced; n would be increased; the prying force (Q) would be reduced; and the design load for the bolt would be reduced.

Structural Steelwork Connections	Subject: Beam to column connection — Bolted end plate.			Chapter Ref. 12
	Design Code: BS 5950 Part 1			Calc. Sheet No. Example 6/4
	Calc. by B.D.C	Date Aug, '87.	Check by G.W.O.	Date Nov, '87

Code Ref.	Calculations	Output
	It is desired to standardize the use of M24 General Grade HSFG bolts throughout the structure.	
	Tension capacity $= \dfrac{0{\cdot}7 \times 827 \times 353 \times 10^{-3}}{1{\cdot}1}$ $= 185{\cdot}8$	
	Maximum allowable prying force $= 185{\cdot}8 - 151{\cdot}7 = 34{\cdot}1$ kN	Use M24 General Grade HSFG bolts.
	Try 35 mm thick plate; $n = 60$ mm (edge distance governs)	
	Moment at toe of weld $= Fb - Qn$ $= 151{\cdot}7 \times 50 - 34{\cdot}1 \times 60$ $= 7585 - 2046$ $= 5539$ Nm	

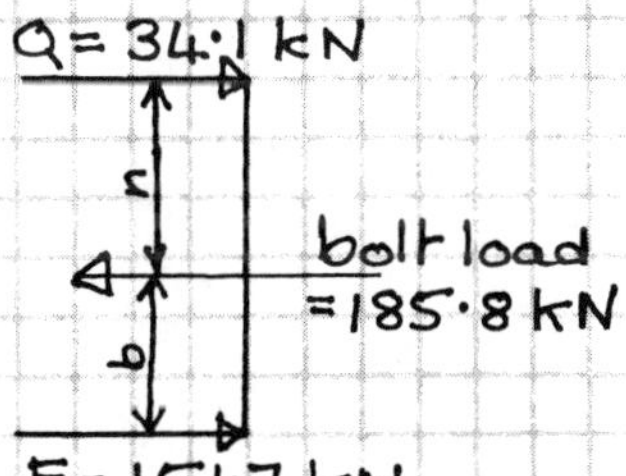

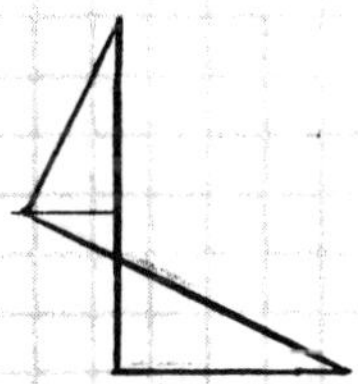

B.M. diagram

	Moment capacity $= \dfrac{265 \times 137.5 \times 35^2 \times 10^{-3}}{1{\cdot}15 \times 4}$ $= 9703$ Nm > 5539 Nm O.K.	
	Check using formula for prying force (Q)	
	$\text{Min}^m\, Q = \dfrac{b}{2n}\left[F - \dfrac{\beta \gamma p_o w T^4}{27 n b^2}\right]$	

Commentary: 6/5

Section 8.3

When checking the combined shear and tension the ratio of applied tension to tension capacity has been taken as 1.0, as assumed in the calculations. In fact, the minimum prying force of 25.9 kN could have been used in checking the plate thickness, and the ratio could have been reduced to (151.7 + 25.9)/185.8 = 0.96 if necessary.

The effective length of the weld is the flange width. There is no deduction for end craters since the weld is returned around the edges of the flange. The same weld size is used on the underside of the flange, although some of the bolt load is carried by the web.

Structural Steelwork Connections	Subject: Beam to column connection – Bolted end plate			Chapter Ref. 12
	Design Code: BS 5950 Part 1			Calc. Sheet No. Example 6/5
	Calc. by B.D.C	Date Aug, '87	Check by G.W.O.	Date Nov, '87

Code Ref.	Calculations	Output
	ie $Min^m Q$	
	$= \frac{50}{2\times 60}\left[151\cdot7 - \frac{2\times1\cdot5\times0\cdot587\times137\cdot5\times35^4}{27\times60\times50^2}\right]$	
	$= 0\cdot417\,[151\cdot7 - 89\cdot7]$	
	$= 25\cdot9$ kN	
	$< 34\cdot1$ kN (= prying force used in calculation) O.k	
	By inspection, bolts and end plate are adequate at other bolt positions.	275 × 35 end plate × 750
6·3·2 Table 32	Shear capacity of M24 General Grade HSFG bolt as ordinary bolt $= P_s$	
	$= 0\cdot48\, U_f A_t$	
	$= 0\cdot48 \times 827 \times 353 \times 10^{-3}$	
	$= 140\cdot1$ kN	
	Applied shear per bolt $= F_s = \frac{254}{8} = 31\cdot8$ kN	
6·3·6·3	$\frac{F_s}{P_s} + \frac{F_t}{P_t} = \frac{31\cdot8}{140\cdot1} + 1\cdot0 = 1\cdot23 < 1\cdot4$ O.k.	
	<u>Welds</u>	
	<u>Top flange to end plate</u>	
	Load from bolts above flange	
	$= 2F_1 = 2\times151\cdot7 = 303\cdot4$ kN	
	Load per mm of weld $= \frac{303\cdot4}{227\cdot6} = 1\cdot333$ kN	
6·6·5	Capacity of 10mm fillet web	
	$= 10\times0\cdot7\times215\times10^{-3}$	
	$= 1\cdot505$ kN/mm	
	O.k. Use	10mm fillet welds to flanges.

Commentary: 6/6

Section 7.7.4

The factor of 0.7 used in determining the effective lengths of the plates allows for the formation of fan yield lines at the corners. For the design of the welds, which have limited ductility, the greater of the loads from a plastic distribution (where the bolt load is distributed in proportion to w/b) and an elastic distribution (where the bolt load is distributed in proportion to w/b^3) is taken. In this case, where the dimensions associated with the flange and the web are nearly the same, a short cut is taken. (The full procedure discussed above is illustrated in the design of the column stiffener welds.)

The load is assumed to be applied to the weld and web on the line of the bolt. The effective length of weld and web is taken as the lesser of $2 \times (\frac{1}{2}$ bolt pitch) or $2b_x$. Taking only the length of weld that is equally balanced about the load as being effective is a quick and economic way of designing line welds with an eccentrically applied load.

(See notes on commentary to Example 1/7 of Chapter 14 for additional limitation of effective length of weld $\not> 2\ (b_y + d)$.)

Structural Steelwork Connections	Subject: Beam to column connection – Bolted end plate			Chapter Ref. 12
	Design Code: BS 5950 Part I			Calc. Sheet No. Example 6/6
	Calc. by B.D.C	Date Aug, '87	Check by ...	Date Nov. '87

Code Ref.	Calculations	Output
	<u>Web to end plate.</u> Consider bolt below flange Load = F_2 = 151·7 kN	
	Centre line of bolt to edge of weld to beam flange = b_x = 75 − 14·8 − 10 = 50·2 mm	
	Centre line of bolt to edge of weld to beam web = $b_y = \frac{130}{2} - \frac{10{\cdot}6}{2} - 6$ = 53·7 mm (Assuming 6 mm fillet welds)	
	Effective length of plate at flange $= W_x = \frac{275 - 130}{2} + 0{\cdot}7 \times 53{\cdot}7 = 110{\cdot}1$ mm (Plate) (Bolt)	
	Effective length of plate at web $= W_Y = \frac{150}{2} + 0{\cdot}7 \times 50{\cdot}2 = 110{\cdot}1$ mm	
	<u>Note</u> $W_x = W_Y$ and b_y is slightly greater than b_x. Therefore slightly less than half of bolt load will be supported from web. Check for $\frac{1}{2}$ × bolt load.	
	Effective length of weld = 2 × 50·2 = 100·4 mm	
6·6·5·2	Load per mm of weld $= \frac{0{\cdot}5 \times 151{\cdot}7}{100{\cdot}4} = 0{\cdot}755$	
6·6·5	Capacity of 6 mm fillet weld $= 6 \times 0{\cdot}7 \times 215 \times 10^{-3}$ = 0·903 kN/mm O.k. Use	6 mm fillet welds to web
	<u>Web of beam</u> Load per mm of web $\simeq \frac{2 \times 0{\cdot}5 \times 151{\cdot}7}{100{\cdot}4}$ = 1·51 kN	
	Tension capacity of web = $p_y t$ $= 275 \times 10{\cdot}6 \times 10^{-3} = 2{\cdot}91$ kN/mm O.k	

Commentary: 6/7

Section 5.3.2

The formula for n which is used to ensure that the bolts do not work loose governs the position of the prying force in this example.

Section 7.7.2

$2(a + 1.4b) = 2(1.7 \times 49.6 + 1.4 \times 49.6) =$ 307.5 mm > pitch of 150 mm. Therefore, pitch of bolts governs in determining w. (*Note*: Limit of $a \ngtr 1.7b$ applied in formula.)

Moment capacity = $P_t/1.15 \times$ plastic modulus.

If M30 or M27 general grade HSFG bolts had been used column stiffeners would not be required at the level of the tension flange of the beam. Retaining the M24 bolts in the example allows the design of tension stiffeners and the effect of longitudinal flange stresses to be illustrated.

Structural Steelwork Connections	Subject: Beam to column connection — Bolted end plate			Chapter Ref. 12
	Design Code: BS 5950 Part1			Calc. Sheet No. Example 6/7
	Calc. by B.D.C	Date Aug, '87	Check by G.W.O.	Date Nov, '87

Code Ref.	Calculations	Output

Column flange.

Distance from centre line of bolt to centre of root fillet = b

$= \frac{130}{2} - \frac{15.7}{2} - \frac{15.2}{2} = 49.6\ mm$

Distance from centre line of bolt to prying force = n = lesser of edge distance and $1.1T\sqrt{\frac{\beta P_o}{p_y}}$

$= 1.1 \times 25.0\sqrt{\frac{2 \times 587}{265}}$

$= 57.9\ mm$

Effective length of flange per bolt

$= W = \frac{75+60}{2} + \frac{150}{2}$

$= 142.5\ mm$ for bolt F_2

From end plate calculations

prying force = Q = 34.1 kN

Moment at root fillet = Fb − Qn

$= 151.7 \times 49.6 - 34.1 \times 57.9$

$= 7524 - 1974$

$= 5550\ Nm$

Moment capacity $= \frac{p_y}{1.15} \times \frac{WT^2}{4}$

$= \frac{265 \times 142.5 \times 25.0^2 \times 10^{-3}}{1.15 \times 4}$

$= 5130\ Nm$

$< 5550\ Nm$

No good. Use stiffeners

Try 125×10 Grade 43 stiffeners with 6mm fillet welds.

Consider bolt F_2

Centre line of bolt to edge of weld to stiffener $= b_x = 75 - 10 - 6 = 59\ mm$

Commentary: 6/8

A longitudinal stress in the column flange of 120 N/mm^2 has been assumed for the example. In practice, the stress would be calculated from the axial load and moment. The flange stress due to the moment should be calculated using the elastic modulus.

Section 7.7.6

The plastic moment capacity of a flat plate with an axial load

$$= P_y\left(1 - \frac{f_a^2}{p_y^2}\right)\frac{wT^2}{4}$$

$$= \mu M_p$$

where f_a = average stress in the plate due to axial load (in the absence of bending). Note that in the case of a beam or column f_a is the stress in the flange due to axial load and bending moment on the member.

Where w is a transverse dimension (of the column flange) it is multiplied by the factor μ, when calculating the plastic distribution of load. w as a longitudinal dimension is assumed not to be affected.

Structural Steelwork Connections	Subject: Beam to column connection – Bolted end plate			Chapter Ref. 12
	Design Code: BS 5950 Part 1			Calc. Sheet No. Example 6/8
	Calc. by B.D.C	Date Aug, '87	Check by G.W.O.	Date Nov, '87

Code Ref.	Calculations	Output

Centre line of bolt to centre of root fillet at column web $= b_Y = \frac{130}{2} - \frac{15.7}{2} - \frac{15.2}{2}$

$= 49.6$ mm

Effective length of flange at stiffener

$= W_X = \frac{310.6 - 130}{2} + 0.7 \times 49.6 = 125.0$ mm

(col Flange) (bolt spac)

Effective length of flange at web

$= W_Y = \frac{150}{2} + 0.7 \times 59 = 116.3$ mm

Longitudinal stress in column flange due to axial load and bending $= f_a = 120$ N/mm²

Correction factor for effective length of transverse yield lines $= \mu = 1 - \frac{f_a^2}{p_y^2}$

$= 1 - \frac{120^2}{265^2} = 0.795$

Bolt load supported by stiffener (plastic distribution) $= F_X$

$$= \left[\frac{\frac{0.795 \times 125.0}{59}}{\frac{0.795 \times 125.0}{59} + \frac{116.3}{49.6}} \right] \times 151.7$$

$= 63.4$ kN

Bolt load supported by web (plastic distribution) $= F_Y$

$= 151.7 - 63.4 = 88.3$ kN

With prying force $= 34.1$ kN

Moment at toe of weld to stiffener

$= Fb - Qn$

$= 63.4 \times 59 - 34.1 \times 57.9$

$= 3741 - 1974 = 1767$ Nm

Moment at bolt line $= Qn = 1974$ Nm

Commentary: 6/9

As explained above, the moment capacity of transverse yield lines is multiplied by the factor μ to allow for longitudinal loads.

The corner detail (a plate loaded by the bolt and supported on two adjacent sides by the stiffener and the web) is being analysed as two separate plates, one supported at the stiffener and one at the web. The load has been split into F_x and F_y, and the corresponding minimum prying forces Q_x and Q_y have been calculated. Whereas the total applied load $(F) = F_x + F_y$, the total prying force (Q) is not necessarily equal to $Q_x + Q_y$. The reason for this is that a prying force located diagonally from the bolt could act as Q_x and Q_y. The general rule would be to take Q equal to the greatest of Q_x or Q_y or $(Q_x + Q_y = P_y T^2/2)$. The third requirement is necessary because the value of the prying force that can be located diagonally is limited by the possibility of the formation of a local yield line.

Tension capacity $= p_y wt$.
Effective width (w) = lesser of $2b_x$ and bolt pitch.
Tension capacity adjacent to other bolts is adequate by inspection.

Structural Steelwork Connections	Subject: Beam to column connection – Bolted end plate			Chapter Ref. 12
	Design Code: BS 5950 Part 1			Calc. Sheet No. Example 6/9
	Calc. by B.D.C	Date Aug, '87	Check by G.W.O	Date Nov, '87

Code Ref.	Calculations	Output
	Moment capacity $= \frac{p_y}{1.15} \times \frac{\mu W T^2}{4}$	
	$= \frac{265 \times 0.795 \times 125.0 \times 25.0^2 \times 10^{-3}}{1.15 \times 4}$	
	$= 3578$ Nm	
	> 1974 Nm O.K.	
	Moment at root fillet to web $= Fb - Qn$	
	$= 88.3 \times 49.6 - 34.1 \times 57.9$	
	$= 4380 - 1974$	
	$= 2405$ Nm	
	Moment capacity $= \frac{p_y}{1.15} \times \frac{W T^2}{4}$	
	$= \frac{265 \times 116.3 \times 25.0^2 \times 10^{-3}}{1.15 \times 4}$	
	$= 4187$ Nm	
	> 2405 Nm O.K.	
	Check minimum prying force	
	$Q_x = \frac{b_x}{2n_x}\left[F_x - \frac{\beta \gamma p_o \mu w_x T^4}{27 n_x b_x^2}\right]$	
	$= \frac{59.0}{2 \times 57.9}\left[63.4 - \frac{2 \times 1.5 \times 0.587 \times 0.795 \times 125 \times 25^4}{27 \times 57.9 \times 59^2}\right]$	
	$= 0.509[63.4 - 12.6] = 25.9$ kN	
	< 34.1 kN O.K.	
	$Q_Y = \frac{b_Y}{2n_Y}\left[F_Y - \frac{\beta \gamma p_o w_Y T^4}{27 n_Y b_Y^2}\right]$	
	$= \frac{49.6}{2 \times 57.9}\left[88.3 - \frac{2 \times 1.5 \times 0.587 \times 116.3 \times 25.0^4}{27 \times 57.9 \times 49.6^2}\right]$	
	$= 0.428[88.3 - 20.8] = 28.9$ kN	
	< 34.1 kN O.K.	
	Column web	
4.6.1	Tension capacity of length of web $(= 2b_x)$	
	$= 265 \times 2 \times 59 \times 15.7 \times 10^{-3}$	
	$= 491$ kN $> 2F_Y$ O.K.	

Commentary: 6/10

The load is assumed to be applied to the stiffener and the welds on the line of the bolt. Only the length of stiffener that is balanced about the bolt line is used for the design. (See commentary on Example 6/6 of this chapter.)

Section 7.7.4

The stiffener design is based on the 'plastic distribution' of load because the stiffener material is assumed to have reasonable ductility. The fillet welds, which have limited ductility, are designed using the greater of the loads from the plastic and elastic distributions. As well as 15 mm for the snipe, a further 6 mm has been deducted to allow for the end crater of a 6 mm fillet weld.

The weld on the underside of the stiffener is being designed to carry the portion of load (F_x) from bolt load F_2. The weld on the topside will carry the portion of bolt load F_1.

Effective length of weld = (Distance between inside of flanges) − (Snipes) − (Allowance for end craters).

Structural Steelwork Connections	Subject: Beam to column connection – Bolted end plate		Chapter Ref. 12
	Design Code: BS 5950 Part 1		Calc. Sheet No. Example 6/10
	Calc. by B.D.C	Date Aug, '87	Check by G.O.O.	Date Nov, '87

Code Ref.	Calculations	Output
	Stiffeners	
	Area of stiffeners required $= \frac{4F_x}{p_{ys}}$	
	$= \frac{4 \times 63.4 \times 10^3}{275} = 922 \text{ mm}^2$	
	Area of 2 No. 125 × 10 stiffeners (with 15mm snipes) $= 2 \times 2\left(\frac{130}{2} - \frac{15.7}{2} - 15\right) \times 10$	
	$= 1686 \text{ mm}^2$	
	O.K. but use 125 × 12.5 stiffeners to standardize with compression stiffeners.	2 No 125 × 12.5 stiffeners
	Bolt load supported by stiffener (elastic distribution $= F_x' = \left[\dfrac{\frac{125.0}{59^3}}{\frac{125.0}{59^3} + \frac{116.3}{49.6^3}}\right] \times 151.7$	
	(annotations: w_x; distance of bolt to prying force; w_y; distance from ℄ of bolt to root fillet.)	
	$= 59.1 \text{ kN}$	
	Plastic distribution governs design of stiffener to flange welds.	
6·6·5·2	Effective length of weld $= 2 \times \left(\frac{130}{2} - \frac{15.7}{2} - 15 - 6\right)$	
	$= 72.3 \text{ mm}$	
	Load per mm of weld $= \frac{63.4}{72.3} = 0.877 \text{ kN}$	
6·6·5	Capacity of 6 mm fillet weld $= 6 \times 0.7 \times 215 \times 10^{-3}$	
	$= 0.903 \text{ kN/mm}$	
	O.K. Use	
	Stiffener to web welds.	
	Effective length of weld	
	$= 327.2 - 2 \times 25 - 2 \times 15 - 2 \times 6$	
	$= 235.2 \text{ mm}$	
	Load per mm of weld $= \frac{63.4}{235.2} = 0.270 \text{ kN}$	
	Use 6mm fillet welds	6 mm fillet welds, stiffeners to column.

Commentary: 6/11

A longitudinal stress in the column flange of 240 N/mm^2 has been assumed for the example. In practice, the stress would be calculated from the axial load and moment. The flange stress due to the moment should be calculated using the elastic modulus.

Bearing is checked on length of column web
= (Beam flange thickness) + (45° dispersion through end plate)
+ (Dispersion through the flange and root radius at a slope of 1:2.5$\sqrt{\mu}$ to the plane of the flange).

Analysis of the effect of a lateral load on the flange of a column indicates that the slope of 1:2.5 should be modified by a factor of $\sqrt{\mu}$ to allow for the effect of longitudinal loads in the flange.

$$\mu = \left(1 - \frac{f_a^2}{p_y^2}\right)$$

Section 7.7.6

The buckling resistance of the stiffeners is based on the compressive strength of a strut, the effective section of which is based on the full area of the stiffeners together with an effective length of web on each side of the centreline of the stiffeners limited to twenty times the web thickness.

The stiffeners are assumed to be fitted to the loaded flange and nominal 6 mm fillet welds are adequate. The 6 mm fillet welds connecting the stiffeners to the web have a capacity greater than the total applied load (F_c). If necessary, it would be reasonable to take some of the applied load directly into the column through the stiff portion of the bearing (i.e. assuming a dispersion of 45° through the end plate and flange).

Structural Steelwork Connections	Subject: Beam to column connection – Bolted end plate			Chapter Ref. 12
	Design Code: BS 5950 Part 1			Calc. Sheet No. Example 6/11
	Calc. by B.D.C	Date Aug. '87	Check by G.W.O.	Date Nov, '87.

Code Ref.	Calculations	Output

<u>Column at compression flange of beam</u>

Longitudinal stress in column flange due to axial load and bending $= f_a = 240 \text{ N/mm}^2$

Correction factor $= \mu = 1 - \dfrac{f_a^2}{p_y^2}$

$= 1 - \dfrac{240^2}{265^2} = 0.180$

Local bearing capacity of the column web for the load applied by the bottom flange

4.5.1.3 of the beam

4.5.3 $= [14.8 + 2 \times 35 + 2 \times 2.5 \times (25.0 + 15.2)\sqrt{0.18}] \times 15.7 \times 265 \times 10^{-3}$

$= 707.6 \text{ kN}$

$< F_c$ ∴ stiffeners required

<u>Stiffeners</u>

Area of stiffeners required (in contact with flange) $= \dfrac{0.8 F_c}{p_{ys}}$

$= \dfrac{0.8 \times 811.8 \times 10^3}{275}$

$= 2362 \text{ mm}^2$

Try 2 No 125×12.5 stiffeners (with 15 mm snipes)

Area in contact $= 2(125 - 15) \times 12.5$

$= 2750 \text{ mm}^2$ o.k. Use

4.5.1.5 Buckling resistance

Area of effective section

$= 2 \times 125 \times 12.5 + 2 \times 20 \times 15.7^2$

$= 12985 \text{ mm}^2$

$I = \dfrac{12.5 \times (2 \times 125 + 15.7)^3}{12} = 19.54 \times 10^6$

$r = \sqrt{\dfrac{19.54 \times 10^6}{12985}} = 38.8 \text{ mm}$

$\dfrac{L}{r} = \dfrac{0.7 \times (327.2 - 2 \times 25.0)}{38.8} = 5.00$

Table 27(c) $p_c = 275 \text{ N/mm}^2$

Commentary: 6/12

Section 12.4.3

For columns with moment capacity connections to only one of the flanges (for example, external columns) it is quite likely to be found that the column web is overstressed by the high shear that occurs over the depth of the beam. If the web is overstressed, the column could be strengthened by a diagonal stiffener or possibly by plating the web. When the web is to be strengthened, check that there is still adequate clearance for installation and tightening of the bolts and that reasonable connections can be detailed for beams and ties connected to the column on the minor axis.

Structural Steelwork Connections	Subject: Beam to column connection — Bolted end plate			Chapter Ref. 12
	Design Code: BS 5950 Part I			Calc. Sheet No. Example 6/12
	Calc. by B.D.C	Date Aug, '87	Check by G.W.O	Date Nov, '87

Code Ref.	Calculations	Output
	Buckling resistance $= 12985 \times 275 \times 10^{-3}$	
	$= 3571$ kN	2 No 125×12.5 stiffeners
	$> F_c$ O.K.	
	Connection of stiffeners to web of column	
	Try 6 mm fillet welds	
	Capacity $= 4 \times 235.2 \times 0.903$	
	$= 849.5$ kN	6mm fillet welds
	$> F_c$ O.K.	
	<u>Column web shear</u>	
	Shear in column above connection = 26 kN	
	Maximum shear in column (at bottom flange of beam)	
	$= 2(151.7 + 151.7 + 95.5 + 18.0) - 26$	
	$= 807.8$ kN	
4.2.3	Shear capacity $= 0.6 \times 265 \times 15.7 \times 327.2 \times 10^{-3}$	
	$= 816.8$ kN	
	> 807.8 kN O.K.	

13

Beam-to-beam connections

13.1 Grillage connections

The structural requirements for grillage connections are similar to those for beam-to-column connections. These are related to the assumptions made in overall analysis, which in turn are a function of the connection type. Thus simple connections are only required to transmit shear from the secondary to the primary beams. Rigid connections are required both to transmit shear from the secondary to the primary beams and also to provide moment continuity between the two secondary beams that meet at the connection. Because of its torsional flexibility, the moment that is transferred into the primary beam as a torsion can usually be ignored.

Because of the similarity of structural action design of the beam, end connection is similar for both beam-to-column and grillage connections. To avoid repetition, this aspect of design is not discussed here and the designer of grillage connections should study the relevant sections of Chapter 12. In what follows discussion is limited to particular aspects of grillage connections.

13.1.1 Simple grillage connections

Figure 13.1 illustrates the common types of simple grillage connection. Either web cleats or flexible end plates are used, because the main girder web is usually too weak to support any form of seating bracket without extensive stiffening. It is frequent practice to position secondary beams on a common line. This means that the bolts through the main girder web are in double shear, but it can lead to erection difficulties. The first beam will have to be propped in some way until after the second beam is placed and the bolts have been inserted. One solution is to provide additional bolts to support the first beam, shown as the left-hand beam in all the examples in Figure 13.1.

Where end plates are used, the resulting inflexibility of beam length can lead to erection difficulties if the main girders will not flex laterally to permit the beam to be swung into position. The problem is greatest with short secondary beams where the corner-to-corner dimension is markedly greater than the beam length. This can be overcome by shortening the secondary beam and welding a pack plate onto one of the main beams. This permits the secondary beam to be swung in and aligned parallel to its final position before being slid into place. The pack plate should be attached to the main beam web with sufficient weld for it to be able to carry all the beam shear, otherwise the bolts could be subject to unacceptable bending moments.

Where the secondary beams may be set to a lower top flange level than the primary ones there will be no interference between flanges. However, if the top flanges are required to be coplanar it will be necessary to notch the secondary beams, as shown in Figures 13.1(c) and (d). This raises two potential problems. If web cleats are used with bolts at minimum pitch a local tearout or 'shear block' failure may occur in the beam web; appropriate design procedures are discussed in Section 7.3.3. In addition, the loss of the top flange will reduce both bending strength and robustness.

Where major and minor beams are of the same depth it is necessary to notch out both flanges, as shown in Figure 13.2. Design procedures for both singly and doubly notched beams are discussed in Section 13.3.

13.1.2 Moment-resisting grillage connections

Figure 13.3 shows typical 'rigid' grillage connec-

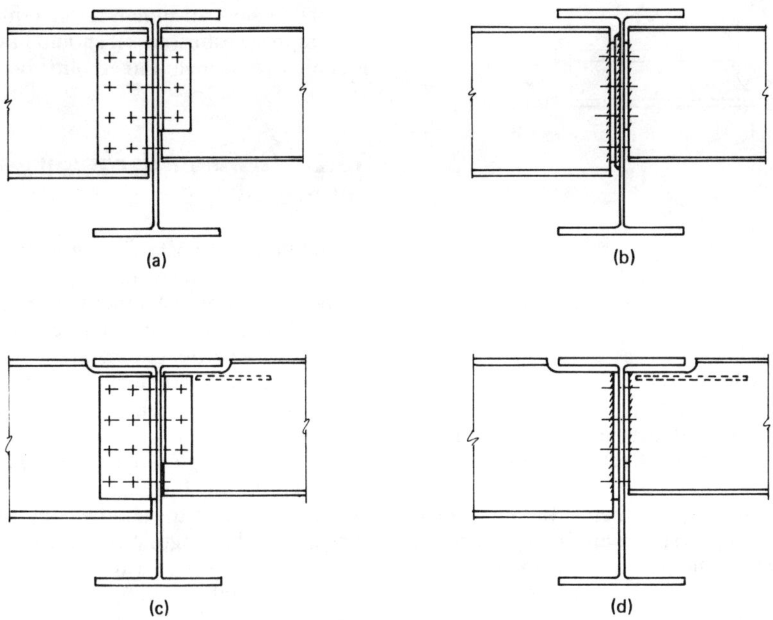

Figure 13.1 Simple beam-to-beam grillage connections

tions. Where the secondary beams may be set below the top flange of the primary ones the most economic connection is likely to utilize end plates. As with simple connections, use of a pack plate is recommended for short, wide secondary beams to ease erection.

If coplanar top flanges are required then the combined splice-plate/end-plate connection shown in Figure 13.3(b) must be used. As with other splice plate moment connections, HSFG bolts must be utilized if local rotation is to be avoided.

13.1.3 Notches in secondary beams

It is possible to carry out straightforward calculations to check the residual capacity of a notched beam. Assuming no connection fixity, the moment $Q_x l_n$ on section XX in Figure 13.4 must be resisted by a Tee section consisting of the web and bottom flange for a single-notched beam or the web alone for a double-notched one. The top fibre stress should be less than the limiting compressive stress; this is generally based on the slenderness $1.5 l_n/(t_w/\sqrt{12})$ for local buckling of the unrestrained web, where t_w is web thickness. The effective length is taken as 1.5 l_n where there is possible out-of-plane

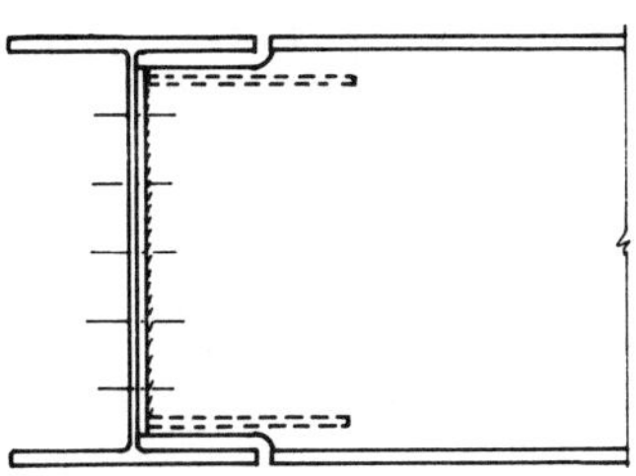

Figure 13.2 Grillage connection between beams of similar depth

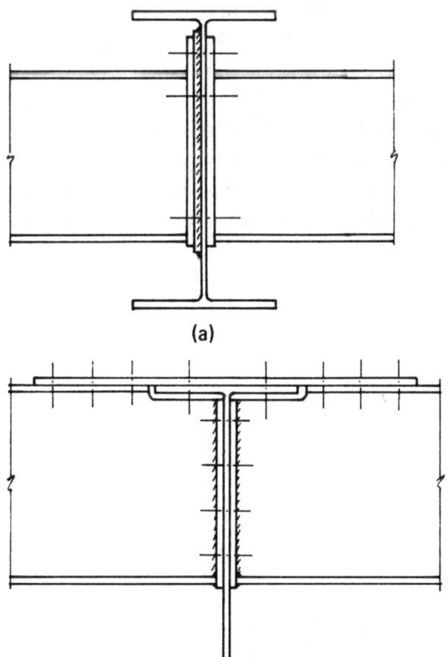

Figure 13.3 Moment-resisting grillage connections

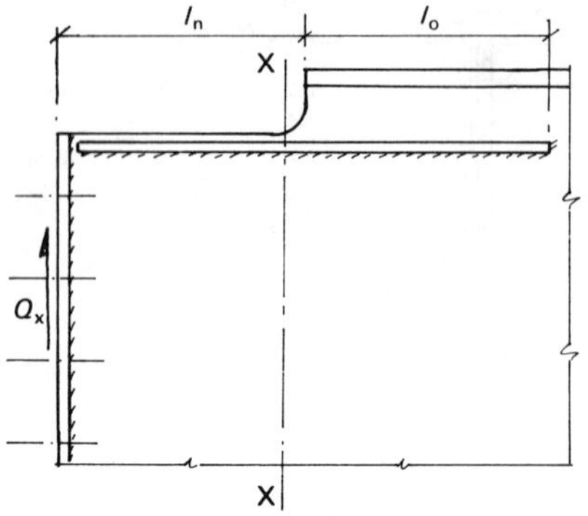

Figure 13.4 Notch reinforcement details

movement of the end of the notch. Where such movement is prevented, an effective length of 0.7 l_n may be adopted.

If section XX is inadequate then stiffener(s) must be provided as shown. Even if there is no overstress, careful consideration should be given to providing stiffening to maintain the robustness of the element. If there is any possibility of:

1. Lateral loading to the beam during erection or service life;
2. Thermal effects imposing overall compression on the beam;
3. Torsion in the main girder increasing the moments on the critical section above the calculated value,

then stiffening should be provided.

The stiffening need only be provided to one face of the web. As with intermediate web stiffeners, it may be curtailed by 3 × t_w from the end plate. Although not necessary to resist local buckling, it will improve robustness if it can have an appreciable overlap l_0 with the flange: ideally, l_0 should not be less than l_n. If provided to overcome an overstress the stiffener size may be optimized by calculation, based on its enhancement to the critical section. If it is to improve robustness it should have a section that is similar to a single-flange outstand.

13.2 Cross-girder/main girder connections

In addition to transferring moment between cross-girders and/or shear into the main girder these connections in bridge structures are usually required to provide lateral–torsional restraint to the main girders. If the cross-girders are at the bottom of the main ones, in some form of through-bridge, this restraint will be required in sagging moment regions. If they are located at the top of the main girders, as in a typical deck bridge, this restraint will be required in hogging moment regions. It is common practice to maintain the same connection throughout the length of the bridge. As can be seen in Figure 13.5, this restraining function alters the connection configuration of these beam end connections. Because the stabilizing restraint is primarily required for the flanges it is necessary to stiffen the web in order to provide sufficient stiffness of the restraint system through to the former elements. As in the previous section, design of this connection is very similar to many aspects of beam-to-column connection design. This section, which only highlights special aspects of behaviour, should be read in conjunction with the earlier part of this chapter.

End-plate connections may be used in conjunction with Tee-section stiffeners. At external girder connections, the restraining moment can generally be developed with a short end plate and no local stiffening to the main stiffener. At internal girders the moments are much greater because of the effective continuity of the cross-girder. Extended end plates are likely to be necessary, together with stiffening to the Tee sections. In both situations

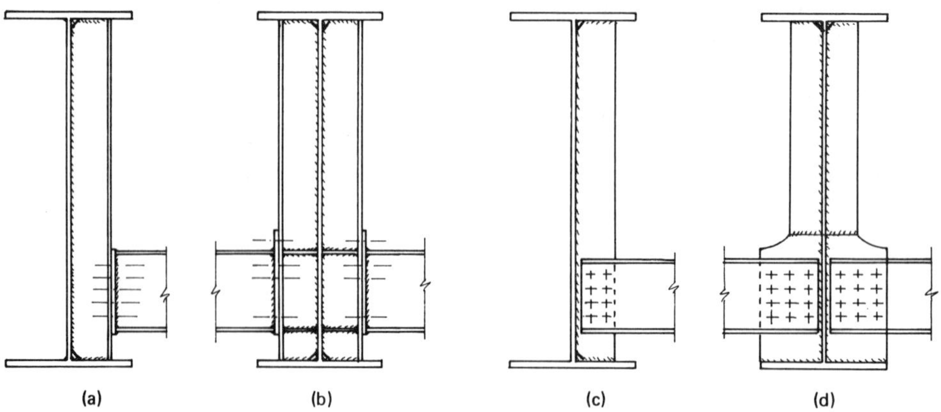

Figure 13.5 Cross-girder/main girder connections for bridges. (a) and (b) End-plate connections; (c) and (d) lapped connections

design procedures are very similar to those for end-plate beam-to-column connections. Note, however, that the greater concern over corrosion in bridges compared to most building structures does have an important influence on workmanship standards for this form of connection. Most bridge specifications will require continuous contact over mating surfaces. This is difficult and costly to achieve in the presence of eccentric welding to the end plate and Tee-section flange.

For this reason, many bridge engineers prefer the lapped connections shown in Figures 13.5(c) and (d). If channel section cross-girders are used it is possible to develop the full section capacity. If UB or UC sections are employed the flange outstands on one side will have to be cut back for the overlaps. At external girders the required moment capacity can be developed by lapping the cross-girder onto a conventional single-sided stiffener (though this should not be treated as a conventional intermediate stiffener but should be welded to the compression flange). At internal girders larger bolt groups will generally be required to carry the combined continuity and restraint moments. Concern over through-thickness tensile loading on the main girder web may lead to the arrangement shown where the main 'gusset plate' is threaded through a slot in the web. A conventional web stiffener can then be welded to the top of this plate to restrain the top flange.

As referred to in Section 6.5.1, considerations of fatigue have an influence on the design of such connections. At external connections the main girder can generally rotate and take up the end slope of the cross-girder in the middle of the span. However, near supports to the main girders these rotations are restrained, and there will be additional moments in the connections. At internal connections there will be large fluctuations in continuity moments as loading varies across the width of the bridge. Such action should, of course, be predicted by an appropriate grillage analysis. Where cross-girders are only used to provide torsional restraint to the main girders, some designers now only brace alternative spaces between girders to minimize transverse bending stiffness.

Note also that, where end-plate connections are used in fatigue situations, the usual simplifying assumption for the distribution of bending stresses cannot be made, nor can any plastic analysis be used for the design of the end plate.

Reference

1. *Engineering for Steel Construction*, AISC, 1984.

Commentary: 1/1

The supported beams are on the same grid line, and the same bolts are needed to connect both beams to the supporting beam. The arrangement used, for the bolts and cleats, is suitable where the larger (left-hand) beam is erected before the smaller one. The larger beam can be held in position by the bottom pair of bolts while the smaller one is erected. If the smaller beam is to be erected first, an arrangement is required so that it can be supported until the larger one is erected (for example, by an erection/seating cleat, or by rearranging the bolts and cleats). 120 mm cross-centres are used so that the local distortion of the cleats can accommodate part of the end rotations of the simply supported beams.

A reasonably full set of calculations is given to illustrate the design procedure. In practice, in this and other design examples the amount of calculation required can be reduced by the use of tables and shear and bearing capacities of bolts.

Structural Steelwork Connections	Subject: Beam to beam connection with bolted web cleats			Chapter Ref. 13
	Design Code: BS 5950 Part 1.			Calc. Sheet No. Example 1/1
	Calc. by B.D.C	Date Aug, '87	Check by G.W.O.	Date Nov, '87

Code Ref.	Calculations	Output

Beam to beam connection with bolted web cleats.

Design bolted web cleat connections for a 610 × 229 × 101 UB (540 kN reaction due to factored loads) and a 356 × 171 × 45 UB (228 kN reaction due to factored loads) connected to the web of a 686 × 254 × 140 UB. All the beams are in Grade 50 steel.

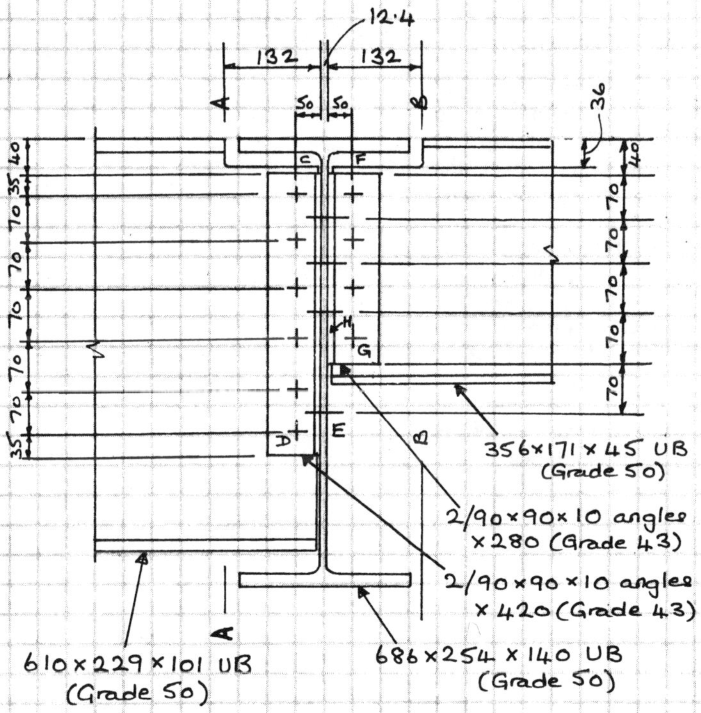

18 bolts M20 grade 8·8
(bolts in outstand legs at 120 mm cross centres.)

Commentary: 1/2

Shear capacity of bolt in single shear = $p_s A_t$

Bearing capacity of web = $P_{bs} dt$

Note that for a simple vertical shear and an end distance of 39 mm the bearing capacity of the web would be limited to $\frac{1}{2} p_{bs} et$ (since $e < 2d$). However, the end distance in the direction of the resultant shear force is greater than $2d$ and the bearing based on the bolt diameter governs.

Section 8.2.2

Horizontal shear force on bolt due to moment due to eccentricity

$$= V_m = \frac{P_x e_x r_i}{\sum_n r_i^2}$$

Vertical shear force per bolt = $V_x = \dfrac{P_x}{n}$

Resultant shear force = $\sqrt{(V_m^2 + V_x^2)}$

For the larger reaction the eccentricity of the load has been taken as the distance from the bolt line to the centre of the supporting web. The bolt capacity is the least of the bolt capacity in double shear, the bearing capacity of the bolt on the web or the cleats, of the web (which governs here) or of the cleats.

In traditional design the effect of eccentricity is ignored when designing the bolts connecting the cleats to the supporting beam. However, with the lower safety factors adopted in current design codes it would be prudent to allow for the effect of eccentricity.

In this example the increase in resultant shear force due to the eccentricity is only 10%; but for the shorter connection on the right-hand beam it is 17%.

Structural Steelwork Connections	Subject: Beam to beam connection with bolted web cleats			Chapter Ref. 13
	Design Code: BS 5950 Part 1			Calc. Sheet No. Example 1/2
	Calc. by: B.D.C	Date: Aug, '87	Check by: G.W.O.	Date: Nov, '87.

Code Ref.	Calculations	Output
	For M20 Grade 8.8 bolts	
6.3.2	Shear capacity of bolt in single shear $= 375 \times 245 \times 10^{-3} = 91.9$ kN	
6.3.2	Shear capacity in double shear $= 2 \times 91.9 = 183.8$ kN	
	<u>Left hand beam</u> (610 × 229 × 101 UB)	
	<u>Connection to web of left hand beam.</u>	
6.3.3.3	Bearing capacity of web of beam $= 550 \times 20 \times 10.6 \times 10^{-3} = 116.6$ kN	
	Try 6 bolts at 70 mm vertical pitch, 50 mm from heel of cleats.	
	Horizontal shear force on bolt due to moment due to eccentricity $= \dfrac{540 \times (50 + 0.5 \times 12.4) \times 175}{2(35^2 + 105^2 + 175^2)} = 61.9$ kN	
	Vertical shear force per bolt $= \dfrac{540}{6} = 90$ kN	
	Resultant shear force $= \sqrt{61.9^2 + 90^2} = 109.2$ kN < bolt capacity O.K	6/M20 bolts Grade 8.8
	<u>Connection to web of supporting beam</u>	
6.3.3.3	Bearing capacity on 10 mm cleat (grade 43) $= 460 \times 20 \times 10 \times 10^{-3} = 92$ kN	
	Single shear capacity of 91.9 kN governs	
	Try 8 bolts, 2 columns at 120 mm c/c (vertical pitch as in sketch)	
	Assume centre of pressure 25 mm below top of cleat.	

Commentary: 1/3

An assessment of the maximum horizontal shear force on the bolts has been made by assuming a compressive reaction between the cleats (25 mm from the top of the cleats) and by assuming that the horizontal loads on the bolts are proportional to the vertical distances from this point.

Hence, horizontal shear force on bottom bolt

$$= V_m = \frac{R}{2} \times e \times \frac{r_i}{\sum_n r_i^2}$$

$$\text{Vertical shear force} = V_x = \frac{R}{2n}$$

(there are n bolts per cleat)

$$\text{Resultant shear force} = \sqrt{(V_m^2 + V_x^2)}$$

The design procedure for the right-hand beam is similar to that adopted for the left-hand one. The one variation is in the eccentricity used to calculate the horizontal shear force on the bolts connecting the cleats to the supported beam.

For the smaller reaction, the eccentricity of the load has been taken as the distance from the bolt line to the face of the supporting web, instead of the distance to the centreline of the web. It is assumed that the difference is balanced by the larger reaction from the left-hand beam.

Structural Steelwork Connections	Subject: Beam to beam connection with bolted web cleats			Chapter Ref. 13
	Design Code: BS 5950 Part 1			Calc. Sheet No. Example 1/3
	Calc. by: B.D.C	Date: Aug. '87	Check by: G.W.O.	Date: Nov, '87.

Code Ref.	Calculations	Output
	Horizontal shear force on bottom bolt due to moment due to eccentricity	
	$= \dfrac{540 \times 0{\cdot}5(120-10{\cdot}6) \times 325}{2(45^2+115^2+185^2+325^2)}$	
	$= 30{\cdot}9$ kN	
	Vertical shear force per bolt $= \dfrac{540}{8} = 67{\cdot}5$ kN	
	Resultant shear force $= \sqrt{30{\cdot}9^2+67{\cdot}5^2}$	
	$= 74{\cdot}2$ kN	
	< bolt capacity O.k	8/M20 bolts Grade 8·8
	Use 2/90×90×10 angle cleats × 420	2/90×90×10 angle cleats × 420 mm
	<u>Right hand beam</u> (356×171×45 UB)	
	<u>Connection to web of right hand beam</u>	
	Bearing capacity of web of beam	
	$= 550 \times 20 \times 6{\cdot}9 \times 10^{-3} = 75{\cdot}9$ kN	
	Try 4 bolts at 70 mm vertical pitch, 50 mm from heel of cleats	
	Horizontal shear force on bolt due to moment due to eccentricity	
	$= \dfrac{228 \times 50 \times 105}{2 \times (35^2+105^2)} = 48{\cdot}9$ kN	
	Vertical shear force per bolt $= \dfrac{228}{4} = 57{\cdot}0$ kN	
	Resultant shear force $= \sqrt{48{\cdot}9^2+57{\cdot}0^2}$	
	$= 75{\cdot}1$ kN	
	< bolt capacity O.k.	4/M20 bolts Grade 8·8
	<u>Connection to web of supporting beam.</u>	
	Try 6 bolts, 2 columns of 3 bolts at 120 mm c/c, 70 mm pitch.	

Commentary: 1/4

For the left-hand beam, the horizontal shear force was calculated to be 30.9 kN. The horizontal shear force on the bolt in the third row is proportional to its distance from the assumed centre of pressure (rotation):

$$V_{m3} = V_{m4} \times \frac{r_3}{r_4}$$

Structural Steelwork Connections	Subject: Beam to beam connection with bolted web cleats		Chapter Ref. 13
	Design Code: BS 5950 Part 1		Calc. Sheet No. Example 1/4
	Calc. by B.D.C	Date Aug. '87	Check by G.W.O.	Date Nov, '87.

Code Ref.	Calculations	Output

Assuming centre of pressure 25 mm below top of cleat, horizontal shear force on bottom bolt due to moment due to eccentricity

$$= \frac{228 \times 0{\cdot}5(120 - 6{\cdot}9) \times 185}{2(45^2 + 115^2 + 185^2)}$$

$= 24{\cdot}1$ kN

Vertical shear force per bolt $= \frac{228}{6} = 38$ kN

Resultant shear force $= \sqrt{24{\cdot}1^2 + 38^2}$

$= 45{\cdot}0$ kN

< bolt capacity O.K — **Output:** 6/M20 bolts Grade 8·8

Use 2/90×90×10 angle cleats × 280 — **Output:** 2/90×90×10 angle cleats × 280 mm

<u>Web of supporting beam</u> Check for combined load from left and right hand beams

Bearing capacity on 12·4 mm web

$= 550 \times 20 \times 12{\cdot}4 \times 10^{-3}$

$= 136{\cdot}4$ kN

Total horizontal shear force due to moment due to eccentricity (on third row)

$$= \frac{30{\cdot}9 \times 185}{325} + 24{\cdot}1 = 41{\cdot}7 \text{ kN}$$

Total vertical shear force $= 67{\cdot}5 + 38{\cdot}0$

$= 105{\cdot}5$ kN

Resultant shear force $= \sqrt{41{\cdot}7^2 + 105{\cdot}5^2}$

$= 113{\cdot}4$ kN

< bearing capacity O.K

<u>Left hand beam.</u>

At end of notch (A–A)

Depth of section $= 602{\cdot}2 - 36 = 566{\cdot}2$ mm

Commentary: 1/5

The beams are checked at the end of the notch (the part cut away to clear the top flange of the supporting beam) for shear and bending capacity.

Shear capacity $= 0.6p_yA_v$
where the shear area $= A_v = 0.9tD$

Structural Steelwork Connections	Subject: Beam to beam connection with bolted web cleats	Chapter Ref. 13
	Design Code: BS 5950 Part 1	Calc. Sheet No. Example 1/5
	Calc. by: B.D.C · Date: Aug. '87 · Check by: G.W.O.	Date: Nov, '87

Code Ref.	Calculations	Output
4.2.3	Shear capacity $= 0.6 \times 355 \times 0.9 \times 566.2 \times 10.6 \times 10^{-3}$ $= 1151$ kN	
	Shear force = 540 kN < shear capacity O.K.	
	Slenderness of compression zone of web $= \dfrac{1.5\,(132-50)}{10.6/\sqrt{12}} = 40.2$	
Table 27(c)	Compressive strength = 300 N/mm²	
	Moment capacity $> 300 \times \dfrac{10.6 \times 566.2^2}{6} \times 10^{-6}$ $= 170$ kNm	
	Moment $= 540 \times \left(132 + \dfrac{12.4}{2}\right) \times 10^{-3}$ $= 74.6$ kNm < moment capacity O.K.	
	Check shear capacity through bolt holes (C–D–E)	
4.2.3 4.6.1 3.3.3	Block shear capacity = shear capacity of C–D + ½(tensile capacity of D–E) $= 0.6 \times 355 \times 0.9 \times 1.1 (5 \times 70 + 39 - 5.5 \times 22) \times 10.6 \times 10^{-3}$ $+ 0.5 \times 355 \times 1.1 (48 - 0.5 \times 22) \times 10.6 \times 10^{-3}$ $= 599.0 + 76.6$ $= 675.6$ kN	
	Shear force = 540 kN < Block shear capacity O.K.	
	<u>Right hand beam.</u> At end of notch (B–B)	
	Depth of section = 352 − 36 = 316 mm	
	Shear capacity $= 0.6 \times 355 \times 0.9 \times 316 \times 6.9 \times 10^{-3}$ $= 418$ kN	

Commentary: 1/6

Section 13.1.3

The compressive strength of the web has been checked at the notch, assuming an effective length of 1.5L (where L has been taken as the distance from the bolt line to the end of the notch). This effective length assumes that there is no restraint to out-of-plane movement of the top flange of the beam.

Radius of gyration = $t\surd 12$

For an initial check, the moment capacity is calculated ignoring the outstand flanges ($Z = tD^2/6$)

Moment capacity = $p_c Z$

The moment capacity, ignoring the outstand flanges, is adequate for the left-hand beam and nearly adequate for the right-hand beam and there is no need to calculate the modulus of the Tee-section. (For the left-hand beam the modulus of the Tee-section is 49% greater than the modulus of the flat plate. For the right-hand beam it is 55% greater.)

As before, the eccentricity assumed in calculating the moment is taken from the centreline of the web for the left-hand beam and the face of the web for the right-hand one.

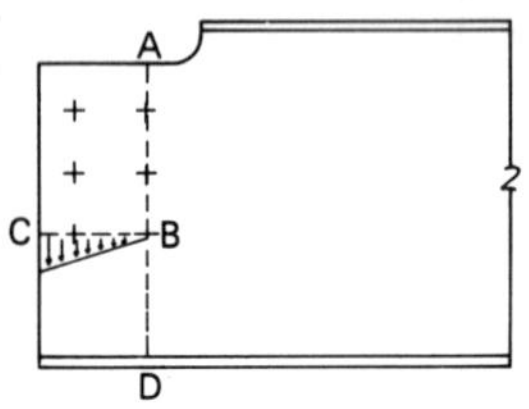

Section 13.1.1: Figure 7.7

When checking the beam web for failure at the bolt holes, two possibilities are considered:

1. Shear failure through the bolt holes on line A–B–D;
2. Shear failure on line A–B combined with tensile failure on linç B–C. This is usually referred to as 'block shear' failure. A triangular distribution is assumed for the tensile stress.

In this example the block shear failure is the more critical of the two possibilities.

Note that for flat plates and coped beams the effective area (A_e) should be used in the design, whereas for beams without copes the bolt holes are ignored when calculating the shear capacity.

Block shear capacity = shear capacity of A–B

+ tensile capacity of B–C assuming a triangular stress distribution

$= 0.6\,p_y \times 0.9A_{e(AB)}$

$+ 0.5p_yA_{e(BC)}$

From BS 5950: Part 1, 1985, Clause 3.3.3:

$A_e = K_e$ (net area) but $\not>$ (gross area)

$K_e = 0.75\,\dfrac{U_s}{Y_s} \not> 1.2$ $\quad K_e = 1.2$ for Grade 43

$K_e = 1.1$ for Grade 50

Structural Steelwork Connections	Subject: Beam to beam connection with bolted web cleats		Chapter Ref. 13
	Design Code: BS 5950 Part 1		Calc. Sheet No. Example 1/6
	Calc. by B.D.C	Date Aug, '87 / Check by G.W.O.	Date Nov, '87

Code Ref.	Calculations	Output
	Shear force = 228 kN < shear capacity O.K	
	Slenderness of compression zone of web $= \dfrac{1.5(132-50)}{6.9/\sqrt{12}} = 61.8$	
	Compression strength = 242 N/mm²	
	Moment capacity $> 242 \times \dfrac{6.9 \times 316^2}{6} \times 10^{-6}$ $= 27.8$ kNm	
	Moment $= 228 \times 132 \times 10^{-3} = 30.1$ kNm	
	This is greater than moment capacity based on modulus of web only, but would be less than the moment capacity based on the full modulus of the Tee section.	
	Check shear capacity through bolt holes (F-G-H)	
4.2.3	Block shear capacity	
4.6.1	= shear capacity of F-G + ½ (tensile capacity of G-H)	
3.3.3	$= 0.6 \times 355 \times 0.9 \times 1.1(3\times70+39-3.5\times22)\times6.9\times10^{-3}$ $+ 0.5\times355\times1.1(48-0.5\times22)\times6.9\times10^{-3}$	
	$= 250.3 + 49.8$ $= 300.1$ kN	
	Shear force = 228 kN < Block shear capacity O.K	

14

Portal frame connections

14.1 Introduction

14.1.1 Practical optimization

The industrial building sector is both the largest and the most competitive part of the structural steelwork market. In the UK, portal frame structures are the most popular structural form, being used for 40% of all steel structures in this sector. Design conditions probably vary less in this type of steelwork than in any other design situation. Frame span and building height remain variable but frame spacings will usually be within the limits of 4,5 to 7,5 m, environmental loadings are almost constant and crane loadings, if any, will generally be standardized for various capacities.

Because of the scale of the market, its competitiveness and the narrow range of design conditions, there has been more practical optimization of connections in portal frame structures than in any other type of structural steelwork. This optimization has been assisted by the widespread use of 'design and construct' contracts, where the designer becomes closely involved with the practicalities of economic fabrication. There is also a tradition of structural tesing to prove designs in this section which has further helped the optimization process. Thus even designers who do not work in this field can benefit considerably from studying the detailed work that has evolved.

14.1.2 Connection location

Originally, site connections were at or near the points of contraflexure in the rafters. However, this led to difficulties with both transportation and erection, so at an early stage in frame development connections moved to the eaves and apex positions.

14.1.3 Design context

In order to have a proper understanding of portal frame connection design it is necessary to appreciate the design context for the frames themselves. Originally, frames were designed by elastic analysis; Figure 14.1(a) shows a typical distribution of moments. The critical moment generally occurred at the eaves, and because of this, both columns and

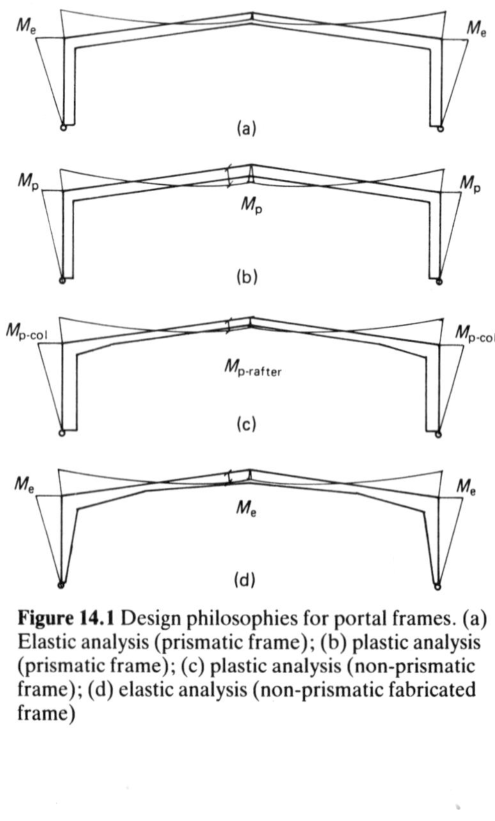

Figure 14.1 Design philosophies for portal frames. (a) Elastic analysis (prismatic frame); (b) plastic analysis (prismatic frame); (c) plastic analysis (non-prismatic frame); (d) elastic analysis (non-prismatic fabricated frame)

rafters had to be of the same section. This configuration did not lead to economy because of the inefficiency of the rafters, which were only stressed to their capacity at the eaves position. In order to improve economy, plastic analysis was introduced, giving the redistribution of moments shown in Figure 14.5(b). Because redistribution was permitted it was possible to make better use of the rafter capacity near the frame apex. A plastically designed, prismatic frame would generally be of lighter weight than its elastically designed equivalent. However, difficulty was experienced in producing sound connection designs at the eaves because of the requirement for rotation capacity at that point. This led to the introduction of haunches, as shown in Figure 14.1(c), and had the added benefit of enabling non-prismatic frames to be designed with rafter sections that were lighter than the columns. This led to further economy, particularly for buildings of low aspect ratio.

The above discussion relates to frames from rolled sections. However, fabricated, tapered sections have also been used for portal frames. Here structural efficiency is achieved, not by redistribution but by fitting the strength envelope as closely as possible to the elastic moment envelope, as shown in Figure 14.1(d). This form of construction had declined in popularity because of the high costs of manual fabrication. However, recent advances in automatic fabrication have led to a resurgence of this type of frame.

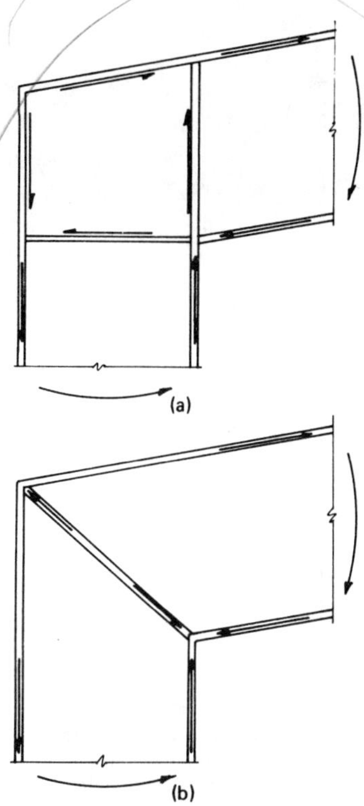

Figure 14.2 Moment transmission at eaves. (a) Corner shear panel; (b) diagonal stiffener

14.2 Eaves connections

14.2.1 General behaviour

Before looking at types of eaves connection in detail it is important to examine the primary modes of behaviour of a typical connection. Generally, moments will predominate; Figure 14.2 shows the two principal ways in which moments are transmitted around the corner. In both cases it is helpful to replace the moments by pairs of forces in each element, i.e. tensile and compressive forces in the flanges of the beam and of the column. In the absence of a diagonal stiffener these pairs of forces achieve equilibrium solely by shear on the corner panel, as shown in Figure 14.2(a). If a diagonal stiffener is added then overall equilibrium may be achieved by the system of forces shown in Figure 14.2(b). In practice, in the presence of a diagonal stiffener, overall equilibrium is obtained by a combination of both modes of behaviour. This is a very good example of the sort of indeterminacy of structural action that was referred to in Chapter 1. In general, designers will not be able to resolve this indeterminacy and should design accordingly. Either they should make one or other force path sufficiently strong to carry all the moment, or, if they partition the forces between the two modes, they must ensure that all the elements concerned can behave in a ductile way, thus permitting any redistribution that may be necessary.

14.2.2 Welded connections

Site welding is very rare in industrial building construction. However, it may sometimes be possible to have a welded eaves connection which is completed in the fabrication shop and to have simple bolted site connections near to the points of contraflexure under symmetric loading in the rafters.

The most efficient form of welded eaves connection is that shown in Figure 14.3(a) with a diagonal stiffening element. If this element is only required to have a thickness of 10–15 mm then it should take the form of a division plate; the crucial tension flange weld should have the butt weld configuration shown in Figure 14.3(b). If a division plate with sound through-thickness properties (i.e. one where there is no possibility of lamellar tearing) is used then it is possible to adopt the weld configuration shown in Figure 14.3(c). In this instance the detail is suitable for use with division plates of any thickness. Above 15 mm and without appropriate material for a

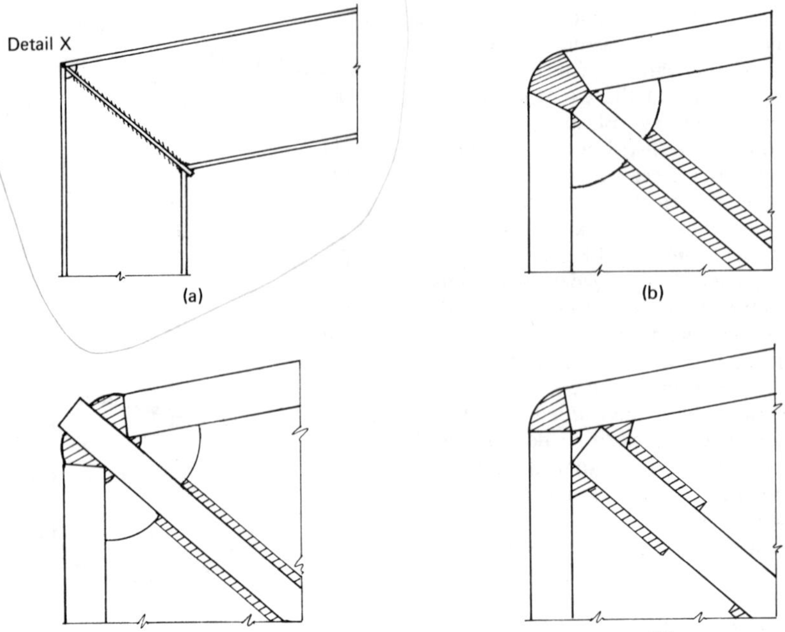

Figure 14.3 Welded eaves connections. (a) General arrangement; (b) detail X with curtailed division plate; (c) detail X with long division plate; (d) detail X with butt welds connecting main sections and subsequently fitted stiffener

division plate the welding detail shown in Figure 14.3(d) should be adopted, where the butt joint is carried out first and stiffeners are subsequently added. Use of the configuration shown in Figure 14.3(b) with thick division plates would result in an excessive volume of weld metal; this would be uneconomic and could create problems of distortion.

14.2.3 Splice plate bolted joints

Figure 14.4 shows two types of eaves connection that make use of splice plates. Because of the requirement for connection rigidity that arises from the overall frame action, such connections have to be made using HSFG bolts. For this reason, these connections are usually expensive and are now rarely used. Care must be taken to ensure that the rafter is not trapped between elements that are welded to the column in order to avoid difficulties with erection. If a bottom splice plate or erection cleat is used then the top splice plate has to be separated from the column, as shown in Figure 14.4(a). This increases the number of bolts and generally makes the connection correspondingly more expensive. The more economical version of this connection is that shown in Figure 14.4(b).

It is customary to provide a column web stiffener to resist the rafter bottom flange compression. This may be horizontal, as shown in Figure 14.4(a), if the

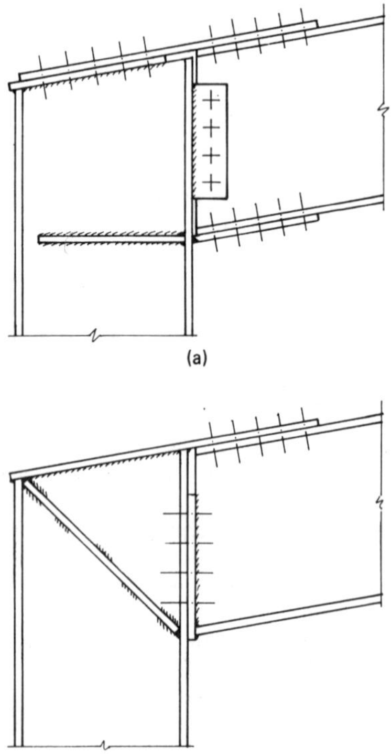

Figure 14.4 Bolted eaves connections with splice plates

column web has sufficient capacity to resist the shear above such a stiffener. Alternatively, a diagonal stiffener may be provided, as shown in Figure 14.4(b).

14.2.4 End-plate connection without haunches

Figure 14.5 shows the various forms of end-plate eaves connections that can be used when the frame is not haunched. Examples (a) and (b) have a vertical plane of connection, but similar connections would result with a horizontal joint and continuity given to the rafter rather than the column. The vertical plane of connection is generally preferred because it can equally well be used for multi- or single-bay frames, thus permitting greater standardization.

The short end plate shown in Figure 14.5(a) has the advantage that a smooth top profile to the frame is obtained. However, because the bolts are operating at a shorter lever arm than the rafter depth it is not possible to develop the full section capacity with this form of connection for rolled sections which have relatively heavy flanges. It is still difficult with minimum-weight fabricated sections which generally have much lighter flanges.

The long end-plate connection of Figure 14.5(b) is much more robust. It is frequently possible to accommodate this extended rafter end plate and extended column within the depth of the purlins without interference to the roof support. In such a case the eaves beam then has to be supported off the face of the column.

Figures 14.5(c) and (d) show the short and extended end-plate connections which have a diagonal plane of connection. Superficially, these appear to be very economical connections: the end plates also act as diagonal stiffeners and no other stiffening is required. However, note that there are particular problems with the stability of the inside corner of this connection, as discussed in Section 14.4. Stabilizing bracing will always be necessary to that inside corner, and this can create severe problems because there is usually no convenient purlin or side rail from which to brace. In addition, it is thought that such connections may cause difficulties with erection. Not many erectors are able to think clearly about three-dimensional geometry in non-orthogonal terms! For these two reasons,

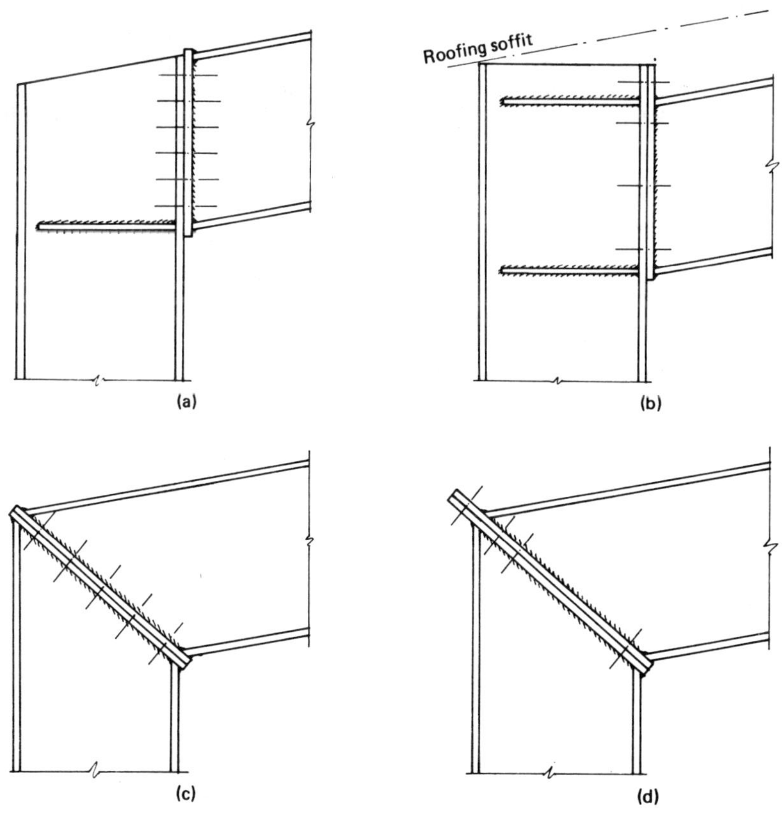

Figure 14.5 Bolted eaves connections with end plates

these connections are not popular, despite their apparent simplicity and economy. If they are used it should be noted that, because of the greater lever arm that arises with the diagonal cut, it is usually possible to develop the full strength of the sections with the short end-plate connection of Figure 14.5(c). It is therefore rarely necessary to extend the end plate in the way shown in Figure 14.5(d).

For all these end-plate connections it is the moment, and in particular the associated tensile forces, that create the greatest design difficulty. However, the following points should be noted:

1. Column web stiffening will generally be provided at the point where the rafter bottom flange bears against the column. As discussed in the previous section, this may be diagonal or horizontal.
2. In extended end-plate connections, separate bolts should be provided to transmit the shear across the joint. Where these bolts are a significant distance from the centre of rotation, they will inevitably attract tensile load; in such circumstances it is essential to ensure that the end plate is attached to the web with adequate welds, as discussed in the commentary to Example 12.4.
3. Irrespective of any requirement for shear bolts, it is essential that a pair of bolts be provided near the bottom of the end plate in order to ensure that the compressive parts of the connection are pulled up into proper bearing contact.
4. The bolts closest to the outside flange must be so positioned that they may readily be inserted and tightened.

14.2.5 End-plate connections with haunches[1]

Figure 14.6 shows a typical end-plate eaves connection for a haunched portal frame. It is generally found that the greatest economy of overall frame design is achieved with a haunch length of approximately one-tenth of the span. This ensures a substantial reduction in rafter weight but enables the haunch section to be cut from the same section as the rafter.

Because of the greater depth available, it is possible to be more flexible with the load paths through the connection. The main tension bolts are moved down the connection as shown, transferring the tension force from the line of the rafter flange into the upper region of the rafter and column webs. This usually means that all tension stiffening to the web can be avoided because of the greater vertical spread of tension in both the column and rafter webs. However, small stiffeners (X) may be necessary to increase the flexural capacity of the column flanges under the local bolt loads, particularly if the column is a light rolling.

The compression load path is assigned to the level of the bottom flange of the haunch. As in the previous section, a pair of bolts is provided to prevent springing of this point. A stiffener is usually attached to the column web, and it is certainly required if a plastic hinge may form at that point in the column. In the same way, if a hinge may form in the rafter at the end of the haunch, a stiffener (Y)

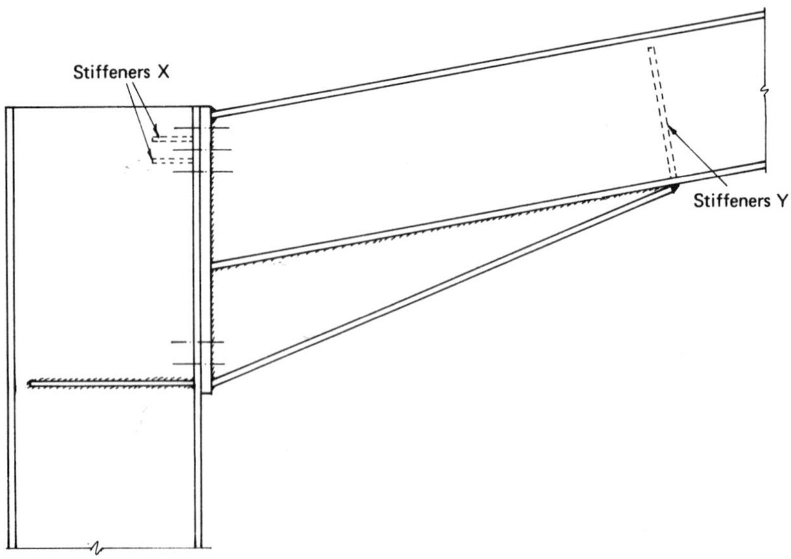

Figure 14.6 A popular eaves connection for plastically designed portal frames

should be provided to resist the local loads on the already plastic section.

Figure 14.7 illustrates various stiffening arrangements for the corner panel which is subject to high shear forces from the moment rotation at the corner that was discussed in Section 14.2.1. A diagonal, tension stiffener (called a Morris stiffener) may be used and this can combine the function of column flange stiffening with web shear stiffening. Full-strength tension welds are required at both ends and it must be used in conjunction with the fitted horizontal compressive stiffener that is shown. While its satisfactory behaviour has been demonstrated by tests for rolled section frames of conventional proportions, it should be used with caution in situations where the effective lever arm (i.e. the distance between the diagonal stiffener and the compressive line of action at the base of the haunch) is significantly less than the column depth. The necessity for maintaining an adequate lever arm around the connection is best demonstrated by considering the stiffening arrangement shown in Figure 14.7(b) for lightweight fabricated sections.[2] Although the slender web panel (with d/t perhaps approaching 200, where d is the panel depth and t is its thickness) may seem to be adequate when checked by traditional design methods, it is necessary to recall that these design approaches rely on a 'tension field' mode of shear resistance. Because of its slenderness, this panel is not acting as a conventional shear panel but is resisting the load by the diagonal band of tension shown, in conjunction with the compressive forces necessary for equilibrium in the horizontal stiffener and vertical column flange and rafter end plate. The lever arm between the tension and compressive forces is thus less than that of the rafter or column within the connection. This reduction in lever arm leads to a local increase in compressive force at the inside corner of the connection, which can result in premature failure of that part of the connection. With a slender corner web panel (say, with d/t of more than 120) where significant tension field action can be expected, the stiffening arrangement shown in Figure 14.7(c) is to be preferred. By providing a compressive diagonal stiffener the tension load path is pushed out away from the inside corner of the connection and a satisfactory lever arm is achieved throughout.

For frames of conventional proportions, no special consideration is necessary for the detailed design of the haunch other than aspects of lateral stability which are beyond the scope of this text. Economy of bolting leads naturally to the use of the greatest depth of haunch that can be obtained from a cutting of the main rafter section. The moment gradient in the rafter ensures that the haunch will have a considerable length, usually of the order of 10% of the frame span. Continuous welds should always be used between the haunch web and both the main rafter and the end plate in order to avoid any local instability. With conventional proportions continuous welds of minimum size will achieve satisfactory connections between the haunch and rafter. Section 12.4.2 discusses the design of haunches of unusual proportions in the context of the short, steep haunches that may be used for rigid beam-to-column connections.

14.3 Apex connections

It is possible to have the same categorization of apex connections as eaves connections.

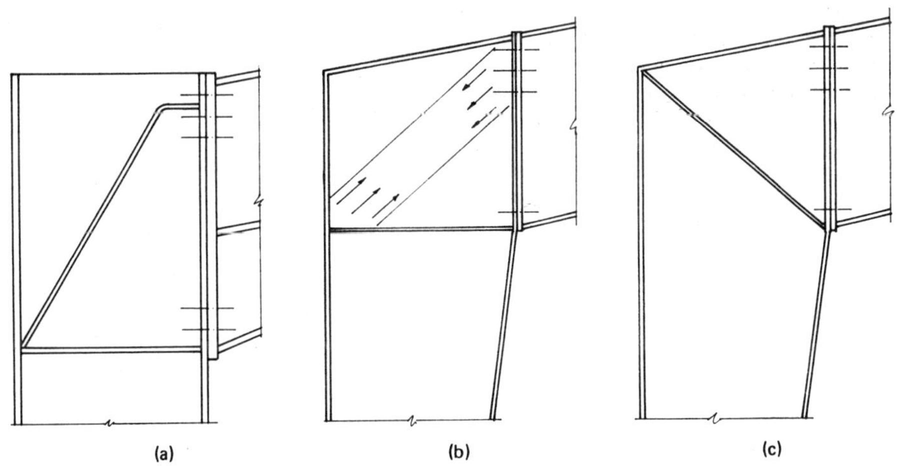

Figure 14.7 Stiffening arrangements for eaves connections. (a) Morris stiffeners for haunched connection; (b) horizontal stiffener for fabricated sections; (c) diagonal stiffener for fabricated sections

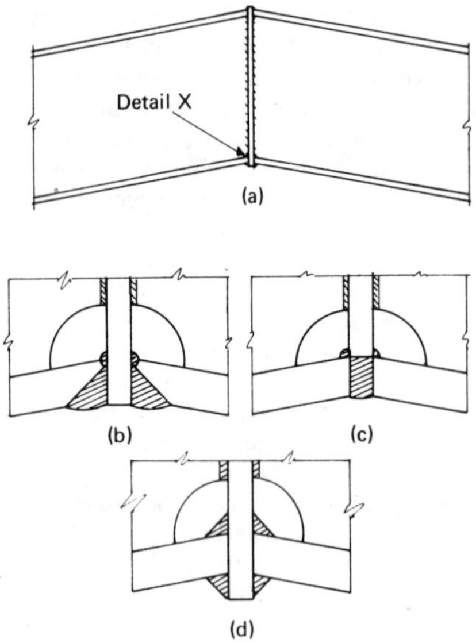

Figure 14.8 Welded apex connections. (a) General arrangement; (b)–(d) different approaches to the tension flange connection

14.3.1 Butt-welded connections

Figure 14.8 shows the various types of butt- and fillet-welded apex connections. These are subject to broadly the same considerations as eaves connections, particularly with regard to the crucial tension flange weld.

14.3.2 Splice plate bolted joints

These connections are not commonly used because they are generally less economic than end-plate connections; Figure 14.9 shows the most efficient type. Because the connection is generally subject to sagging moments it is necessary to provide haunches to overcome out-of-plane effects on the tensile splice plates. On the compressive load path bent splice plates would be most inefficient because they would have to be designed for out-of-plane bending; load transfer should therefore be by bearing.

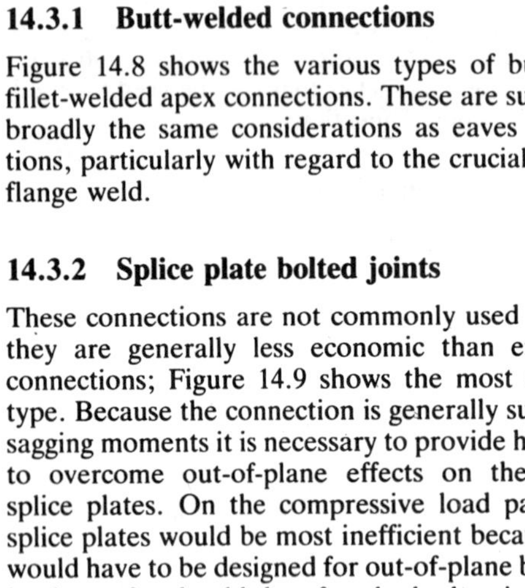

Figure 14.9 Bolted apex connection with splice plates. (This is unlikely to be an economic connection in practice)

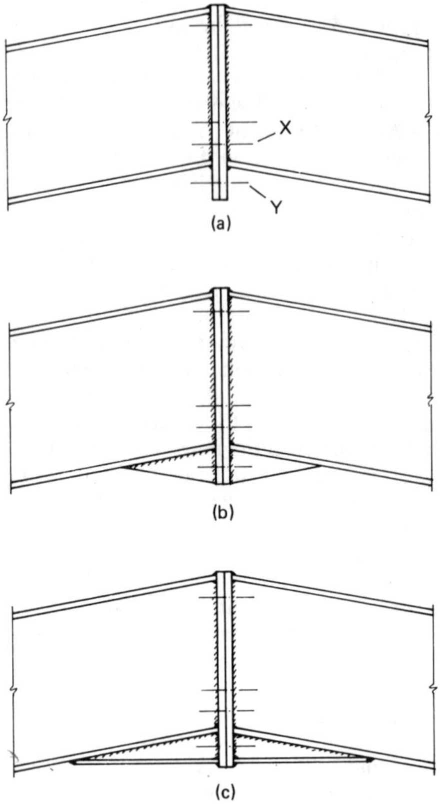

Figure 14.10 Bolted apex connections with end plates

14.3.3 End-plate connections

Figure 14.10 shows the basic types of extended end-plate apex joint. In Figure 14.10(a) the end plate is simply extended downwards, in Figure 14.10(b) it has a simple gusset reinforcement to the end-plate extension and in Figure 14.10(c) full haunches are provided.

The most straightforward connection (a) will behave satisfactorily if a reasonably conservative approach is taken to the end-plate thickness, i.e. that it is designed elastically. However, in plastically designed portal frames it is becoming increasingly common to use plastic methods of design for end plates. This reduces the end-plate thickness to the same order as the bolt diameter and leads to connections where local deformations of the end plate significantly influence bolt force distributions. In this simple connection this could cause bolts X to be carrying significantly greater forces than bolts Y, because the former are loaded by stiffer parts of the end plate. This can seriously reduce the efficiency of this type of connection. The solution is either to use elastic design for the end plates (which will increase their thickness) or to reinforce the extension to the end plate, as shown in Figures 4.10(b) and (c). Note that this is less of a problem for eaves connections because the column provides additional in-plane restraint to the end plate.) Where the frame is made of fabricated sections of variable depth it is possible to use short end-plate connections, because of the greater lever arm and correspondingly reduced tensile forces in such construction.

14.4 Stability in portal frame connections

All portal frames are made with sections that have a high ratio of I_x:I_y. The elements of these frames are therefore prone to lateral–torsional buckling. There is reasonable guidance for the provision of stabilizing bracing along the lengths of these members, but less about the stabilizing of the connections between these elements.

The apex connection creates the lesser problem because the compression flange is at the top, and this is generally stabilized at intervals along its length by the purlins. However, it should be ensured that reasonable continuity of stiffness of the compression flange is achieved through the apex connection and that the ridge purlins are not more than 0.5 m or so from the connection.

The eaves connection presents a much greater problem, because the compression flange is on the inside of the frame and therefore is not stabilized by the purlin or eaves beam. Three classes of situation should be recognized:

1. Situations where there is inadequate continuity of stiffness in any direction through the inside corner of the connection, as shown in Figure 14.11(a). In this instance, whether elastic or plastic design is used, a direct restraint to the inside rafter or column flange must be provided within 0.5 m of the corner (position U).
2. Situations where elastic design is used and where stability of the inside corner of the connection is only required up to the point where the compression flange reaches nominal yield stress. In these instances continuity of the rafter compression flange, as in Figure 14.11(b), through to the tension load path will be sufficient to ensure stability, and the elements concerned may be designed on the basis of effective lengths equal to the distance between the tension flange and the first braced position of the inside flange.
3. Frames designed plastically where rotation capacity is required at a hinge position in the immediate vicinity of the connection. In such situations direct bracing must be provided close to the inside corner. If a hinge is anticipated in either the rafter at XX in Figure 14.11(b) or in the column at YY in Figure 14.11(c) then bracing must be provided within 0.5 m of the inside corner points V or W, respectively.

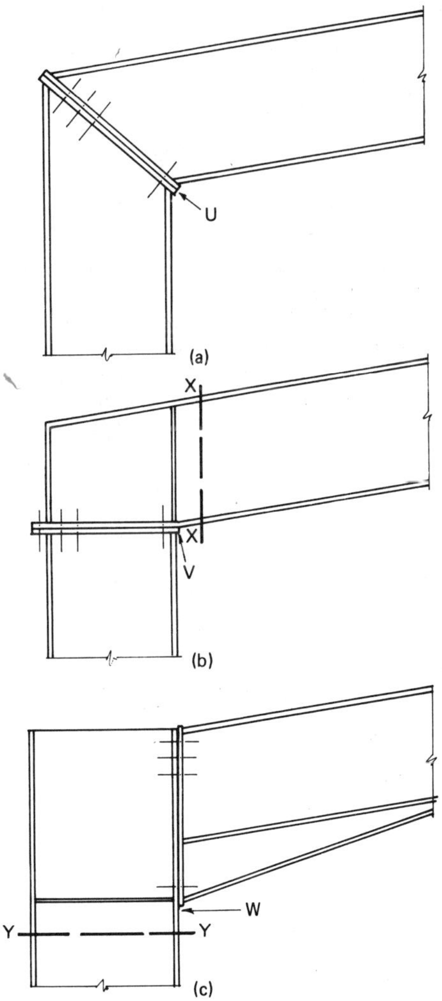

Figure 14.11 Stability of eaves connections

References

1. Horne, M. R. and Morris, L. J., *Plastic Design of Low-rise Frames*, Granada, St Albans, 1981.
2. Dowling, P. J., Mears, T. F., Owens, G. W. and Raven, G. R., 'A development in the automated design and fabrication of portal framed industrial buildings', *The Structural Engineer*, **60A**, No. 10 (1982).
3. Dowling, P. J., Mears, T. F., Owens, G. W. and Raven, G. R., 'Discussion', *The Structural Engineer*, **61A**, No. 12 (1983).

Commentary: 1/1

Section 14.2.5

The example illustrates the design of a bolted eaves connection where the end plate and the column are extended above the level of the top flange of the rafter. If the alternative detail, finishing the end plate and column at the level of the top flange, were adopted it would be necessary to stiffen the column in the tension region or increase the depth of the haunch. However, in this example the depth of haunch required, to avoid stiffening the column, would be excessive. After sketching the proposed eaves connection and before making any calculations it is advisable to check that the lower purlin and the gutter can be accommodated. With the extended end plate connection, the top of the outside flange and part of the column web could be cut away to accommodate the gutter.

The example illustrates the design of the various parts of the connection, but in practice the complete design of all the parts may not be necessary. For example, here the beam web and the 6 mm fillet welds could be seen to be adequate without doing the load partition calculation.

Section 5.3.2 (equation 5.2.1)

HSFG bolts, which have a heavy series head and nut, are chosen as being suitable for use in tension. Neither the shear capacity nor slip of the joint are critical, therefore the bolts can be classed as 'non-preloaded' when applying the formulae for the minimum prying force (if required) and the distance n at which the prying force acts.

Some preloading of the bolts is advisable, as it increases the stiffness of the connection (and therefore the frame). If power-operated tools are to be used for tightening, make sure that adequate clearance is provided. (The clearances in this example would not be adequate for some power-operated tools.)

The formula for n, which is used to ensure that the bolt strain is acceptable, governs the position of the prying force in this example.

Load paths

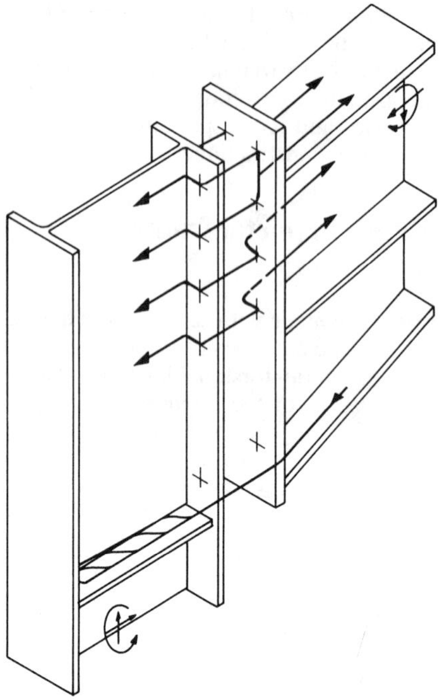

For clarity only the nearside load paths are shown. The moment in the beam is carried by tension in the top flange and upper portion of the web; bending in the end plate; tension in the bolts; bending in the column flange and tension/shear in the column web, together with compression in the bottom (haunch) flange and compression/shear in the stiffeners and column web.

Structural Steelwork Connections	Subject: Bolted end plate splice for eaves connection of pitched portal frame		Chapter Ref. 14
	Design Code: BS 5950 Part 1		Calc. Sheet No. Example 1/1
	Calc. by B.D.C	Date Aug, '87	Check by G.W.O.	Date Nov, '87

Code Ref.	Calculations	Output

Bolted end plate splice for eaves connection of pitched portal frame.

Design a connection between a 356×171×51 UB rafter (with haunch) in Grade 50 steel and a 406×178×60 UB column in Grade 43 steel.

The connection is to carry the following factored loads at the end of the rafter:

Bending moment	314 kNm
Axial load in rafter	64 kN (compression)
Shear at end of rafter	112 kN

406 × 178 × 60 UB Column.
356 × 171 × 51 UB (Grade 50) rafter.
Haunch out of 356×171×51 UB (Grade 50)

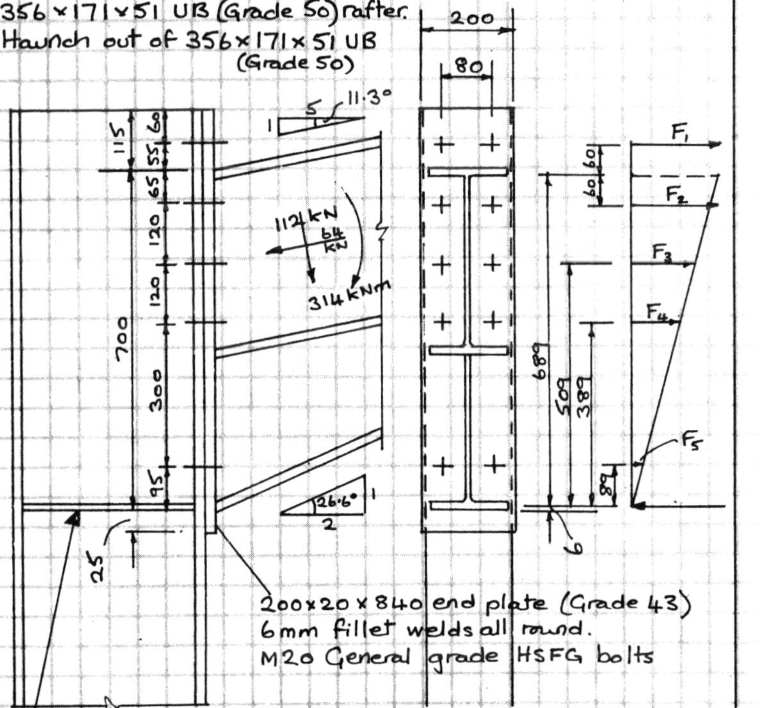

Commentary: 1/2

Section 8.3

The connection is assumed to 'pivot' about the 'hard spot' at the bottom flange (haunch) and the loads in the bolts are assumed to be proportional to their distance from the centre of the bottom flange. However, some account is taken of the greater flexibility of the cantilever end plate supporting bolts F_1, compared with the portion of plate supporting bolts F_2, which is stiffened by the beam web, by assuming that the loads in bolts F_1 and F_2 are equal. Note that the column flange is checked before the design of the end plate, since the column section is already specified. (If the first check is unsuccessful, the setting out and/or the design concept could be modified to justify the column section.)

Structural Steelwork Connections	Subject: Bolted end plate splice for eaves connection of pitched portal frame			Chapter Ref. 14
	Design Code: BS 5950 Part 1			Calc. Sheet No. Example 1/2
	Calc. by B.D.C	Date Aug, '87	Check by G.W.O.	Date Nov, '87

Code Ref.	Calculations	Output

Distribution of load

Try bolt spacing in sketch

By moments about bottom flange

$$314 \times 10^3 - (64 \cos 11{\cdot}3 - 112 \sin 11{\cdot}3) \times 513$$

$$= (2F_1 + 2F_2) \times 689 + 2F_3 \times 509 + 2F_4 \times 389 + 2F_5 \times 89$$

$$= \frac{2F_1}{689}\left[2 \times 689^2 + 509^2 + 389^2 + 89^2\right]$$

$F_1 = F_2 = 73{\cdot}8$ kN

$F_3 = 54{\cdot}5$ kN

$F_4 = 41{\cdot}7$ kN

$F_5 = 9{\cdot}5$ kN

Reaction at bottom flange of haunch $= F_c$

$$= 2(73{\cdot}8 + 73{\cdot}8 + 54{\cdot}5 + 41{\cdot}7 + 9{\cdot}5) + 64 \cos 11{\cdot}3 - 112 \sin 11{\cdot}3$$

$$= 547{\cdot}4 \text{ kN}$$

Column flange

Use HSFG General Grade bolts and classify bolts as "non-preloaded".

Distance from centre line of bolt to centre line of root fillet $= b = \frac{80}{2} - \frac{7{\cdot}8}{2} - \frac{10{\cdot}2}{2}$

$= 31{\cdot}0$ mm

Distance from centre line of bolt to prying force $= n =$ lesser of edge distance and $1{\cdot}1T\sqrt{\frac{\beta p_o}{p_y}}$

$$= 1{\cdot}1 \times 12{\cdot}8 \sqrt{\frac{2 \times 587}{275}}$$

$= 29{\cdot}1$ mm

Effective length of flange per bolt $= w$

$= 2(b+d) = 2(31+20) = 102$ mm

(assuming M20 bolts)

Commentary: 1/3

Section 7.7.2

As discussed in section 7.7.2, the traditional approach to determining w, i.e. assuming a spread of 60° each side of the bolt line, becomes unsafe when $b > 3a$. There are no such difficulties with the American approach of spreading at $b + d$ either side of the bolt line and this is adopted here, even though $b \not> 3a$, to demonstrate its use. A maximum spread of $(a + 1.4b)$ could be used instead of $\sqrt{(3)}b$ by assuming the yield line pattern in Figure 7.17(g). It would then be necessary to design the end plate in single curvature to maintain compatibility of the yield lines.

Moment capacity = p_y × plastic modulus. $w > b$, therefore the bolt holes are ignored. Tension capacity of web = $p_y wt$.

In determing w for the web design a maximum load spread of d + 45° from bolt line to centreline of web would be reasonable.

Section 7.4.1

Local bearing capacity = (Effective bearing length at root line of column) × (web thickness) × P_y

Effective bearing length = stiff bearing (beam flange) + spread at 45° through end plate + spread at 1 in 2.5 through column flange and root radius.

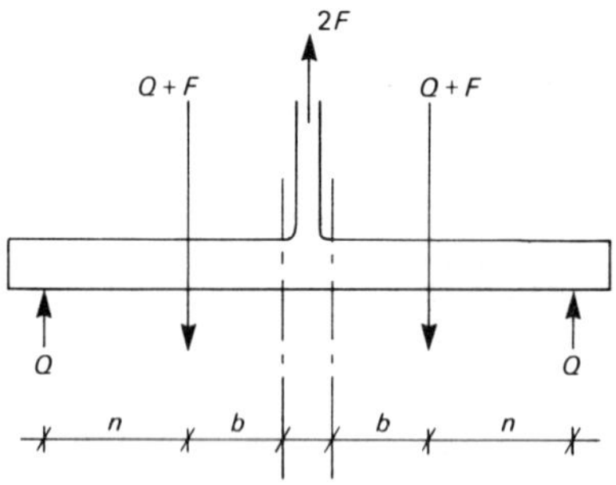

Structural Steelwork Connections	Subject: Bolted end plate splice for eaves connection of pitched portal frame			Chapter Ref. 14
	Design Code: BS 5950 Part 1			Calc. Sheet No. Example 1/3
	Calc. by B.D.C	Date Aug, '87	Check by G.W.O.	Date Nov, '87

Code Ref.	Calculations	Output
	Using "minimum thickness/maximum prying force" design for bolt F_1	
	$M = \frac{F_1 b}{2} = \frac{73 \cdot 8 \times 31}{2} = 1144$ Nm	
	Moment capacity $= p_y \times \frac{wT^2}{4}$	
	$= \frac{275 \times 102 \times 12 \cdot 8^2 \times 10^{-3}}{4}$	
	$= 1149$ Nm	
	$>$ Design moment (M) o.k.	
	Prying force $= \frac{M}{n} = \frac{1144}{29 \cdot 1} = 39 \cdot 3$ kN	
	Bolt load $= 73 \cdot 8 + 39 \cdot 3 = 113 \cdot 1$ kN	
	Column web.	
4·6·1	Tension capacity of 120mm length of web	
	$= 275 \times 120 \times 7 \cdot 8 \times 10^{-3}$	
	$= 257$ kN	
	$> 2F_1$ o.k.	
	Assuming 20mm thick end plate, local bearing capacity of the column web for the load applied by the bottom flange of the haunched beam	
4·5·1·3 and 4·5·3	$= (11 \cdot 5 + 2 \times 20 + 2 \times 2 \cdot 5 \times (12 \cdot 8 + 10 \cdot 2)) \times 7 \cdot 8 \times 275 \times 10^{-3}$	
	$= 357 \cdot 1$ kN	
	$< F_c$ Stiffeners required	
	Stiffeners	
4·5·4·2	Area of stiffeners required (in contact with flange) $= \frac{0 \cdot 8 F_c}{p_{ys}}$	
	$= \frac{0 \cdot 8 \times 547 \cdot 4 \times 10^3}{275}$	
	$= 1592$ mm^2	

Commentary: 1/4

Area of effective section = (Area of stiffeners) + (Area of web, up to 20 × web thickness each side of stiffener)

I = second moment of area of stiffeners about centreline of web

Buckling resistance = (effective area) × p_c

The stiffeners are assumed to be fitted to the loaded flange and nominal 6 mm fillet welds are adequate. (See commentary on Example 6/12 of Chapter 12 for note on stiffener to column web welds.)

Shear capacity = 0.6 p_y × (web thickness) × (overall depth of section).

Structural Steelwork Connections	Subject Bolted end plate splice for eaves connection of pitched portal frame			Chapter Ref. 14
	Design Code BS 5950 Part 1.			Calc. Sheet No. Example 1/4
	Calc. by BDC	Date Aug, '87	Check by G.W.O	Date Nov, '87

Code Ref.	Calculations	Output
	Try 2 No 80 × 12·5 stiffeners (with 15 mm snipes)	
	Area in contact $= 2(80-15) \times 12{\cdot}5$	
	$= 1625\ mm^2$	2 No 80 × 12·5 stiffeners (15 mm snipes)
	O.k. Use	
4·5·1·5	Buckling resistance	
	Area of effective section	
	$= 2 \times 80 \times 12{\cdot}5 + 2 \times 20 \times 7{\cdot}8^2$	
	$= 4434\ mm^2$	
	$I = \frac{12{\cdot}5 \times 167{\cdot}8^3}{12} = 4{\cdot}92 \times 10^6$	
	$r = \sqrt{\frac{4{\cdot}92 \times 10^6}{4434}} = 33{\cdot}3\ mm$	
	$\frac{L}{r} = \frac{0{\cdot}7 \times (406{\cdot}4 - 2 \times 12{\cdot}8)}{33{\cdot}3} = 8{\cdot}0$	
Table 27(c)	$p_c = 275\ N/mm^2$	
	Buckling resistance $= 4\,434 \times 275 \times 10^{-3}$	
	$= 1219\ kN$	
	$> F_c$ o.k.	
	Connection of stiffeners to web of column	
	Effective length of weld	
	$= 406{\cdot}4 - 2 \times (12{\cdot}8 + 15 + 6)$	
	$= 338{\cdot}8\ mm$	
	Load per mm $= \frac{547{\cdot}4}{4 \times 338{\cdot}8} = 0{\cdot}404\ kN$	Stiffeners to column, 6 mm fillet welds
	Use 6mm fillet welds	
	Column web shear	
	Maximum shear in column (at bottom flange of haunch) $= 2(73{\cdot}8 + 73{\cdot}8 + 54{\cdot}5 + 41{\cdot}7 + 9{\cdot}5)$	
	$= 506{\cdot}6\ kN$	
4·2·3	Shear capacity $= 0{\cdot}6 \times 275 \times 7{\cdot}8 \times 406{\cdot}4 \times 10^{-3}$	
	$= 523{\cdot}0\ kN > 506{\cdot}6\ kN$. O.k	

Commentary: 1/5

Distance b is measured to the toe of a fillet weld but the half radius for the root fillet of a rolled section.

$b + d = 49 + 20 = 69\,\text{mm} > \frac{1}{2} \times$ pitch of 80 mm
and > edge distance of 60 mm

Therefore

$\frac{1}{2} \times 80 + 60 = 100\,\text{mm}$ governs w

By inspection, it can be seen that the end plate at other bolt positions is also adequate. (F_2 is a similar distance from the flange to F_1 and is also supported by the web. F_3 has a smaller distance b and a greater length w than F_1.)

It has been assumed that a section of the column flange and the top section of the end plate act as short Tee stubs in double curvature, with the webs of the Tees at right angles. If it were necessary to use an effective length of yield line greater than $2(b + d)$ to justify the column flange the possibility of incompatibility between the yield lines in the column flange and the end plate arises. For example, if the column flange deformed in a manner similar to that illustrated in Figure 7.17(e) it would tend to prevent the yield lines forming through the bolt holes in the extended portion of the end plate. Where this could arise, the end plate should be designed in single curvature bending, i.e. $M = F_b$.

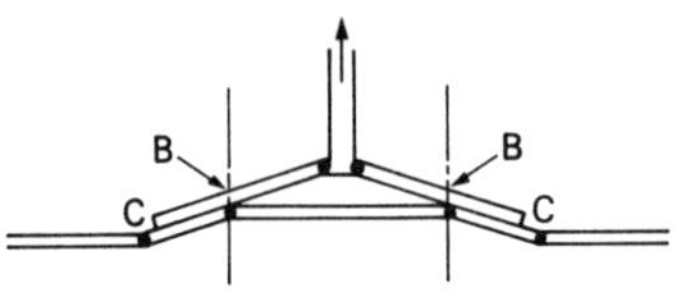

Section through bolt lines, showing Tee section resting on deformed shape of long flange. Flanges of Tee are not bent to form hinges at B.

The bolt is designed using the higher of the prying forces from the column flange and end-plate calculations. The bolts are designed for the bolt load including the prying force, regardless of whether the bolts are designed as ordinary bolts or 'friction grip fasteners'. The procedure in BS 5950: Part 1, Clause 6.3.6, where prying action is ignored, is not used as it could lead to an unsafe connection. The tension capacity for design using factored loads is taken as:

$$\frac{Y_f A_t}{1.1\gamma_m} \text{ but } \not> \frac{0.7 U_f A_t}{1.1\gamma_m}$$

(In BS 5950, the material factor γ_m is equal to 1.0.) For general grade HSFG bolts, $0.7U_f A_t/1.1$ is virtually the same as $0.9P_0$, the tension capacity in BS 5950: Part 1, Clause 6.4.4.2.

Since the prying force is greater than 10% of the applied load (F), there is no need to check the bolts for compliance with BS 5950: Part 1, Clause 6.3.6.

Note that if the rafter and haunch had been in Grade 43 steel, the haunch flange would have been inadequate. Options available to overcome the problem would be:

1. Make the haunch out of plate material;
2. Use a heavier section for the haunch;
3. Use the same section but in Grade 50;
4. Deepen the haunch (but the haunch required would be too deep to use a cutting from the rafter section).

Structural Steelwork Connections	Subject: Bolted end plate splice for eaves connection of pitched portal frame		Chapter Ref. 14
	Design Code: BS 5950 Part 1		Calc. Sheet No. Example 1/5
	Calc. by B.D.C	Date Aug, '87	Check by G.W.O.	Date Nov, '87

Code Ref.	Calculations	Output
	End plate	
	Consider portion above top flange (assume 6 mm fillet welds to flange and 200 mm wide end plate.)	
	Distance from centre line of bolt to toe of fillet weld $= b = 55 - 6 = 49$ mm	
	Distance from centre line of bolt to prying force $= n =$ lesser of edge distance and $1.1T\sqrt{\frac{\beta p_o}{p_y}}$	
	$= 1.1 \times 20\sqrt{\frac{2 \times 587}{265}}$	
	$= 46.3$ mm	
	Effective length of end plate per bolt $= w = \frac{1}{2} \times 200 = 100$ mm	
	$M = \frac{F_1 b}{2} = \frac{73.8 \times 49}{2} = 1808$ Nm	
	Moment capacity $= p_y \times \frac{wT^2}{4}$	
	Minimum $T = \sqrt{\frac{4 \times 1808 \times 10^3}{265 \times 100}}$	
	$= 16.5$ mm	200 × 20
	Use 20 mm	end plate × 840 Grade 43
	Prying force $= \frac{M}{n} = \frac{1808}{46.3} = 39.0$ kN	
	The prying force from the column flange being greater governs the design of the bolts.	
	Bolts	
	Tension capacity of M20 General Grade HSFG bolt $= \frac{0.7 U_f A_t}{1.1} = \frac{0.7 \times 827 \times 245 \times 10^{-3}}{1.1}$	
	$= 128.9$ kN $>$ Bolt load $= 113.1$ kN	
6.3.6.3	By inspection, bolt group can also carry the shear load of $(64 \sin 11.3 + 112 \cos 11.3) = 122$ kN	M 20 General Grade HSFG bolts.

Commentary: 1/6

$$\text{Load per mm of weld} = \frac{\text{Load from adjacent bolts}}{\text{Flange width}}$$

No deduction is required for end craters, as the weld is returned round the edge of the flange.

The design rules for symmetrical fillet welds, in BS 5950: Part 1, Clause 6.6.5.1 are not used for two reasons:

1. The welds and loads are not strictly symmetrical;
2. Due to their limited ductility, it is important that the welds do not form a weak link in the connection.

Section 7.7.4

The factor of 0.7 used in determining the effective lengths of the plates allows for the formation of fan yield lines in the corners.

Structural Steelwork Connections	Subject: Bolted end plate splice for eaves connection of pitched portal frame			Chapter Ref. 14
	Design Code: BS 5950 Part I			Calc. Sheet No. Example 1/6
	Calc. by B.D.C	Date Aug, '87	Check by G.W.O.	Date Nov, '87

Code Ref.	Calculations	Output
	Beam flanges.	
	Capacity of flange $= p_y A$	
	$= 355 \times 171.5 \times 11.5 \times 10^{-3}$	
	$= 700.1$ kN	
	Tension in top flange $= \dfrac{2F_1 + 2F_2}{\cos 11.3}$	
	$= \dfrac{4 \times 73.8}{\cos 11.3} = 301.0$ kN	
	< 700.1 kN O.K.	
	Compression in haunch flange $= \dfrac{F_c}{\cos 26.6}$	
	$= \dfrac{547.4}{\cos 26.6} = 612.2$ kN	
	< 700.1 KN O.K.	
	Welds.	
	Top flange to end plate	
	Load from bolts above flange $= 2F_1$	
	$= 2 \times 73.8 = 147.6$ kN	
6·6·5·2	Load per mm of weld $= \dfrac{147.6}{171.5} = 0.861$ kN	
6·6·5	Capacity of 6mm fillet weld	
	$= 6 \times 0.7 \times 215 \times 10^{-3}$	
	$= 0.903$ kN/mm O.K. Use	
	Web to end plate	
	Consider bolt below flange.	
	Load $= F_2 = 73.8$ kN	
	Centre line of bolt to edge of weld to beam flange $= b_x = 65 - 11.5 - 6 = 47.5$ mm	
	Centre line of bolt to edge of weld to beam web $= b_y = \dfrac{80}{2} - \dfrac{7.3}{2} - 6 = 30.4$ mm	
	Effective length of plate at flange	
	$= W_x = \dfrac{171.5 - 80}{2} + 0.7 \times 30.4 = 67.0$ mm	

Commentary: 1/7

The bolt load is shared between the web and the flange in proportion to w/b when assuming a plastic distribution and w/b^3 when assuming an elastic one.

In this example it can be seen by inspection that the level of loading on the beam web and fillet welds adjacent to bolt F_3, etc. is less than that calculated adjacent to bolt F_2.

Section 7.7.4

In recognitiion of the limited ductility of fillet welds the greater of the loads calculated by assuming plastic or elastic distribution of load is used for their design.

The load is assumed to be applied on the line of the bolt. The effective length of weld is taken as the lesser of 2 × ($\frac{1}{2}$ bolt pitch) or $2b_x$. Taking only the length of weld that is equally balanced about the load as being effective is a quick and economic way of designing line welds with an eccentrically applied load. Although for some details an effective length of up to $2(a_y + 1.4b_y)$ could be assumed for the design of the plate, it is likely that there would be a concentration of reaction opposite the bolt. In view of this it would be prudent to add a further limitation of effective length of weld $\not> 2(b_y + d)$ to avoid problems due to lack of ductility in the fillet welds.

b_y = distance of bolt from fillet weld
d = bolt diameter

Structural Steelwork Connections	Subject: Bolted end plate splice for eaves connection of pitched portal frame			Chapter Ref. 14
	Design Code: BS 5950 Part 1			Calc. Sheet No. Example 1/7
	Calc. by B.D.C	Date Aug, '87	Check by G.W.O.	Date Nov, '87.

Code Ref.	Calculations	Output

Effective length of plate at web

$$= W_Y = \frac{120}{2} + 0{\cdot}7 \times 47{\cdot}5 = 93{\cdot}2 \text{ mm}$$

Bolt load supported by web (plastic distribution) =

$$\left[\frac{\frac{93{\cdot}2}{30{\cdot}4}}{\frac{67{\cdot}0}{47{\cdot}5} + \frac{93{\cdot}2}{30{\cdot}4}}\right] \times 73{\cdot}8 = 50{\cdot}5 \text{ kN}$$

Bolt load supported by web (elastic distribution) =

$$\left[\frac{\frac{93{\cdot}2}{30{\cdot}4^3}}{\frac{67{\cdot}0}{47{\cdot}5^3} + \frac{93{\cdot}2}{30{\cdot}4^3}}\right] \times 73{\cdot}8 = 62{\cdot}1 \text{ kN}$$

Elastic distribution governs design of welds

Effective length of weld = $2 \times 47{\cdot}5$
$= 95{\cdot}0$ mm

Load per mm of weld $= \frac{62{\cdot}1}{95{\cdot}0} = 0{\cdot}65$ kN

Use 6 mm fillet weld

Bottom (haunch) flange to end plate

Horizontal component of flange force
$= F_c = 547{\cdot}4$ kN

Vertical (shear) component of flange force
$= \frac{1}{2} \times 547{\cdot}4 = 273{\cdot}7$ kN

"Fit" end plate to flange and assume that horizontal compression is taken in direct bearing and that vertical shear is carried by fillet welds.

Load per mm of weld $= \frac{273{\cdot}7}{2 \times 171{\cdot}5} = 0{\cdot}798$ kN

Use 6 mm fillet welds

Output: Weld end plate to beam and haunch with 6 mm fillet welds all round

Commentary: 1/8

When checking the web of the beam, as with the weld, the load is assumed to be applied on the line of the bolt (and the effective length is $2b_x$). As a final check, look at the connection to make sure that there are no incompatibilities in the assumptions made for the design of the column flange and the end plate. In particular, compare bolt loads, prying forces and yield line patterns. In this example the bolt loads have already been considered. A possible problem with the compatibility of the yield lines has also been dealt with.

Structural Steelwork Connections	Subject: Bolted end plate splice for eaves connection of pitched portal frame			Chapter Ref. 14
	Design Code: BS 5950 Part 1			Calc. Sheet No. Example 1/8
	Calc. by B.D.C	Date Aug, '87	Check by G.W.O	Date Nov, '87

Code Ref.	Calculations	Output
	Web of beam	
	Load per mm of web $= \frac{2 \times 50.5}{95.0} = 1.06$ kN	
	Tension capacity of web $= p_y t$	
	$= 355 \times 7.3 \times 10^{-3}$	
	$= 2.59$ kN/mm	
	O.K.	

Commentary: 2/1

The example illustrates the design of an apex connection for a pitched portal frame. A haunch is used to increase the lever arm of the bolts and at the same time reduce the compressive reaction on the top flange. The compression in the top flange is close to the flange capacity. The bending moment of 262 kNm is 82% of the plastic moment capacity of the rafter. If the proportions of the frame required the full moment capacity of the rafter to be developed at the connection it would be necessary to extend the end plate and add an extra row of bolts below the haunch flange or increase the depth of the haunch.

Section 8.3

The connection is assumed to 'pivot' about the 'hard spot' at the top flange and the loads in the bolts are assumed to be proportional to their distance from the centre of the flange.

Load paths

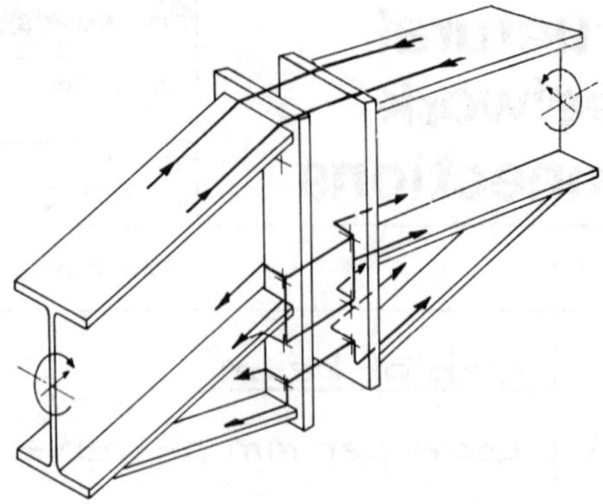

For clarity only the nearside load paths are shown. The moments in the beams are carried by tension in the haunches and bottom flanges; bending in the end plates and tension in the bolts, together with compression in the top flanges.

Structural Steelwork Connections	Subject: Bolted end plate splice for apex connection of pitched portal frame			Chapter Ref. 14
	Design Code: BS 5950 Part 1			Calc. Sheet No. Example 2/1
	Calc. by B.D.C.	Date Aug, '87	Check by G.W.O.	Date Nov,'87

Code Ref.	Calculations	Output

<u>Bolted end plate splice for apex connection of pitched portal frame.</u>

Design an apex connection between 356 × 171 × 51 UB rafters in Grade 50 steel.
The connection is to carry the following factored loads:

Bending moment 262 kNm
Axial load in rafter 49 kN (compression)

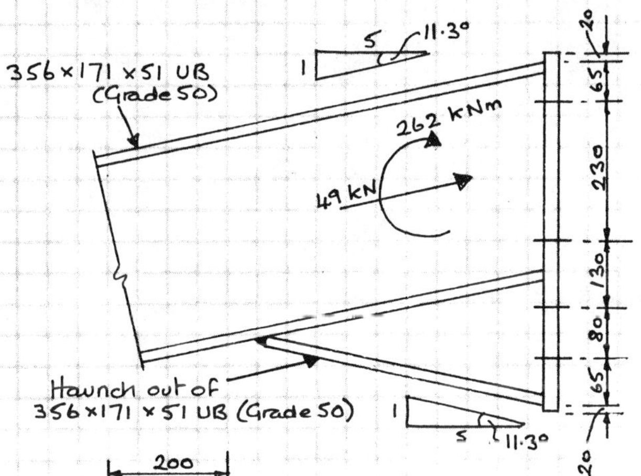

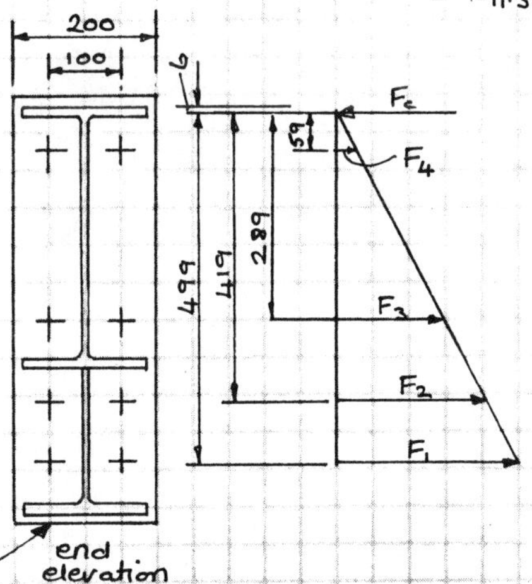

200 × 20 × 610 end plate (Grade 43)
6 mm fillet welds all round rafter
10 mm fillet welds all round haunch flange and web
M22 General grade HSFG bolts

Commentary: 2/2

The loads from bolts F_1, F_2 and F_3 are partly carried by the flanges and partly by the webs. However, since $2F_1 = 247.4$ kN and $2F_2 + 2F_3 = 350.8$ kN are less than the flange capacity the flanges must be adequate.

Section 7.7.2

Edge distance $< a + 1.4b$, therefore edge distance governs.

$\frac{1}{2} \times$ bolt pitch $< a + 1.4b$, therefore $\frac{1}{2} \times$ pitch governs.

Structural Steelwork Connections	Subject: Bolted end plate splice for apex connection of pitched portal frame			Chapter Ref. 14
	Design Code: BS 5950 Part 1			Calc. Sheet No. Example 2/2
	Calc. by B.D.C	Date Aug, '87.	Check by G.W.O.	Date Nov, '87

Code Ref.	Calculations	Output

Distribution of load.

Try bolt spacing in sketch

By moments about top flange

$$262 \times 10^3 - (49 \cos 11{\cdot}3) \times 175$$

$$= 2F_1 \times 499 + 2F_2 \times 419 + 2F_3 \times 289 + 2F_4 \times 59$$

$$= \frac{2F_1}{499}\left[499^2 + 419^2 + 289^2 + 59^2\right]$$

$F_1 = 123{\cdot}7$ kN
$F_2 = 103{\cdot}8$ kN
$F_3 = 71{\cdot}6$ kN
$F_4 = 14{\cdot}6$ kN

Reaction at bottom flange of haunch $= F_c$

$$= 2(123{\cdot}7 + 103{\cdot}8 + 71{\cdot}6 + 14{\cdot}6) + 49 \cos 11{\cdot}3°$$

$$= 675{\cdot}5 \text{ kN}$$

Beam flanges.

Capacity of flange $= p_y A$

$$= 355 \times 171{\cdot}5 \times 11{\cdot}5 \times 10^{-3}$$

$$= 700{\cdot}1 \text{ kN}$$

Compression in top flange $= \frac{F_c}{\cos 11{\cdot}3°} = \frac{675{\cdot}5}{\cos 11{\cdot}3°}$

$$= 688{\cdot}9 \text{ kN}$$

$$< 700{\cdot}1 \text{ kN}$$ O.K.

Bottom flange and haunch flange O.K. by inspection

End plate.

Consider portion associated with bolt F_1 (assume 200 mm wide end plate and 6 mm fillet weld around beam, but 10 mm fillet weld around haunch flange and web.)

Centre line of bolt to edge of weld to haunch flange $= b_x = 65 - 11{\cdot}5 - 10 = 43{\cdot}5$ mm

Centre line of bolt to edge of weld to haunch web $= b_y = \frac{100}{2} - \frac{7{\cdot}3}{2} - 10 = 36{\cdot}3$ mm

Commentary: 2/3

Section 7.7.4

Due to the 'pin' support at the flange, the load supported from the flange = $F_x = M_x/b_x$; whereas the load supported from the web (where two hinge lines form) = $F_y = 2M_y/b_y$.

B.M. diagrams

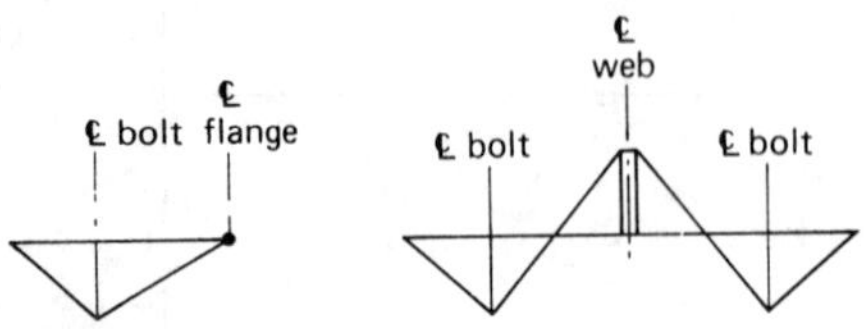

From flange

From web

The factor of 2 by which the calculations differ is the reason for the 0.5 that appears in the calculation for the distribution of the bolt load.

Section 7.7.2

Moment capacity = p_y × plastic modulus
$w > b$, therefore the bolt holes are ignored.

It follows from the procedure for distributing the bolt load that the minimum thickness (T) calculated in (a) and (b) will be the same.

By inspection, it can be seen that the end plate at other bolt positions is also adequate. The bolt load is less, the bolt spacing is not too different and, since the end plate is continuous between bolts F_2 and F_3 each side of the flange, there are additional plastic hinges.

Structural Steelwork Connections	Subject: Bolted end plate splice for apex connection of pitched portal frame			Chapter Ref. 14
	Design Code: BS 5950 Part1			Calc. Sheet No. Example 2/3
	Calc. by B.D.C	Date Aug, '87	Check by G.W.O.	Date Nov, '87

Code Ref.	Calculations	Output
	Effective length of plate at flange $= W_X = \frac{200-100}{2} + 0.7 \times 36.3 = 75.4$ mm	
	Effective length of plate at web $= W_Y = \frac{80}{2} + 0.7 \times 43.5 = 70.4$ mm	
	Assume a "pin" support at the haunch flange.	
	Bolt load supported by flange (plastic distribution) $= F_{1X}$ $= \left\{\dfrac{\frac{0.5 \times 75.4}{43.5}}{\frac{0.5 \times 75.4}{43.5} + \frac{70.4}{36.3}}\right\} \times 123.7 = 38.2$ kN	
	Bolt load supported by web (plastic distribution) $= F_{1Y} = 123.7 - 38.2 = 85.5$ kN	
	Using "minimum thickness" design	
	a) Plate supported from flange	
	$M_X = F_{1X} b_X = 38.2 \times 43.5 = 1662$ Nm	
	Moment capacity $= p_y \times \frac{WT^2}{4}$	
	$T = \sqrt{\frac{4 \times 1662 \times 10^3}{265 \times 75.4}} = 18.2$ mm	
	Use 20 mm	200 × 20 end plate × 610 (Grade 43)
	b) Plate supported from web	
	$M_Y = \frac{F_{1Y} b_Y}{2} = \frac{85.5 \times 36.3}{2} = 1552$ Nm	
	$T = \sqrt{\frac{4 \times 1552 \times 10^3}{265 \times 70.4}} = 18.2$ mm	

Commentary: 2/4

Section 5.3.2

See Commentary on Example 1/2 of this chapter on 'non-preloaded' bolts.

The corner detail (the portion of end plate loaded by the bolt and supported on two adjacent sides by the flange and the web) is being analysed as two separate plates supported at the flange and the web. Whereas the total applied load = $F = F_x + F_y$, the total prying force (Q) is not necessarily equal to $Q_x + Q_y$. The reason for this is that a prying force located diagonally from the bolt could act as Q_x and Q_y. The general rule would be Q equal to the greatest of Q_x or Q_y or $(Q_x + Q_y - p_y T^2/2)$.

The third requirement is necessary because the value of the prying force that can be located diagonally is limited by the possibility of the formation of a local yield line.

Tension capacity of bolts. See commentary on Example 1/5 of this chapter.

If the apex connection were part of the same frame as the haunch detail in Example 14/1 it would be good practice to use M22 general grade HSFG bolts for both connections.

The factors of 0.25 allow for the reduction in elastic stiffness due to the assumed pin support at the flange relative to the continuous support at the web.

In recognition of the limited ductility of fillet welds, the greater of the loads calculated by assuming plastic or elastic distribution of load is used for their design. The load is assumed to be applied on the line of the bolt.

Structural Steelwork Connections	Subject: Bolted end plate splice for apex connection of pitched portal frame		Chapter Ref. 14
	Design Code: BS 5950 Part 1		Calc. Sheet No. Example 2/4
	Calc. by B.D.C	Date Aug, '87.	Check by G.W.O.	Date Nov, '87

Code Ref.	Calculations	Output

Classify bolts as non-preloaded

Distance from centre line of bolt to prying force = n = lesser of edge distance and $1.1T\sqrt{\frac{\beta p_o}{p_y}}$

$$1.1T\sqrt{\frac{\beta p_o}{p_y}} = 1.1 \times 20\sqrt{\frac{2 \times 587}{265}} = 46.3 \text{ mm}$$

Plate supported from flange $n_x = 46.3$ mm

$$\text{Prying force} = Q_x = \frac{m_x}{n_x} = \frac{1662}{46.3} = 35.9 \text{ kN}$$

Plate supported from web $n_Y = 46.3$ mm

$$\text{Prying force} = Q_Y = \frac{m_Y}{n_Y} = \frac{1552}{46.3} = 33.5 \text{ kN}$$

The prying force as plate supported from the flange governs

Bolt load = 123.7 + 35.9 = 159.6 kN

Bolts.

Tension capacity of M22 General Grade HSFG bolt

$$= \frac{0.7\,U_f A_t}{1.1} = \frac{0.7 \times 827 \times 303 \times 10^{-3}}{1.1}$$

= 159.5 kN = bolt load. O.k. Use

Output: M22 General Grade HSFG bolts.

Welds.

Bolt load supported by flange (elastic distribution)

$$= \left[\frac{\dfrac{0.25 \times 75.4}{43.5^3}}{\dfrac{0.25 \times 75.4}{43.5^3} + \dfrac{70.4}{36.3^3}}\right] \times 123.7 = 16.7 \text{ kN}$$

Bolt load supported by web (elastic distribution) = 123.7 − 16.7

= 107.0 kN

Commentary: 2/5

The effective length of weld is taken as the lesser of $2 \times (\frac{1}{2}$ bolt pitch) or $2 \times b_x$. Taking only the length of weld that is equally balanced about the load as being effective is a quick and economic way of designing line welds with an eccentrically applied load. Although a calculation for the end plate to flange weld using the load of 38.2 kN indicates that a 6 mm fillet weld would be adequate, it is considered good practice to carry the 10 mm fillet weld round the flange. One reason for this is that a pinned support from the flange was assumed in the calculations, whereas in practice there would be some restraint from the flange which would increase the share of the bolt load carried by the flange. For other notes on welds see the commentary on Examples 1/6 and 1/7 in this chapter.

Section 7.7.4

Assuming a plastic distribution, the bolt load is shared between the supports in proportion to w/b. Assuming an elastic distribution, the load is shared in proportion to w/b^3.

Structural Steelwork Connections	Subject Bolted end plate splice for apex connection of pitched portal frame			Chapter Ref. 14
	Design Code BS 5950 Part 1			Calc. Sheet No. Example 2/5
	Calc. by B.D.C	Date Aug, '87	Check by G.W.O.	Date Nov, '87.

Code Ref.	Calculations	Output

Elastic distribution governs design of web welds.

Effective length of weld $= 2 \times \frac{80}{2} = 80$ mm

Load per mm of weld $= \frac{107.0}{80.0} = 1.34$ kN

6.6.5 Capacity of 10 mm fillet weld

$= 10 \times 0.7 \times 215 \times 10^{-3} = 1.50$ kN/mm

O.K. USE.

For bolt F_2

$b_x = 62.3 - 6 = 56.3$ mm $\quad b_y = 36.3$ mm

$W_x = 75.4$ mm $\quad W_y = \frac{80}{2} + 0.7 \times 56.3$

$= 79.4$ mm

Bolt load supported by flange (plastic distribution)

$$= \left[\frac{\frac{75.4}{56.3}}{\frac{75.4}{56.3} + \frac{79.4}{36.3}} \right] \times 103.8 = 39.4 \text{ kN}$$

Bolt load supported by web $= 103.8 - 39.4$

$= 64.4$ kN

Bolt load supported by flange (elastic distribution)

$$= \left[\frac{\frac{75.4}{56.3^3}}{\frac{75.4}{56.3^3} + \frac{79.4}{36.3^3}} \right] \times 103.8 = 21.1 \text{ kN}$$

Bolt load supported by web $= 103.8 - 21.1$

$= 82.7$ kN

Plastic distribution governs design of flange welds.

Commentary: 2/6

Since the web should have adequate ductility only the load derived assuming a plastic distribution of load is considered. The local capacity of the web is checked using a dispersion of 1:2.5.

Structural Steelwork Connections	Subject: Bolted end plate splice for apex connection of pitched portal frame			Chapter Ref. 14
	Design Code: BS 5950 Part 1			Calc. Sheet No. Example 2/6
	Calc. by B.D.C	Date Aug, '87	Check by G.W.O.	Date Nov, '87.

Code Ref.	Calculations	Output
	Effective length of weld $= \frac{171.5 - 100}{2} \times 2$	
	$= 71.5$ mm	
	Load per mm of weld $= \frac{39.4}{71.5} = 0.551$ kN	Weld end plate to beam with 6mm fillet welds all round and to haunch flange and web with 10mm fillet welds all round.
	Use 6mm fillet welds (capacity 0.90 kN/mm)	
	By inspection 6mm fillet welds are adequate for other beam to end plate welds.	
	<u>Web of haunch</u>	
	Load per mm of web $= \frac{2 \times 85.5}{80} = 2.14$ kN	
	Tension capacity of web $= p_y t = 355 \times 7.3 \times 10^{-3}$	
	$= 2.59$ kN/mm	
	> 2.14 kN/mm O.k	
	<u>Connection of haunch flange to rafter flange</u>	
	Use full strength butt weld for flange connection	
	If rafter flange at this point is fully stressed, flange force $= p_y A = 700.1$ kN	
	Component of force perpendicular to rafter $= 700.1 \tan(2 \times 11.3) = 291.4$ kN	
	Assume stiff bearing length = flange thickness $= 11.5$ mm	
	Local "bearing" capacity of web	
	$= (11.5 + 2 \times 2.5 \times (11.5 + 10.2)) \times 7.3 \times 355 \times 10^{-3}$	
	$= 311.0$ kN > 291.4 kN	
	No stiffeners required	

15

Other industrial building connections

15.1 Column brackets

Figure 15.1 shows typical examples of column bracket connections. Different structural forms are necessary for major and minor column axis eccentricity. In the former case the moment imposes local horizontal tensile and compressive forces on the column flange and these have to be transmitted into the column web. In the latter the moment is transmitted into the column flanges by shear of the welds or bolts. Of course, in both cases the bracket must transmit the shear into the column. Welded connections will generally be used for shop attachment and bolted brackets for site attachment. Bearing bolts may generally be used for statically loaded major axis brackets and torqued high-strength bolts for fatigue loading. Bearing bolts may be used for minor axis brackets if the overall design can accommodate the possible bracket rotation as the bolts slip into bearing. If the bracket bearing surface has to be held level then HSFG bolts should be used.

Many elementary design texts use brackets to illustrate methods of connection analysis. Such examples frequently consider only the forces on the welds and fasteners. This may beguile the unwary designer into thinking that bracket design is trivial. In reality, sound design of this form of connection requires the same attention to detail as other types. Because there are important differences in behaviour between the various types of bracket connection they are discussed separately below.

15.1.1 Major axis welded bracket

The traditional design of such brackets was based on the analysis of the weld group. Elastic analysis of the Tee-section weld was carried out and the deforma-

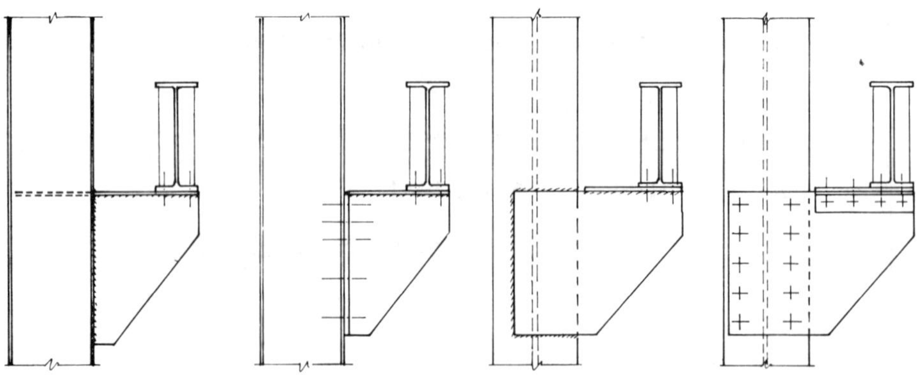

Figure 15.1 Column brackets. (a) and (b) Major axis eccentricity; (c) and (d) minor axis eccentricity

tion of other connection components was ignored. This was acceptable provided that:

1. This distribution of bending stresses was maintained when checking other connection components;
2. Weld sizes were sufficient for ductile behaviour and elements were of compact proportions so that any discrepancies between this rather simplistic analysis and true behaviour could be accommodated safely by redistribution.

For many brackets, the critical component will be the stiffening gusset, and it would be more relevant for the design to concentrate initially on this element. Provided that the top bracket flange can resist the tensile force, a triangular distribution of compressive stress would be an appropriate idealization for the gusset plate, as shown in Figure 15.2. Note that the maximum elastic compressive stress could be up to 33% greater than that calculated by this simplified procedure. However, this potential lack of conservatism is accommodated by the underestimate of compressive strength that arises from the simplified procedures of the following paragraph. (In checking the top flange tensile capacity the designer must ensure that its anchorage into the column web, via the dispersing action of the column flange, is satisfactory. The effective length of column web and the effective breadth of bracket flange should be determined from Section 7.4.2 and their corresponding tensile capacities checked.) If the top flange is not adequate in tension then either a stiffener should be provided or the tensile zone extended down the gusset plate with an appropriate modification to the distribution of compressive stresses.

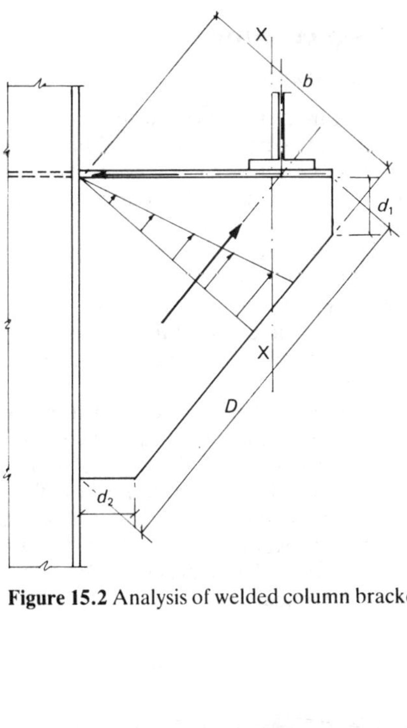

Figure 15.2 Analysis of welded column bracket

In reality, the gusset plate acts partly as an outstand and partly as a column. For design purposes its adequacy to resist the applied compression is checked by assuming only one mode of behaviour. The resulting conservatism should overcome any elastic overstress arising from the simplifications of the overall analytical model presented above.

The proportions of the gusset plate will determine which is the more beneficial restraint. The slenderness for the local column check is based on a column length of D, built in at both ends, and the gusset plate thickness t_g: it takes the value $0.7D/(t_g\sqrt{12})$ or $2.4\ D/t_g$. Its stability as a slender outstand should be checked in accordance with Section 7.2.3, using the outstanding dimension b shown in Figure 15.2.

While it is acceptable to make the gusset plate shape convex, there are clearly limits to dimensions d_1 and d_2 above which it would be unreasonable to assume full effectiveness. In the absence of more detailed guidance it would seem sensible to limit d_1/t_g and d_2/t_g to less than 5.

Another critical section exists at XX, where shear stresses are at a maximum. This should be checked in accordance with conventional practice. With this second approach to design the weld sizes follow directly from the stress flows. It is assumed that the shear is distributed parabolically through the gusset plate, in accordance with conventional bending theory.

Note that much of the above complexity disappears if the bracket is made from a full UB section with both top and bottom flanges intact. It may be designed as a conventional cantilever and the connection treated as a welded beam-to-column connection.

Finally, the compressive action of the lower portion of the bracket on the column web should be checked (see Section 7.5). If the web is overstressed it may be stiffened, although it would probably be more economic to increase the bracket depth.

15.1.2 Major axis bolted bracket

Traditional analysis of major axis bolted brackets concentrated on the distribution of bolt forces. A neutral axis position was postulated that was one-sixth of the bracket depth up from the bottom and an elastic analysis carried out. This would seem to be acceptable in that it gave a distribution of bolt forces that was in reasonable agreement with experimental observations. It is worth noting that the magnitude of the critical top bolt force is not sensitive to reasonable variations in neutral axis position. It was necessary to check that this analysis did not imply a compressive overstress either in the gusset plate or the column web. If an overstress did occur then either the bracket depth had to be increased or stiffeners added. If the latter course

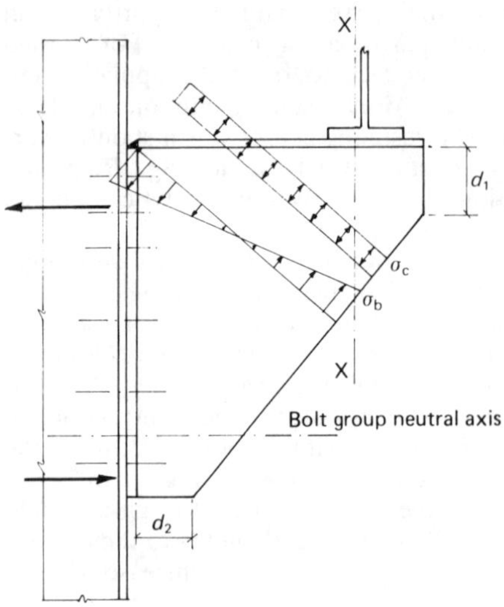

Figure 15.3 Analysis of bolted, major axis column bracket

were adopted then this would create a 'hard spot' on the compression load path. It would then be preferable to assume that the compressive force was concentrated at that level with a centre of rotation immediately above.

However, a complete design must also examine the other components of the bracket. If an extended end plate is used it is possible to assign the tensile force to the top flange and proceed in a similar way to the welded connection described above. However if, as is shown in Figure 15.3, a short end plate is used, it is no longer practicable to make this assumption because of the vertical eccentricity between this element and the line of action of the bolts. Rigorous analysis of the true tensile force distribution is not feasible in a design office. In its place the bending stress distribution (σ_b), shown in Figure 15.3, while clearly conservative would seem to offer reasonable continuity to both compression and tensile load paths. The conservatism arises because no account is taken of the contribution of the end plate or top flange to bending resistance. However, because of the influence of shear lag near the discontinuity at the corner and the coincidence of the critical section with this discontinuity the contribution is unlikely to be very significant. To this bending stress distribution should be added the compressive stress (σ_c) arising from the normal component of the eccentric load. The shear on this critical cross-section from the tangential component of the eccentric load should also be determined, though it is most unlikely to govern the design. With this total stress distribution, other checks on the bracket may proceed in a way similar to those for the welded major axis bracket described in the previous section, including the shear check on the critical section adjacent to the point of application of the load. It will also be necessary to check the adequacy of the tension load path in the vicinity of the upper bolts, and the column web in shear, both in a way similar to a bolted beam-to-column connection.

15.1.3 Minor axis column brackets

These connections present much less difficulty than their major axis counterparts. The main connections to the column are straightforward and are most unlikely to lead to any local points of weakness because the connection is directly into the flanges, which provide the minor axis bending strength of the column.

The gussets may be checked in a way similar to the bolted major axis bracket with a stress distribution based on the diagonal cross-section of the gusset plate alone. The only variations relate to its stability. If bolted, the outstand (b/t) should be based on the distance from the bolts to the free edge; if welded, it should only be based on the distance to the near edge of the column range if the gusset plate is welded to that edge. Where the design strength is based on column behaviour of the free edge it is important to consider the overall behaviour of the bracket in determining effective length.

Only if a suitable top flange prevents lateral displacement of the bracket tips should an effective length of 0.7 × unsupported length be used. In the absence of any top flange, an effective length of 2 × unsupported length should be adopted.

15.2 Built-up columns

15.2.1 Battened columns

The battens are required to provide a shear transfer between the elements of the built-up column with sufficient stiffness to ensure that the entire fabrication acts compositely. Because the battening does not form a triangulated network this can only be achieved by rigid connections. Historically, rivets performed satisfactorily in such connections but have now fallen into disuse. Bearing bolt connections would be too flexible and HSFG connections are generally too expensive. Thus almost all modern battened columns are welded.

Empirical rules have been developed for the arrangement of such connections, and these are illustrated in Figure 15.4. With the three intermittent welds shown, l_c and l_c should be not less than $l_b/6$ and l_r should be not less than four times the batten thickness. The welds should, of course, be designed for the shears and moments specified in the relevant Code of Practice.

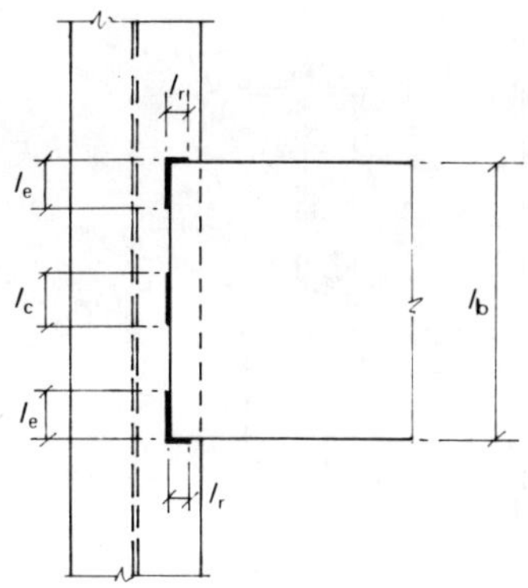

Figure 15.4 Typical weld details for a battened column

15.2.2 Laced columns

Lacing fulfils the same function of shear transfer between elements but, because it is a triangulated system, it is more tolerant of connection flexibility. Thus bearing bolts could be used for site-assembled built-up columns. However, most assembly is carried out in the fabrication shop, and there economics dictate that welded connections should be used. Once again, connection details are based on traditional empirical rules; the connections should be designed for the shears and moments specified in the relevant Code of Practice. A typical arrangement is shown in Figure 15.5. The overlap (l) should not be less than four times the bar (or angle) thickness, or four times the thickness of the flange of the component to which it is attached, whichever is the less. The bar should be welded along the whole length of lap on both sides and the weld should be returned for an aggregate distance of not less than four times the bar thickness, where this product is less than the bar width.

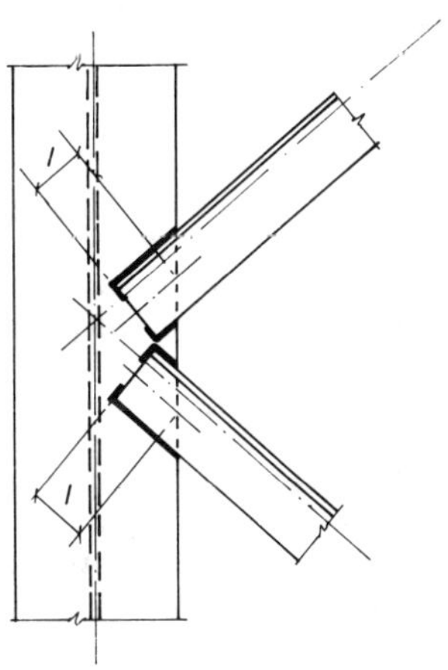

Figure 15.5 Typical weld details for a laced column

15.3 Crane beam connections

Collectively, cranes and their supporting structures have the reputation of being troublesome in service. Most of the problems arise because of the failure of the designer to appreciate the great difference in load history between these and other structural components. Apart from maintenance and other infrequently used installations, cranes are exposed to a significant proportion of their design loading more frequently than almost any other structural system. As extreme examples, cranes in steel mills or in bulk-handling units will be lifting their design capacity up to twenty times an hour, and cranes in heavy workshops will be handling loads of 50% and upwards of their rated capacity at high speeds constantly during working periods. In addition, cranes are subject to more abuse than almost any other structural system; loads may exceed rated capacity; they can swing like a pendulum; the crane may be used to drag loads both across and down the workshop; or the crane may be rammed against the stops. All these misuses will subject both the crane and its supporting structure to severe horizontal and vertical loads.

Conventional fatigue calculations can be used for the main elements of the system, and it is essential that the same attention is devoted to the connections. It is particularly important that the implications of repeated deflection of the main elements are considered in the design of the secondary and restraining elements. For example, simple connections in, say, multistorey frame buildings that utilize cleats or light end plates are quite satisfactory, even though these elements will become plastic locally under working loads, due to the rotation of the beam ends. This is satisfactory because such loading will occur only rarely during the life of the structure. Such details would fail rapidly in fatigue if used on a heavy-duty crane beam. The prudent designer will minimize such problems by limiting the flexibility of the crane beams. A figure of $l/1000$ for live load deflection is commonly used. Even so, this can still lead to significant horizontal movement at the top of a deep crane girder. Such movements are quantified in the following section.

There are many different types of crane and associated support system. While they may have particular problems, most of these have the common origins outlined above.

The most widely used type of crane, and the one that impinges most on its supporting structure, is the overhead travelling crane. The connections used for the associated crane beams or girders have therefore been discussed in detail in the following sections. However, the principles illustrated there may be applied with equal validity to other crane systems.

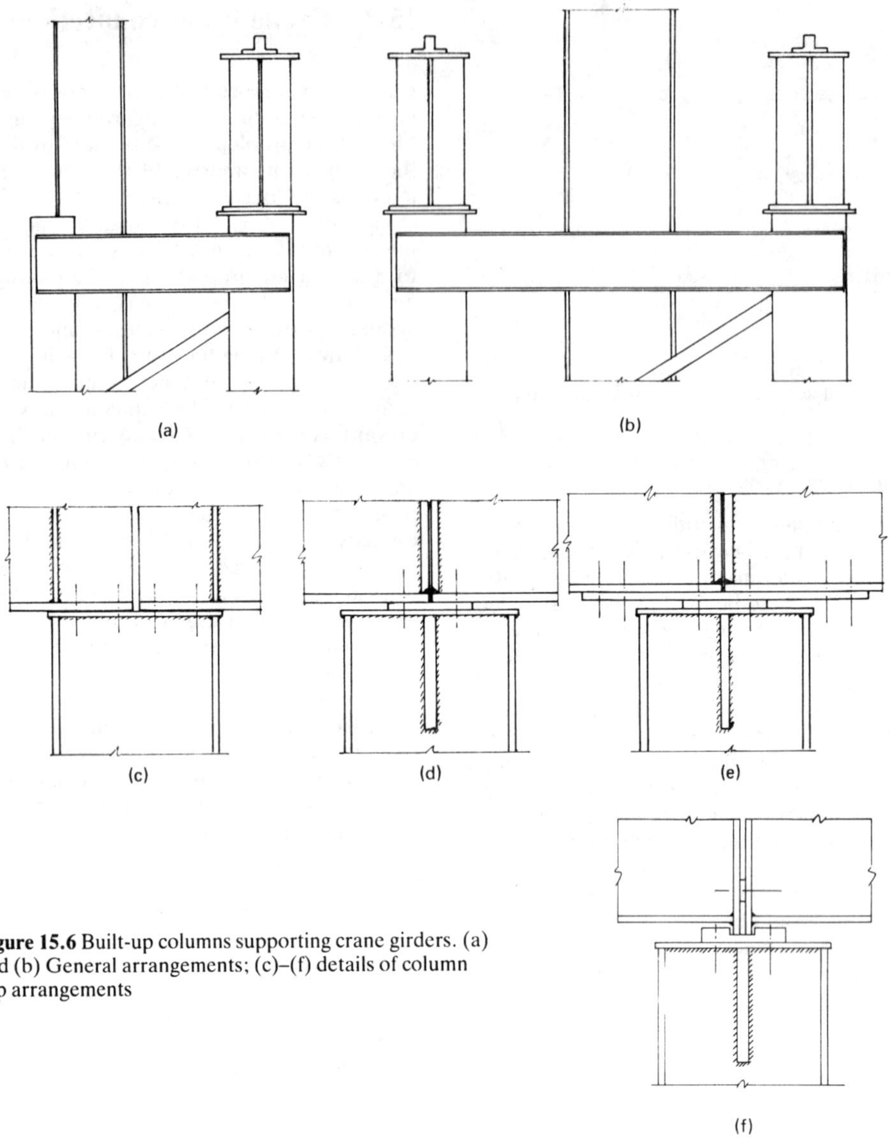

Figure 15.6 Built-up columns supporting crane girders. (a) and (b) General arrangements; (c)–(f) details of column cap arrangements

15.3.1 Beam vertical supports

Figures 15.6(a) and (b) show the general arrangements where built-up columns are used to support crane beams. This would be an appropriate arrangement for heavy cranage; the column brackets discussed earlier would be more economic for light crane systems (say, up to 10–15^{T} capacity). Irrespective of the support system, the most important aspect of design is the accommodation of the beam end rotations which occur when, as is usually the case, simply supported beams are used. Even with a deflection limitation of $l/1000$, beam end rotations of 0.2° will be occurring repeatedly.

If load-bearing stiffeners are positioned directly over the column flanges, as shown in Figure 15.6(c), the bolts and column cap plate will have to accommodate the associated uplift. A thin cap plate should be selected and the bolts should be positioned as far as possible from the column and beam webs, to maximize vertical flexibility.

Alternatively, a local stiffening can be provided to the column web, as shown in Figures 15.6(d)–(f), with some provision for minimizing eccentricity. In Figures 15.6(d) and (e) a simple pack plate is used; in (f) extended end plates slot into a recessed 'keep' plate. In many crane systems it is necessary to transmit surge loadings between girders, along the building, because horizontal load resistance is confined to braced bays. Figure 15.6(e) shows an arrangement for tying the beams together securely without restricting the beam end rotations; bolts X fulfil a similar function in Figure 15.6(f).

Whatever girder support system is used, if either the designer lacks experience of the structural detail or the crane is subject to particularly onerous service it would be prudent to carry out fatigue calculations to ensure that the connection can safely sustain the local deflections imposed by the beam end rotation of 0.2°.

15.3.2 Beam lateral supports

Crane beams have to resist considerable horizontal loads, normal to their longitudinal axes, and different details for resisting these loads are shown in Figure 15.7. These loads are transmitted to the beams at crane rail level, i.e. above the level of the top flange. Thus the most effective restraints are those which can be positioned close to crane rail level. This is satisfactorily achieved in Figures 15.7(a) and (c). However, in (b) the downstanding flanges of the channel have forced the restraint system down, introducing a significant moment into the system. The design should take account of this moment; alternatively, the channel flanges should be notched, to allow a higher position of the restraint system.

As with the vertical support system described in the previous section, the lateral support system must be able to accommodate the repeated end rotation of the beams. A careful balance needs to be struck between this requirement for flexibility and that for compressive capacity to resist the horizontal surges. (Once again, if there is any doubt about the fatigue lives of the details, they should be investigated by calculation.) Some designers specify slotted holes, but these are unlikely to permit movement in practice. For heavy-duty installations the expense of link arms and ball joints is probably justified.

Most of the lateral bending resistance of the crane beam is concentrated in the top flange, and it is important to ensure that there is proper continuity of the restraint system through to this element. If, as in Figure 15.7(b) and (c), the lateral support is attached to a web stiffener, it is essential to weld this stiffener directly to the top flange. Failure to do this will produce a severe stress concentration at X and Y, leading to premature cracking.

Whatever detail is adopted, it is very important to make provision for subsequent adjustment of both line and level of the crane rail and girder by packs or other means. This is likely to be necessary to correct any settlement of the main structure.

15.4 Crane gantry end stops

Figure 15.8 shows typical examples of crane stops. Example (a) would be suitable if it were permissible to stop the crane short of the end column; example (b) shows a compact arrangement where the crane bogie has to extend up to the end column. Geometric details will depend on individual performance specifications. Note that the crane rail is stopped short of the end stop to provide an erection and fabrication tolerance.

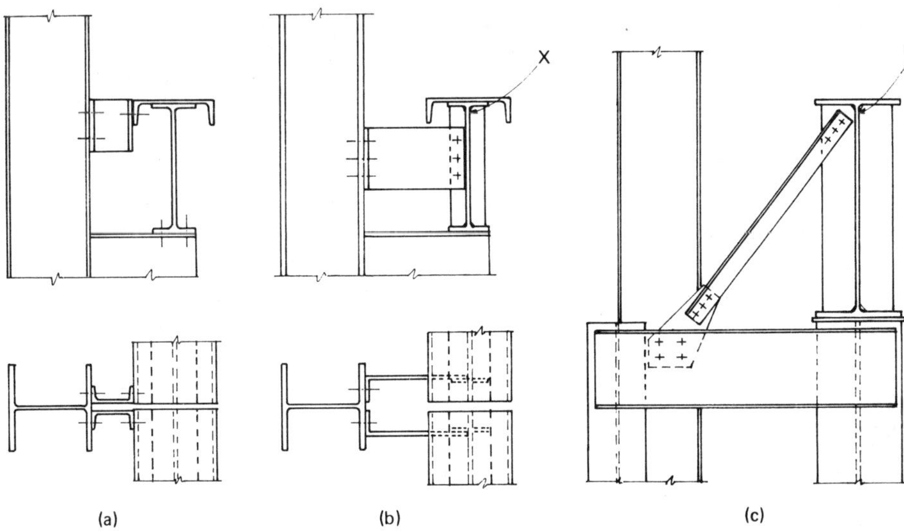

Figure 15.7 Lateral restraint to crane girders

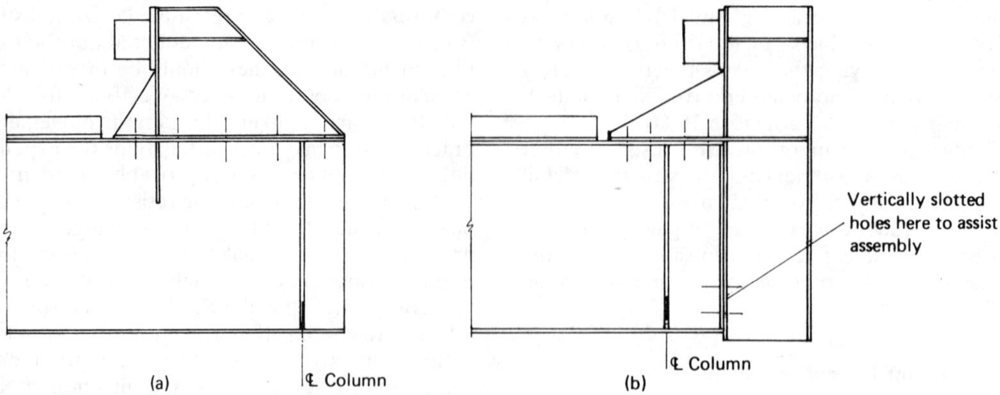

Figure 15.8 Crane gantry end stops. (a) Fabricated from welded plate; (b) fabricated from rolled sections

The design load (F) for the end stop is a function of the spring or plunger design:

$$F = \frac{WV^2}{gT}$$

where W is crane vertical reaction excluding lifted load which will swing in an impact,
V is crane speed,
T is length of travel of spring or plunger.

Unlike other aspects of crane design, it is not necessary to consider fatigue in the design of the end stops because of the relatively low number of occurrences of significant loading. Indeed, it would be acceptable to use bearing bolts in shear if required. However, it is appropriate to have a robust design, that is, one that can absorb energy by plastic deformation if necessary. If this requirement is satisfied then, even if the plungers malfunction, a catastrophe will be avoided. This requirement can usually be achieved by ensuring that:

1. Full-strength welds are used on tension and shear load paths, and
2. Bolts are not the weakest link on any tension or shear path.

Reference

1. Crane rail fastening systems, Supplement to *Steel Construction*, 7, No. 3 (1983), SAISC.

16

Truss connections

16.1 Introduction

In most structural steelwork design it is necessary to consider the connections during the initial conception if economy is to be achieved. With trusses this is essential. A very high proportion of the total cost is attributable to the connections, and it is a false economy to select members that are themselves efficient if they cannot be connected economically.

Figure 16.1 illustrates this important point in relation to welded truss design. Figure 16.1(a) shows the connection that might result if the members only were considered in the initial design. Universal Columns or Universal Beams make very efficient truss chords, particularly if they are subject to local loads between node points. They are better able to resist this local bending if they are orientated with their major bending planes vertical. If similar members are chosen for the post and diagonal members of the truss it is clearly sensible to orientate them as shown, with all the element webs in the same plane. Without any stiffening this is already an expensive connection to fabricate. The web elements have to be precisely cut to length, with a double-angled cut to the diagonal; weld preparations will probably be necessary, at least to the diagonal member. Assembly costs will be high, because of the close control that will be required on weld root gaps. It can be seen that there is no continuity to the flanges of the post and diagonal members. If, as is usually the case, these elements are highly stressed then substantial stiffening must be added. Much of this stiffening will have to be fitted, adding considerably to the cost of what is already an expensive connection.

The requirement for stiffening could be overcome if the truss elements are reorientated and chosen to have the same depth, as shown in Figure 16.1(b). However, all the requirements for precise cutting and fitting remain and butt welds are now required throughout the flange connections. There are also problems with the post and diagonal webs if these cannot be curtailed. Finally, and most importantly, this type of connection severely restricts the choice of truss elements. Only those whose actual, not nominal, depths are such that the flanges can be made coplanar can be used.

However, if economy of the entire truss were considered, including fabrication costs, then the arrangement shown in Figure 16.1(c) might have resulted. The chord is now a pair of battened channels and web members of the same true depth are used. The material cost of all the elements has probably increased significantly and the chords have to be battened, but the total cost is reduced because of the economy of the connections. There is now no precise cutting to length or cutting at an angle (a local eccentricity has been introduced into the connection, but this can readily be accommodated in the element design). All the welds can be fillet welds, with no preparation. Because of the considerable weld length available these are likely to be single-run, 6 mm welds – a further economy.

Another economical alternative is shown in Figure 16.1(d). The truss posts must all have the same depth but the tension members can be sized separately. Because these members are arranged on opposite sides of the chords, it is possible to avoid eccentricity at the nodes. Once again, all cutting to precise length and at an angle has been avoided and there is no butt welding.

16.2 Single-plane trusses

Single-plane trusses, as the name implies, are those where all the connections are (more or less) in the same plane. In traditional, riveted construction each

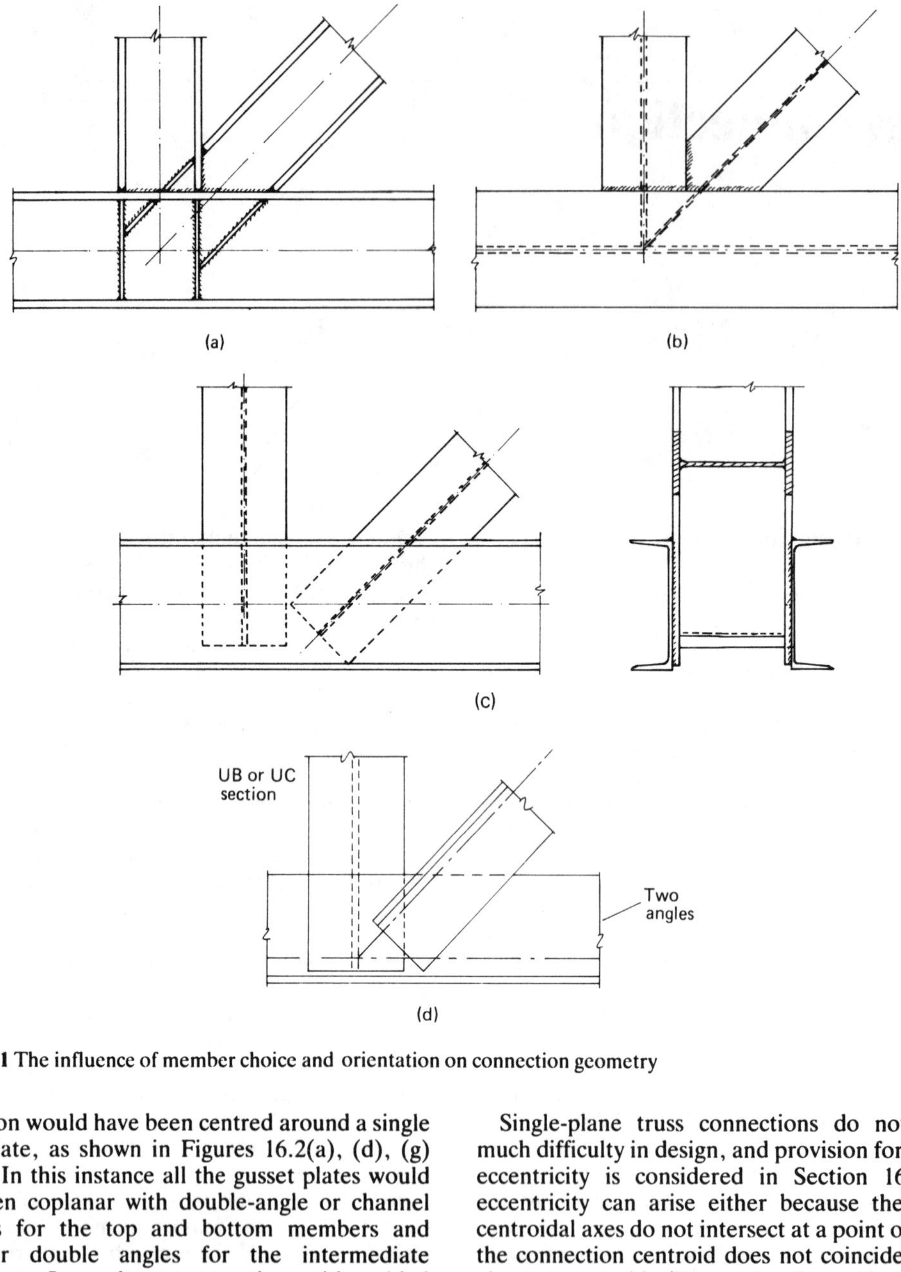

Figure 16.1 The influence of member choice and orientation on connection geometry

connection would have been centred around a single gusset plate, as shown in Figures 16.2(a), (d), (g) and (h). In this instance all the gusset plates would have been coplanar with double-angle or channel members for the top and bottom members and single or double angles for the intermediate components. In modern construction, with welded connections it is frequently possible to omit the gusset plates. The top and bottom members are now likely to be Tees or single angles with unequal legs. The intermediate angles are then welded directly to the vertical elements of the top and bottom members, as shown in Figures 16.2(b), (e), (f) and (i). Although most trusses are now welded, there are occasions when bolted trusses are used, usually where piecemeal transportation is required with on-site assembly prior to erection.

Single-plane truss connections do not present much difficulty in design, and provision for any local eccentricity is considered in Section 16.4. Such eccentricity can arise either because the element centroidal axes do not intersect at a point or because the connection centroid does not coincide with the element centroid. (For angles the setting-out lines are usually used instead of the centroidal axes.) Design of gusset plates is discussed in Section 16.5 and the influence of partial connection on element strength is covered in Section 16.6. However, the examples in Figure 16.2 have been chosen to illustrate particular aspects of truss connection design, and these are discussed below.

The bolted apex connection shown in Figure 16.2(a) is usually designed on the assumption that all the element forces are transmitted through the

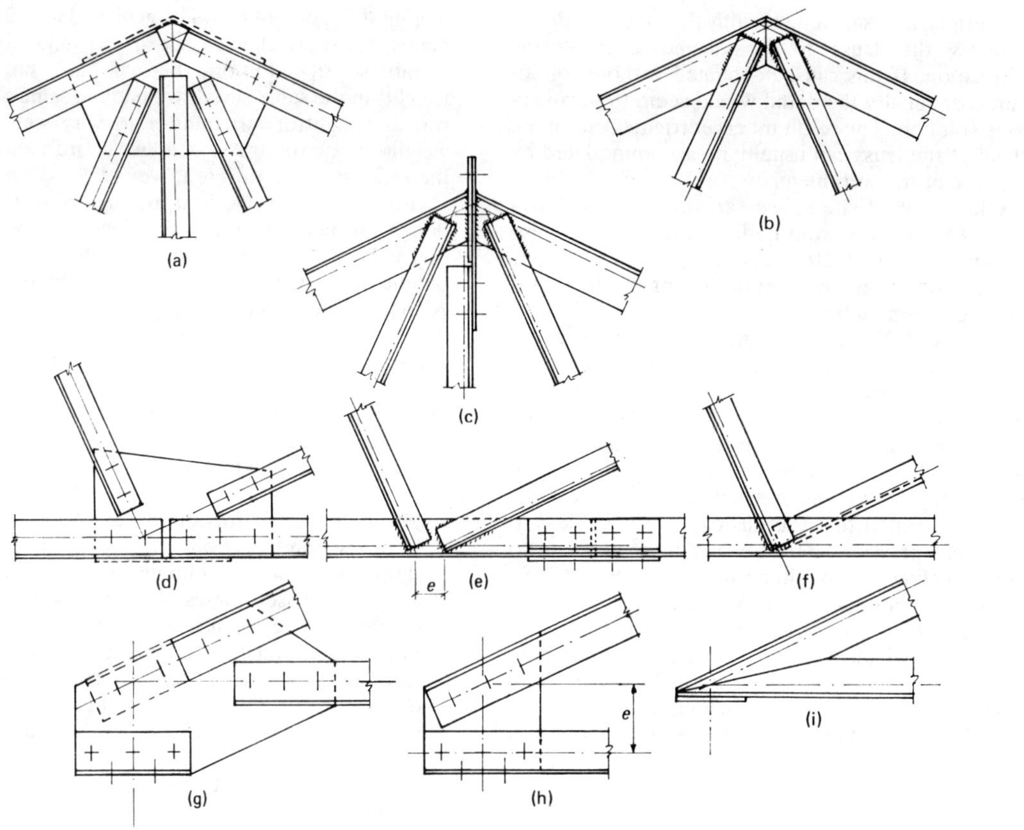

Figure 16.2 Typical connections for single-plane truss construction

central gusset plate. This is satisfactory provided that all the nodes of the truss are held in plane by some means. If, either during erection or in the completed structure, there is any possibility of some overall instability of the truss and associated folding-up of the gusset plate then a splice plate should be added to the top of the rafter members to ensure adequate continuity of out-of-plane bending stiffness.

The welded apex connection shown in Figure 16.2(b) does not have any such discontinuity in out-of-plane bending stiffness. The two rafters may be butt welded together as shown; alternatively, a small division plate may be introduced, allowing fillet welds to be used. If long-stalk Tees are used for the rafters it is usually possible to weld the intermediate members directly to the stalk. (Though more concerned with design of the truss than the connections, it should be noted that the *b*/*t* of the stalk will not generally comply with limiting geometric criteria for compression members. A reduced, effective, stalk should be used in the rafter design; this is usually equated with the maximum permitted outstand.) If other rafter sections are used, it may be necessary to extend the downstand as shown in order to develop sufficient weld length.

To facilitate transportation, site splices are often introduced into welded trusses. Figure 16.2(c) shows a suitable arrangement for a bolted site splice at an apex. An end-plate connection is used, one plate being extended downwards to provide an anchorage for the central post.

Away from the apex, bolted truss connections will generally take the form shown in Figure 16.2(d). Once again, if there is any requirement for continuity of out-of-plane bending stiffness or strength, an additional splice plate should be provided.

A typical welded truss detail is shown in Figure 16.2(e). If a gusset plate is to be avoided it is frequently necessary to space out the intermediate members, introducing local eccentricity to the connection. This will generally lead to the most economic truss, with connection simplicity more than offsetting any penalty in element sizes. An alternative solution, removing the eccentricity, is shown in Figure 16.2(f). Here the intermediate members are arranged on opposite sides of the truss.

The principal disadvantage with this detail is that it requires the truss to be turned over during fabrication. If this can be arranged without significant cost penalty the detail would seem to be a very good solution. The resultant eccentricities out of the plane of the truss can usually be accommodated by bending of the web members.

Where bolted site splices are introduced into a welded truss away from nodal connections, as also shown in Figure 16.2(e), it is again necessary to pay proper attention to continuity, particularly for compression members.

Figures 16.2(g) and (h) are alternative arrangements for the eaves connection of bolted roof trusses. For most practical roof geometries, if eccentricity is avoided, large gusset plates will be necessary, as shown in Figure 16.2(g). It may well be more satisfactory to permit eccentricity and reduce the gusset plate size, as shown in (h). In both cases it should be noted that the gusset plate is subject to direct compression. Because of this, it is prone to local buckling and should be sized accordingly. This is discussed in detail in Section 16.5.

Finally, Figure 16.2(i) shows a welded truss eaves connection which has been arranged to eliminate eccentricity. For roofs of low pitch a very long connection is produced. In such circumstances it is probably preferable to curtail the connection and design for the ensuing eccentricity.

16.3 Double-plane trusses

Double-plane trusses have two connection planes, as shown in Figure 16.3. They were a very popular form of construction in the days of riveted construction when rolled sections had actual sizes that were the same as serial sizes. Thus, if 24 in chord members were to be used with 20 in web members, a good fit could be achieved with 2 in pack plates between the gusset plates and the web members. In addition, because all the rolling weights in the same serial size had the same depth, it was a straightforward matter to vary web section weights along the truss, or splice chord members of the same size and different weight.

However, with modern rolling mills it is the dimension between the inside faces of the flanges that is kept constant. Different rolling weights are obtained by varying the overall section depth. Actual section depths may vary from nominal depths by considerable margins.

It is now rare for actual section sizes for a practical combination of web and chord truss members to differ by an amount which is convenient for packing. Thus for modern heavy truss design one of the following must be adopted:

1. For bolted construction make the entire truss from the same section. This is the simplest but rarely the most economic solution. If it is adopted, gusset plates may be used with the overall configuration shown in Figure 16.3.
2. For welded construction make the entire truss from the same serial size. Because the inside dimensions of all universal sections of the same serial size are the same, there will always be full continuity of the lighter section. However, such designs are subject to the reservations discussed in Section 16.1.
3. Make the truss with rolled section web members and fabricated chords. The chord members will usually be deep channels or boxes fabricated from heavy plate, although for light trusses the solution shown in Figures 16.1(c) and (d) would be very economical. As shown in Figure 16.4, this enables the chord webs to be extended to act as gussets and to be attached directly to the web

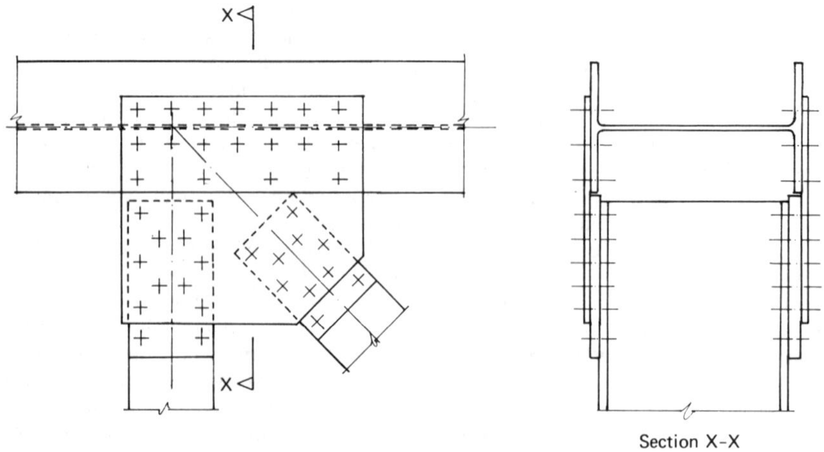

Figure 16.3 Typical connections for double-plane truss construction

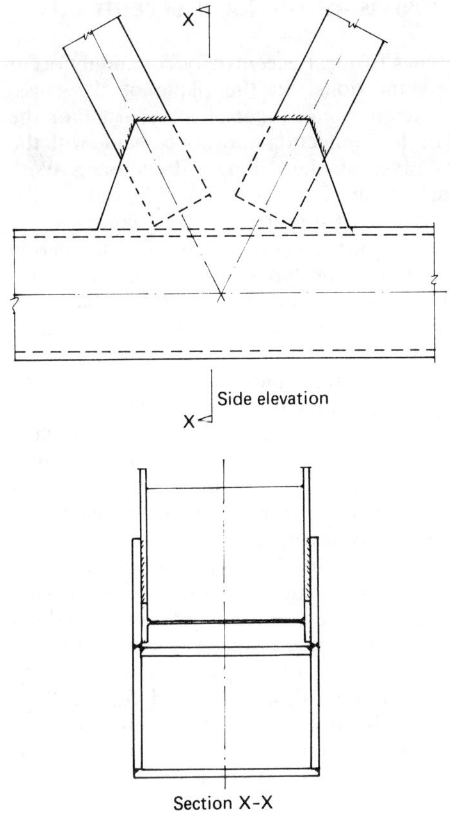

Figure 16.4 Typical connections in a truss with fabricated chords

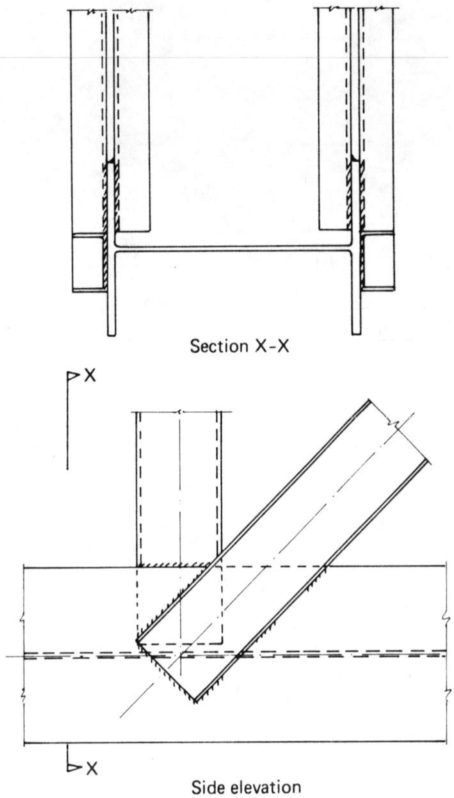

Figure 16.5 Typical truss connection with twin planes of webs

members. The resulting truss will have a pleasing appearance but there are still severe restrictions on the variation in web members along the span. In addition, considerable care in welding sequence will need to be exercised if good fit is to be achieved.

4. Alternatively, the truss may have rolled section chords and fabricated web members. Laced or battened web members can be fabricated so that their depths are compatible with the chords. Connections could then take the general form shown in Figure 16.3, but without any necessity for packs.
5. Finally, it is possible to have two separate planes of web members, one centred around each of the chord flanges, as shown in Figure 16.5. Here the connections can be very similar to those adopted in single-plane truss construction that were illustrated in Figure 16.2.

Whatever type of double-plane truss is adopted, design is generally straightforward; due account should be taken of Sections 16.4–16.6, where appropriate.

16.4 Gusset plate design

Rigorous analysis of gusset plates would be most complex because of their proportions and the presence of local loads. In practice, design is based on very simple analysis to determine stress distribution and straightforward geometric criteria to ensure that buckling does not occur. Because of this simplicity, design calculations should be interpreted with common sense; in critical situations calculated thicknesses should be rounded up. The marginal cost of increasing gusset plate thickness is small and a change from, say, 12 to 15 mm thickness will have little effect on overall economy.

Local stresses arising from the load input from an individual member may be checked on the basis of a 30° dispersal over the length of the connection, as shown in Figure 16.6(a). In addition, critical sections (for example, a–a in Figure 16.6(b)) should be checked under combined direct, bending and shear stresses. Calculations are generally based on engineers' bending theory, which assumes that plane sections remain plane, notwithstanding the unusual proportions of the 'beam'.

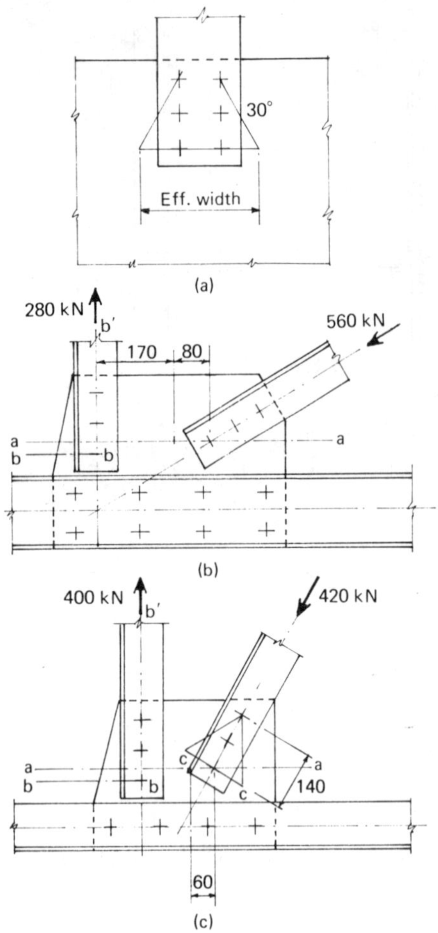

Figure 16.6 Critical sections in gusset plates

It should be remembered that shear stresses in a flat plate have a parabolic distribution with a maximum value of 1.5 times the average. Where a member joins the connection close to a gusset plate edge it may be necessary to check non-linear critical sections (for example b–b–b′ in Figure 16.6(b)). In this instance it would be sufficient to assume that b–b is carrying tension and bb′ is carrying shear.

Traditional geometric criteria for gusset plates are presented in Chapter 7. Note that these are much more slender than those permitted for other plate elements such as stiffeners. It follows that they are only suitable for situations where the elements being connected effectively stabilize the gusset plate. If this is not the case (for example, if there are high compressive stresses near a free boundary or if there is a region of gusset plate that is entirely unsupported (as in Figure 16.2(g) if the dotted portion of the rafter is curtailed), then more stocky criteria should be used. Semi-compact criteria would seem appropriate in such circumstances.

16.5 Provision for local eccentricity

Several types of local eccentricity commonly occur in truss connections. In the plane of the truss, member axes may not meet at a point and/or the centroid of the connection may not coincide with the centroidal axis of the connected member. Web members, notably single angles, will lie outside the principal plane of the truss which contains the centroidal axis of the chord members. This section discusses the first two types of eccentricity. Out-of-plane effects are closely related to problems of partial connection and are therefore discussed in Section 16.6.

Figure 16.7 shows typical examples of in-plane eccentricity. The moment $P.e$ is resisted by all the members framing into the connection. It is usually distributed between them in proportion to their bending stiffnesses (I/l). The member connections should be designed to resist these moments in addition to the axial forces.

Member connections are subject to additional moments if the connection centroids do not coincide with the member axes. For example, in welded lap connections at the ends of angles the longer welds may be on the edges of the legs, as shown in Figure 16.2(c). Traditional design texts differ in their advice about the necessity of designing for these latter moments. Certainly, if designed for, they add considerably to the tedium of connection calculations.

A rational and time-saving way of allowing for both moments is to base the design on axial load capacity and then use the interaction diagrams for weld and bolt groups subject to combined loading in

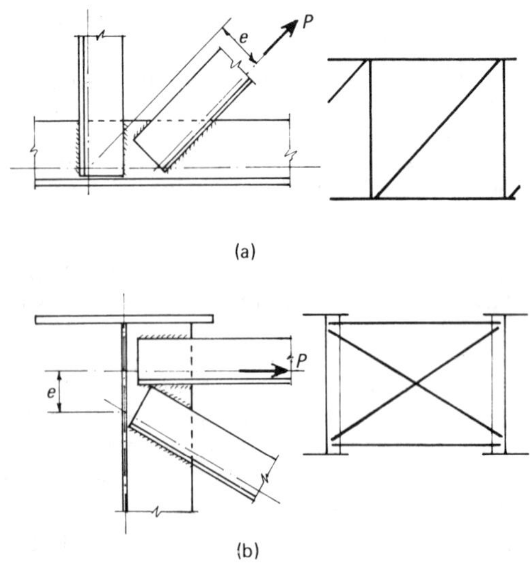

Figure 16.7 Eccentricities in truss connections. (a) Pratt truss; (b) cross-bracing between plate girders

Chapter 8 to determine the increase in weld or bolt group size that is necessary in order to accommodate the coincident moments. If this procedure is adopted it can be seen that modest moments only reduce axial load capacity by 10–15%. With the greater factors of safety inherent in traditional connector design strength it is then clear why these moments could be disregarded with impunity.

16.6 Partial connection

In many truss connections it is only feasible to connect to part of the element cross-sections. In all such cases this raises questions about how the load is dispersed into the unconnected parts of the cross-section. In cases where there is an out-of-plane eccentricity (for example, where a single angle is connected to a gusset plate) there are additional questions about the effect of this eccentricity on connection and element design.

In the first instance let us examine this eccentricity. In the previous section any in-plane moment arising from an eccentricity between the member and its connection was assumed to be resisted by the gusset plate, and the connection was designed to transmit that moment into the gusset plate. That was appropriate, because the in-plane rotational stiffness of the gusset plate was considerably greater than that of the member. When out-of-plane moments are considered the situation is reversed. The gusset plate now has a lower rotational stiffness than the member and the moment must therefore be resisted by the member. (This is allowed for in traditional codes by requiring unconnected elements to be considered as partially effective: in more rigorous codes it may be necessary to consider the element under combined axial load and moment.) Because the moment is resisted by the member, there is no need to take account of any out-of-plane eccentricity moment on the connection. These empirical rules also take general account of load dispersal in such members, though it may still be necessary to check critical net sections in the member.

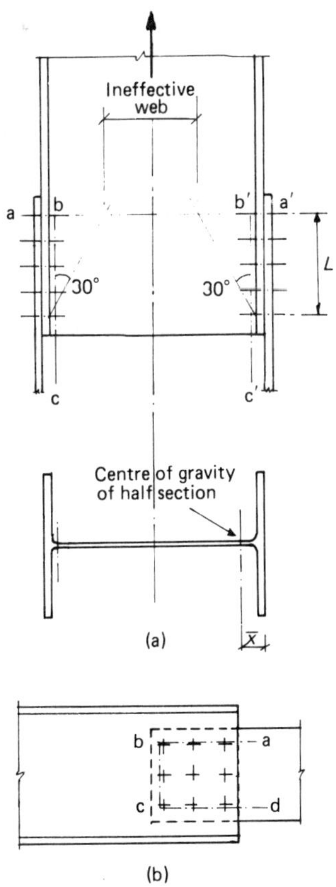

Figure 16.8 Partial connections of tension members (a) to flanges and (b) to web

Where there is no eccentricity it is still necessary to consider partial connection effects in the types of situation shown in Figure 16.8(a). In these cases there is generally little or no codified guidance, and the designer is left to his own devices. One approach that has been suggested incorporates an empirical efficiency factor (η) into the calculation of the effective net section to take account of shear lag in the unconnected web. Thus the effective net section is given by η times the actual net section:

$$\eta = \left(1 - \frac{\bar{x}}{L}\right)$$

where $\bar{x}$ is the distance of the centre of gravity of the half section from the connection plane (as shown), and
L is the connection length.

The two disadvantages of this approach are that:

1. It suggests that it is impossible to achieve full efficiency on the net section irrespective of connection length; this is clearly nonsense.
2. Very low efficiencies are achieved for short connections, i.e. it is not possible to mobilize the flanges in such circumstances. It is difficult to see how shear lag in the web can reduce flange capacity significantly. This does not create difficulties with practical connections if an overstress is permitted on the net section because yielding of the gross section will govern design. However, it can lead to a significant loss in design capacity if net section overstress is not allowed.

One rational alternative would be to base design for short connections on the non-linear critical sections a–b–c shown in Figure 16.8(a). Section a–b is acting in tension and b–c is acting in shear. For long connections the normal a–b–b′–a would govern, at full efficiency.

An even more straightforward approach would be to use a permitted dispersion concept for the unconnected web. The appropriate dispersion angle in the presence of the free boundary is 30°, and this can be used to determine the proportion of the web that may be considered to be effective.

Figure 16.8(b) illustrates another situation where partial connection requires special consideration. The plate strength in the immediate vicinity of the bolts must be checked to ensure that resistance to local tear-out is adequate. The net sections a–b and c–d are in shear and the net section b–c is in tension. Design strength of the connection may be taken as the sum of the net sections' strengths.

Reference

1. *Engineering for Steel Construction*, AISC, 1984.

Commentary: 1/1

The design example is for a welded node for a medium span N-girder truss. The members and their layout have been chosen to avoid the use of gusset plates. Although the aim is to avoid the use of gusset plates, it is feasible (if found to be necessary at the most heavily loaded nodes) to extend the vertical legs of the boom angles locally by butt welding gusset plates to the toes of the angles.

Structural Steelwork Connections	Subject: Welded truss connection			Chapter Ref. 16
	Design Code: BS 5950 Part 1			Calc. Sheet No. Example 1/1
	Calc. by B.D.C	Date Aug, '87	Check by G.W.O.	Date Nov, '87

Code Ref.	Calculations	Output

Welded truss connection.

Design a welded connection for the bottom boom node of an N-girder truss. The forces (due to factored loads) and the member sizes are shown in the sketch.

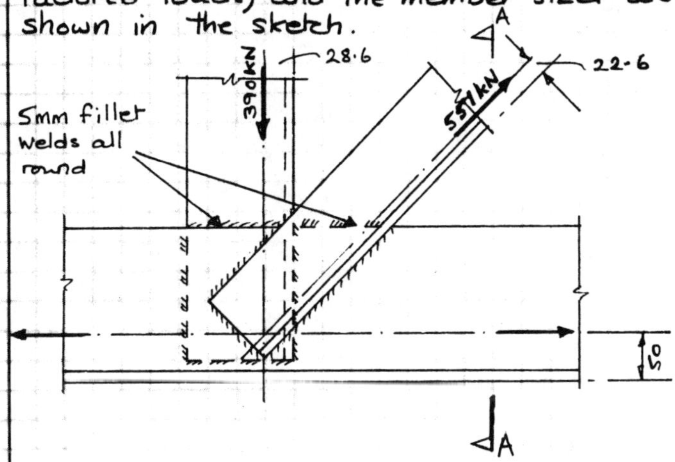

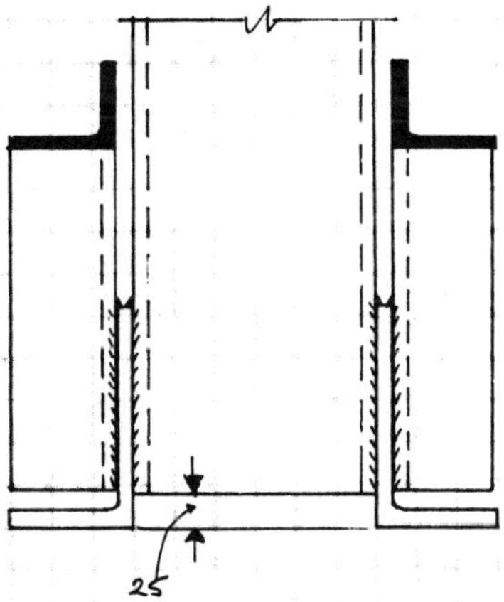

Section A-A.

Vertical 152 × 89 [23·84

Diagonal 2 No. 80 × 80 × 8 angles.

Tension beam 2 No. 150 × 90 × 10 angles

Commentary: 1/2

The member load is shared equally between the welds to the two boom angles.

Section 8.4.1

$$\int r^2 ds = \int x^2 ds + \int y^2 ds$$

The terms $\int x^2 ds$ and $\int y^2 ds$ are the second moments of area of the weld group about the YY and XX axes for a weld of unit width (throat thickness).

The interaction curve for a square weld group in Figure 8.16 could be used for a quick design:

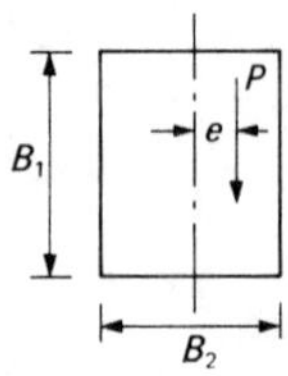

$$\text{Maximum resultant shear on weld} = \frac{\text{Average shear ignoring eccentricity}}{P/P_0 \text{ from curve}}$$

$\frac{e}{B}$ must be calculated taking B = lesser of B_1 and B_2:

For this example,

$$\frac{e}{B} = \frac{15.9}{89} = 0.179 \cdot \frac{P}{P_0} = 0.77 \text{ from curve}$$

$$\text{Maximum resultant shear on weld} = \frac{195}{428 \times 0.77}$$

$$= 0.592\,\text{kN/mm}$$

(*Note*: This approximate procedure is conservative, but not unduly so for the proportions of normal end-connection details.)

Structural Steelwork Connections	Subject: Welded truss connection			Chapter Ref. 16
	Design Code: BS 5950 Part 1			Calc. Sheet No. Example 1/2
	Calc. by: B.D.C	Date: Aug, '87	Check by: G.W.O.	Date: Nov, '87.

Code Ref.	Calculations	Output

Connection for vertical member.

Assume load to be applied at centroid of member.

Load per weld group $= \frac{390}{2} = 195$ kN

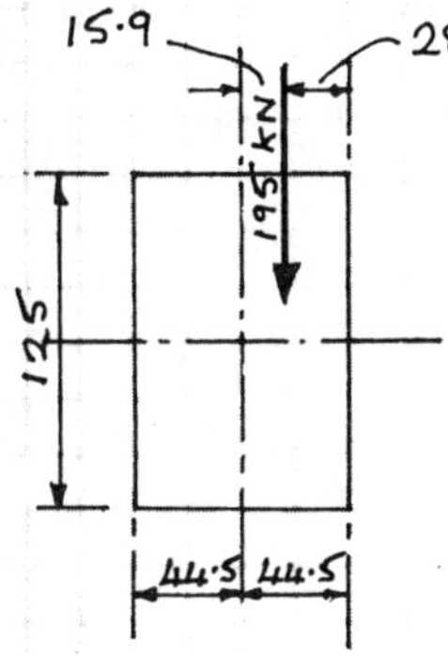

Total length of weld
$= 2[125 + 89]$
$= 428$ mm

$$\int r^2 ds = 2\left[\frac{125^3}{12} + 89 \times 62{\cdot}5^2 + \frac{89^3}{12} + 125 \times 44{\cdot}5^2\right]$$

$$= 1{\cdot}633 \times 10^6 \text{ mm}^3$$

Maximum resultant shear on weld

$$= \sqrt{\left[\frac{195}{428} + \frac{195 \times 15{\cdot}9 \times 44{\cdot}5}{1{\cdot}633 \times 10^6}\right]^2 + \left[\frac{195 \times 15{\cdot}9 \times 62{\cdot}5}{1{\cdot}633 \times 10^6}\right]^2}$$

$$= \sqrt{[0{\cdot}456 + 0{\cdot}084]^2 + [0{\cdot}1187]^2}$$

$= 0{\cdot}553$ kN/mm

Connection for diagonal member.

Assume load is applied at centroid of member

Load per weld group $= \frac{551}{2} = 275{\cdot}5$ kN

Diagram overleaf

Commentary: 1/3

Section 8.4.1

(See notes in commentary on previous page.)

Capacity of fillet weld = 0.7 × (leg length) × P_w

The calculations to find the position of the centroid, $\int r^2 ds$ and the resultant shear can be rather lengthy and a simplified design procedure is often adopted.

The simplified procedure used in this example is to ignore part of the weld group so that the load passes through the centroid of the remainder of the weld group (i.e. the part used in the calculations). With the load passing through the centroid there are no moment calculations, which were somewhat lengthy in the full elastic analysis.

The aim is to incorporate as much weld as possible while keeping the calculations simple. In this example the eccentricity is relatively small so that it is possible to use all the weld on sides AB, BC and CD with part of the weld on side AD to give the correct balance about the line of action of the load.

Structural Steelwork Connections	Subject: Welded truss connection			Chapter Ref. 16
	Design Code: BS 5950 Part 1			Calc. Sheet No. Example 1/3
	Calc. by B.D.C	Date Aug, '87	Check by G.W.O	Date Nov, '87

Code Ref.	Calculations	Output

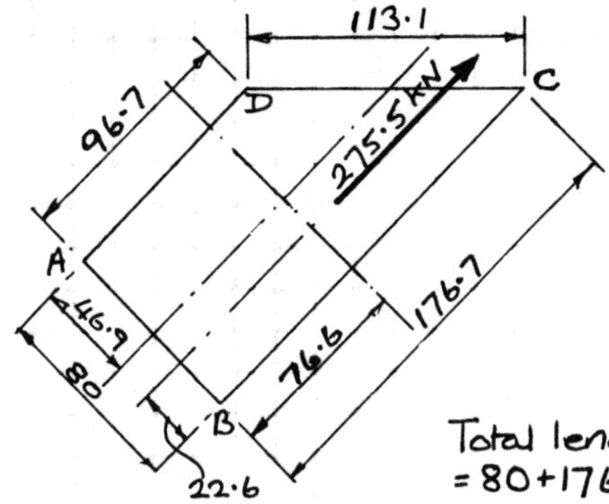

Total length of weld
$= 80 + 176{\cdot}7 + 113{\cdot}1 + 96{\cdot}7$
$= 466{\cdot}5$ mm

Find centroid

By moments about AD

$$\bar{x} = \frac{1}{466{\cdot}5}\left[80 \times \left(\frac{80}{2}\right) + 176{\cdot}7 \times 80 + 113{\cdot}1 \times \left(\frac{80}{2}\right)\right]$$

$$= 46{\cdot}9 \text{ mm}$$

By moments about AB

$$\bar{y} = \frac{1}{466{\cdot}5}\left[176{\cdot}7 \times \left(\frac{176{\cdot}7}{2}\right) + 113{\cdot}1 \times \left(\frac{176{\cdot}7 + 96{\cdot}7}{2}\right) + 96{\cdot}7 \times \left(\frac{96{\cdot}7}{2}\right)\right]$$

$$= 76{\cdot}6 \text{ mm}$$

$$\int r^2 ds = 80 \times 76{\cdot}6^2 + \frac{176{\cdot}7^3}{12} + 176{\cdot}7 \times \left(\frac{176{\cdot}7}{2} - 76{\cdot}6\right)^2$$
$$+ 113{\cdot}1 \times \frac{80^2}{12} + 113{\cdot}1\left(\frac{176{\cdot}7 + 96{\cdot}7}{2} - 76{\cdot}6\right)^2$$
$$+ \frac{96{\cdot}7^3}{12} + 96{\cdot}7 \times \left(76{\cdot}6 - \frac{96{\cdot}7}{2}\right)^2$$
$$+ \frac{80^3}{12} + 80\left(46{\cdot}9 - \frac{80}{2}\right)^2 + 176{\cdot}7(80 - 46{\cdot}9)^2$$
$$+ 113{\cdot}1 \times \frac{80^2}{12} + 113{\cdot}1\left(46{\cdot}9 - \frac{80}{2}\right)^2$$
$$+ 96{\cdot}7 \times 46{\cdot}9^2$$
$$= 2{\cdot}093 \times 10^6 \text{ mm}^3$$

Eccentricity of load from centroid of weld group $= 80 - 46{\cdot}9 - 22{\cdot}6 = 10{\cdot}5$ mm

Structural Steelwork Connections	Subject: Welded truss connection	Chapter Ref.: 16
	Design Code: BS 5950 Part 1	Calc. Sheet No.: Example 1/4
	Calc. by: B.D.C — Date: Aug, '87 — Check by: G.W.O.	Date: Nov, '87

Code Ref.	Calculations	Output
	Maximum resultant shear on weld	
	$= \sqrt{\left[\frac{275.5}{466.5} + \frac{275.5 \times 10.5 \times 33.1}{2.093 \times 10^6}\right]^2 + \left[\frac{275.5 \times 10.5 \times 100.1}{2.093 \times 10^6}\right]^2}$	
	$= \sqrt{[0.591 + 0.046]^2 + [0.138]^2}$	
	$= 0.652$ kN/mm	
6.6.5	Capacity of 5mm fillet weld	
	$= 0.7 \times 5 \times 215 \times 10^{-3}$	
	$= 0.753$ kN/mm O.k. Use	5mm fillet welds all round.

Simplified procedure.

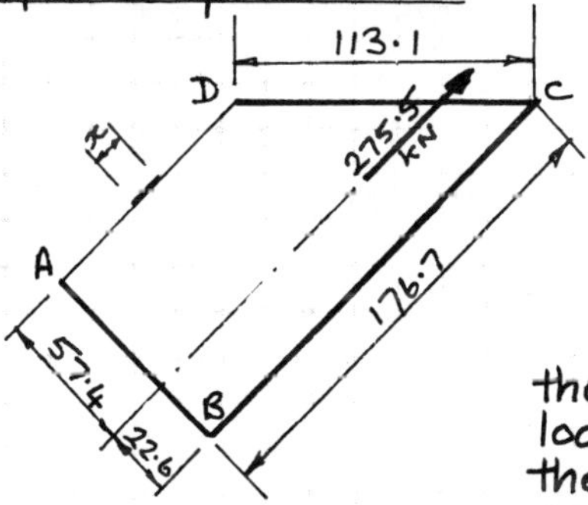

Using only the weld indicated for the design, so that the line of action of the load passes through the centroid of the selected weld group.

By moments about line of load

$x \times 57.4 + 80 \times 17.4 + 113.1 \times 17.4 = 176.7 \times 22.6$

$x = 11.0$ mm

Effective length of weld $= 176.7 + 80 + 113.1 + 11.0$

$= 380.8$ mm

Shear on weld $= \frac{275.5}{380.8} = 0.723$ kN/mm

Use 5mm fillet weld — **Output:** 5mm fillet weld all round.

Commentary: 2/1

The example illustrates the design of a bolted connection for the first node from the end of the top chord of a warren girder. The majority of lattice girders are shop welded into transportable sections, with site-bolted joints between the sections. However, fully bolted girders are still fabricated, usually for reasons associated with shipping or erection.

Grade 8.8 ordinary bolts have been used in the example. The connection is part of a deep lattice girder with a relatively small number of nodes and a depth-to-span ratio of 1/6. The additional deflection due to any slippage of the bolts in their clearance holes is acceptable. In general, if there is any doubt about the additional deflection due to slippage being acceptable then HSFG bolts should be used. (Note that with M20 General Grade HSFG bolts the capacity would be only 143 kN per bolt in double shear.)

The usual practice has been adopted of using the bolt lines, instead of the centroids, for setting out the members.

Shear capacity = $p_s A_s$
Bearing capacity = $P_{bs} = dtp_{bs}$

Therefore, $t = \dfrac{P_{bs}}{dp_{bs}}$

With Grade 8.8 bolts and Grade 43 steel the bearing capacity of the connected ply is less than that of the bolt.

Structural Steelwork Connections	Subject: Bolted gusset plate connection for truss		Chapter Ref. 16
	Design Code: BS 5950 Part 1		Calc. Sheet No. Example 2/1
	Calc. by B.D.C	Date Aug, '87	Check by G.W.O. / Date Nov, '87

Code Ref.	Calculations	Output

<u>Bolted gusset plate connection for truss</u>

Design a bolted gusset plate connection for the members and factored loads shown in the sketch.

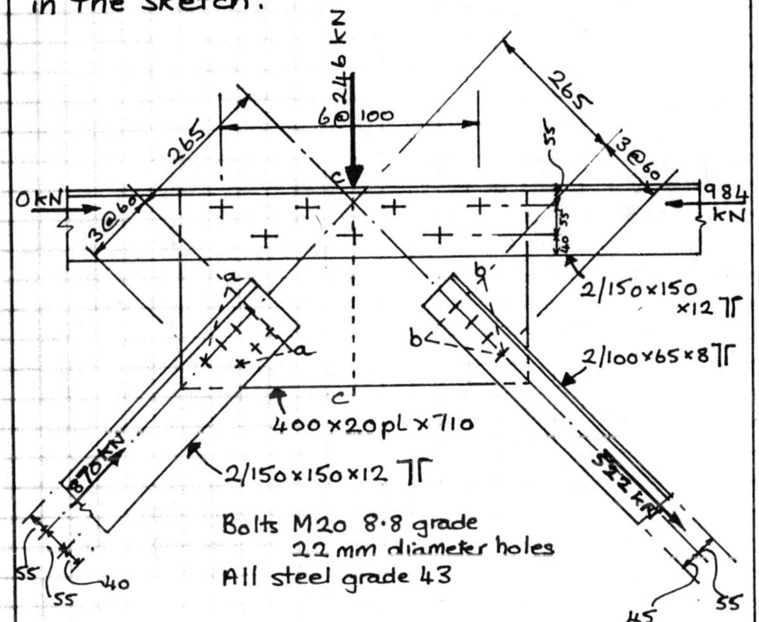

<u>Bolts.</u>

Use M20 8·8 bolts in double shear.

6·3·2 Table 32 — Shear capacity (per bolt)

$= 2 \times 375 \times 245 \times 10^{-3} = 183 \cdot 7\ \text{kN}$

6·3·3 Table 33 — Minimum thickness of plate in bearing

$$= \frac{183 \cdot 7 \times 10^3}{20 \times 460} = 20 \cdot 0\ \text{mm}$$

Use 20 mm thick gusset — Output: 20 mm thick gusset plate

Commentary: 2/2

Section 8.2.6

It has been common practice to ignore eccentricity when calculating the number of bolts required. However, with the two lines of bolts the eccentricity can have a relatively large effect on the bolt load (an increase of 46% in this example). With higher allowable design stresses in modern standards, it would appear to prudent to include the more significant eccentricities in the design. If the interaction curves in Figure 8.6 are used:

If the interaction curves in Figure 8.6 are used:

$a = 55\,\text{mm} \quad b = 180\,\text{mm} \quad D = \sqrt{(55^2 + 180^2)}$
$= 188\,\text{mm}$

$\dfrac{a}{b} = \dfrac{55}{180} = 0.31 \quad e = (55 - 41.2 + 23.6)$
$= 37.4\,\text{mm}$

$$\text{Estimate of } \alpha = \tan^{-1}\left(\frac{D}{4e}\right) = \tan^{-1}\left(\frac{188}{4 \times 37.4}\right) = 51.5°$$

From the interaction curve $\dfrac{P}{P_0} \simeq 0.7$

Load per bolt ignoring eccentricity $= \dfrac{870}{7}$

$= 124.3\,\text{kN}$
$< 183.7\,\text{kN}$
Use seven bolts

Maximum bolt load including eccentricity

$$\doteqdot \frac{124.3}{0.7} = 178\,\text{kN}$$

Still use 7 bolts

Structural Steelwork Connections	Subject Bolted gusset plate connection for truss.			Chapter Ref. 16
	Design Code BS 5950 Part 1.			Calc. Sheet No. Example 2/2
	Calc. by B.D.C	Date Aug, '87	Check by G.W.O.	Date Nov, '87

Code Ref.	Calculations	Output
6.3.3.3 Table 33	Minimum end distance for gusset plate $= \frac{2 \times 183.7 \times 10^3}{20 \times 460} = 39.9$ mm	

Combined thickness of attached legs of the tension diagonal $= 2 \times 8 = 16 < 20$ mm
Therefore, capacity per bolt for tension diagonal (bearing capacity of connected plies)

$$= 2 \times 20 \times 8 \times 460 \times 10^{-3} = 147.2 \text{ kN}$$

Compression diagonal (2/150 × 150 × 12 ⌐⌐)
Number of bolts required ignoring eccentricity

$$= \frac{870}{183.7} = 4.7$$

Try 7 bolts

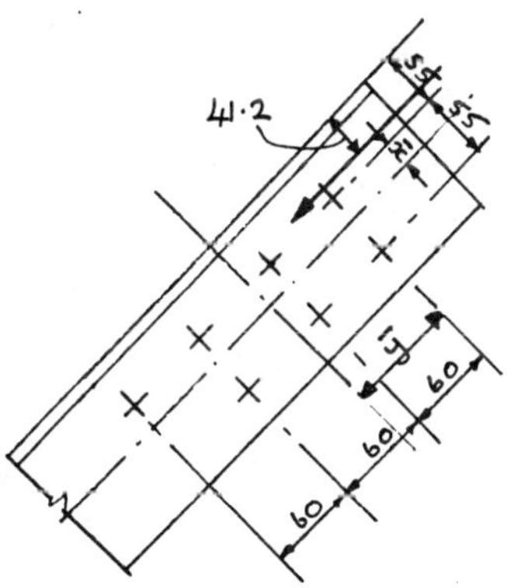

$$\bar{x} = \frac{3 \times 55}{7} = 23.6 \text{ mm}$$

$$\bar{y} = \frac{2 \times 60 + 2 \times 120 + 1 \times 180}{7} = 77.1 \text{ mm}$$

$$\Sigma(x^2+y^2) = 4 \times 23.6^2 + 3 \times 31.4^2 + 2 \times 77.1^2 + 2 \times 17.1^2 + 2 \times 42.9^2 + 1 \times 102.9^2$$

$$= 31{,}929 \text{ mm}^2$$

Moment $= 870 \times (55 - 41.2 + 23.6) = 32538$ Nm

Commentary: 2/3

For the tension diagonal, the maximum increase in shear on a bolt due to the small eccentricity is less than 10% and does not affect the number of bolts required.

Tension capacity = $A_c p_y$

For the top chord, the maximum increase in shear on a bolt due to eccentricity is again less than 10% and does not affect the number of bolts required.

Section 16.4 (see Figure 7.8)

The increases in bolt shear due to eccentricity can be estimated from the interaction curves (Figures 8.7 and 8.6).

Tension capacity = (effective width) × d × p_y

Structural Steelwork Connections	Subject: Bolted gusset plate connection for truss.			Chapter Ref. 16
	Design Code: BS 5950 Part 1			Calc. Sheet No. Example 2/3
	Calc. by B.D.C	Date Aug, '87	Check by G.W.O.	Date Nov, '87.

Code Ref.	Calculations	Output
	Maximum resultant shear on bolt	
	$= \sqrt{\left[\frac{870}{7} + \frac{32538 \times 23.6}{31929}\right]^2 + \left[\frac{32538 \times 102.9}{31929}\right]^2}$	
	$= \sqrt{[124.29 + 24.05]^2 + [104.86]^2}$	
	$= 181.7$ kN < 183.7 kN — Use 7 bolts	7/M20 8.8 bolts in comp^n diagonal
	Tension diagonal (2/100 × 65 × 8 ⅂⌈)	
	Number of bolts required ignoring eccentricity $= \frac{522}{147.2} = 3.55$ — Use 4 bolts	4/M20 8.8 bolts in tension diagonal
	Check tension capacity	
4.6.3.3	Effective area $= 2(1270 - 22 \times 8)$ $= 2188$ mm^2	
4.6.1	Tension capacity $= 2188 \times 275 \times 10^{-3}$ $= 601.7$ kN > 522 kN — o.k.	
	Top chord (2/150 × 150 × 12 ⅂⌈)	
	Number of bolts required ignoring eccentricity $= \frac{984}{183.7} = 5.4$ — Use 7 bolts.	7/M20 8.8 bolts in chord
	Gusset plate.	
	Check section a-a	
	Taking 30° spread from extreme bolts, effective width $= 180 \tan 30 + 55 + 120 \tan 30 - 2 \times 22$ $= 184.2$ mm	
	Tension capacity $= 184.2 \times 20 \times 265 \times 10^{-3}$ $= 976$ kN > 870 kN — o.k	

Commentary: 2/4

Half of the chord load of 984 kN has been assumed to be transferred to the gusset plate on each side of section c–c.

Horizontal force in gusset place at section c–c

$$\doteq \text{(Horizontal component of 870 kN)} + 0 - \frac{984}{2}$$

Shear force = vertical component of 870 kN

Shear capacity = $0.6\ p_y \times 0.9\ tD$

Structural Steelwork Connections	Subject: Bolted gusset plate connection for truss			Chapter Ref. 16
	Design Code: BS 5950 Part 1			Calc. Sheet No. Example 2/4
	Calc. by B.D.C	Date Aug, '87	Check by YGO	Date Nov '87

Code Ref.	Calculations	Output

Section b-b O.K by inspection

Check section c-c

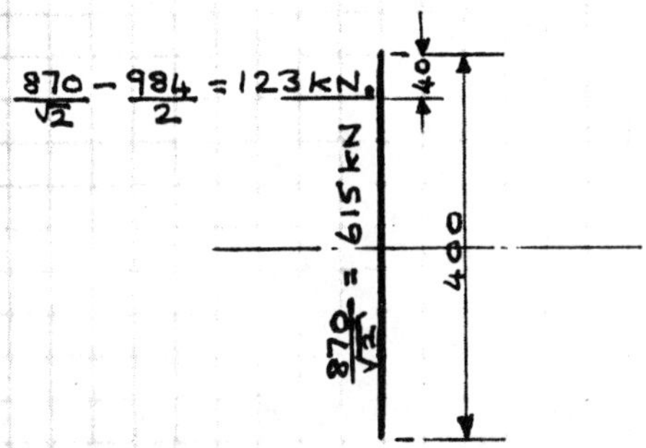

4.2.3 Shear capacity $= 0.6 \times 265 \times 0.9 \times 20 \times 400 \times 10^{-3}$

$= 1145$ kN

> 615 kN O.K.

4.8.2 Tension and moment

$$\frac{F}{A_e p_y} + \frac{M_x}{p_y Z_x} = \frac{123 \times 10^3}{400 \times 20 \times 265} + \frac{123 \times (200-40) \times 10^3}{265 \times \frac{20 \times 400^2}{6}}$$

$$= 0.058 + 0.139$$

$$= 0.197 < 1 \quad \text{O.K.}$$

Index

HSFG is used as an abbreviation for high strength friction grip